DON                          30119 021 205 457                          ST

Chemistry, Society and Environment
A New History of the British Chemical Industry

*Archibald Cochrane, 9th Earl of Dundonald (1749–1831): a founder of the British chemical industry*

# Chemistry, Society and Environment

## A New History of the British Chemical Industry

Edited by

**Colin A. Russell**

*Department of History of Science and Technology, The Open University, Milton Keynes, UK*

**RS•C**

ROYAL SOCIETY OF CHEMISTRY

ISBN 0-85404-599-6

A catalogue record for this book is available from the British Library.

Published by The Royal Society of Chemistry,
Thomas Graham House, Science Park, Milton Road, Cambridge CB4 0WF, UK

For further information see our web site at www.rsc.org

Typeset by Paston PrePress Ltd, Beccles, Suffolk, NR34 9QG
Printed and bound by Redwood Books Ltd., Trowbridge, Wiltshire

# *Preface*

The authors hope that this book will be of value to all those concerned with the social and environmental impacts of the British Chemical Industry. Whether academic chemists, industrialists, or politicians, many people today need an accurate assessment of what that industry has done to Britain. Such an assessment cannot be complete without at least some understanding of how we got to the present position. And that is the story we are trying to tell.

There have been several distinguished attempts to write the history of Britain's chemical industry as a whole, and a much greater number concentrating on individual companies. Some are chiefly concerned with technical detail, and others with the issues of economics facing the big firms. Most offer more than a nodding acquaintance with the people who figure prominently in the unfolding events. Yet, so far as we can discover, few have attempted to analyse the effects of the industry on society as a whole. These range from the varying conditions of employment of chemical workers in Victorian times to the transformation of civilised life in the 20th century. To be sure many chemical authors have written books purporting to show their science as entirely benevolent, and that the chemical industry deserved public support and encouragement. However, many of these works of mild propaganda originated at a time when science, chemistry and the chemical industry were undergoing nothing like the public resentment or suspicion that marks the end of the present century. Moreover, few of them attempted any serious historical analysis but were largely content to dwell on the delights conferred by the industry in the modern period. It therefore seemed to us that one of the primary emphases of any new history ought to be the direct effects the chemical industry has had, and is having, on British society.

That was one reason for writing this book. A second, much more powerful, reason lay in another shortcoming that we perceived in our predecessors. None of them has offered any serious analysis of the industry's impact on the environment. This is surprising in that much of the blame for current environmental troubles has been placed firmly at the door of the chemical industry and its products. Was that fair? Or did the beleaguered industry protest too much? Granted that no one much before 1960 talked about 'the environment' this did not mean that the issues conveyed by that word were never being considered. Recent research has shown the opposite to have been the case. So it appeared to be fairly important that someone should take a close look at the environmental history of the chemical industry in the UK.

We must, however, make one matter crystal clear. Although we are all concerned with some fairly sensitive issues, this is not a book of apologetics for the chemical industry. On occasions we have felt it right to be highly critical. What we have tried to do is to give as objective and fair an assessment of the whole history of the industry as we possibly can. At no point have we been constrained by the Royal Society of Chemistry or any other chemical interest group, and we thank our publishers for the freedom we have been accorded. It may be a matter of some interest that, with our attempts to depict things as they really were, 'warts and all', the industry emerges with a far better image than it popularly 'enjoys' today. Also, history suggests that everyone still has some important lessons to learn.

Because we envisage that most of our readership will have some chemical knowledge we have not hesitated to use chemical terminology or give chemical reactions where appropriate. We believe they actually help in understanding the issues being discussed. The notion that the history of any science can be evacuated of most of its technical content seems to us one of the more absurd ideas still occasionally encountered today. Equally, however, we must point out to non-chemists that, though helpful, such technicalities are not essential to the overall argument. We simply ask that our conclusions be assessed by the normal canons of historical judgement.

Although I have had the final editorial reponsibilities, the book has been produced by four authors acting as a team, being in frequent correspondence with each other, with a few lengthy meetings interspersed. Each of us has seen, and commented on, everything our colleagues have written. Because we recognise that some people will have a selective interest in the material, and may not wish to read systematically through the whole book (though that is by far the best way!), we have intentionally allowed a small amount of overlap between some chapters, so that each is complete in itself.

All the authors have been members of the History of Chemistry Research Group of the Open University and we must thank that University for facilities ranging from continuous use of the Library to a Leverhulme Research Fellowship (for SW). Three of us also have associations with other universities, at Newcastle (WAC) and Cambridge (SW, CAR), and we owe much to them also. We would all wish to acknowledge the help received from the Royal Society of Chemistry not only for undertaking the publication of this book but also for the splendid facilities we all enjoy at Burlington House.

Above all I am grateful for the pleasure of working so closely with such congenial colleagues, whose time and expertise have been given unstintingly to the project. They deserve much more than formal editorial thanks.

As we go to press it is with deep sadness that I have to record the recent death of one of my fellow-authors, Dr. W. A. Campbell. His obituary has appeared in the national press and also in the *Newsletter* of the Historical Group of the Royal Society of Chemistry. The community of history of chemistry is greatly impoverished by his loss, and we shall long remember his meticulous scholarship, his impeccable experimental skills and his infectious sense of humour. It is some consolation to record that he had completed all work on his chapters up to, but just missing, the final stage of page proofs. The rest of us hope that the book may to some degree constitute a memorial to this much valued colleague.

Colin A. Russell

In memory of

**Archie Clow,**

valued friend, colleague and pioneer historian
of the revolution in British chemical industry

# *Authorship team*

*All the authors have been members of the History of Chemistry Research Group in the Department of History of Science and Technology at the Open University*

**W.A. Campbell**[†]

Formerly Senior Lecturer in General Chemistry at the University of Newcastle-upon-Tyne

**N.G. Coley**

Senior Research Fellow in History of Science and Technology at the Open University

**C.A. Russell**

Emeritus and Visiting Research Professor, Department of History of Science and Technology, the Open University

Affiliated Research Scholar, Department of History and Philosophy of Science, University of Cambridge, and former Visiting Fellow, Wolfson College, Cambridge

**S.A.H. Wilmot**

Senior Research Associate, the Darwin Correspondence Project, University of Cambridge

Formerly Leverhulme Research Fellow, Department of History of Science and Technology, the Open University

# Contents

Contents

**Chapter 11**    **Chemical Industry and the Quality of Life**    **319**
*N.G. Coley and S.A.H. Wilmot*

# Acknowledgements

Many of the illustrations in this book are owned by the individual authors. We are glad to acknowledge permission to reproduce others from the following:

Albright & Wilson plc, for the illustrations on pp. 172, 173 and 325 (from *100 years of phosphorus making* by R. E. Threlfall, 1951) and p. 189 (from *Fine chemicals*)

Alcan Smelting and Power, UK, for the illustrations on p. 318 (from *A guide to the Lynemouth smelter*), p. 316 (from *Aluminium in the Highlands*), p. 314 (from *Aluminium in the Scottish Highlands*), p. 317 (from *A commitment to continual environmental improvement*), and p. 315 (from *Foyers, 1896–1996*)

BASF AG, Germany, for the illustration on p. 132 (from *In the realm of chemistry*)

BP Amoco plc, for the illustrations on p. 24 (from *Chemistry and the petroleum industry* by A. E. Dunstan), p. 251 (from *The Geon story* by British Geon Ltd.), p. 260, lower right (from *Our industry* by BP, 1947), and p. 263 (from *British Hydrocarbon Chemicals* by the BHC Group).

Ciba Speciality Chemicals plc, for the illustrations on pp. 233 and 237 (from *Clayton Aniline Co. 1876–1976* by E. N. Abrahart, 1976)

Mrs. Jill Davis (for the late A. and N. Clow), for the illustrations on pp. 49, 54, 60 and 307 (from their *The Chemical Revolution*, Batchworth Press, London, 1952).

The 15th Earl of Dundonald, for the Frontispiece (Photo. by M. Levers)

Enka Technico, GmbH & Co. KG, Germany, for the illustration on p. 342 (from a fashion advertisement,1952, from American Enka Corporation)

Esso Petroleum Co. Ltd., for the illustrations on pp. 260 (top left), 267 and 268 (from *Esso refineries in Britain*)

Frank Cass & Co. Ltd., for the illustration on p. 280 (from *Early iron industry of Furness* by A. Fell)

Gateshead Public Library, for the illustration on p. 17 (from their own collection)

GlaxoWellcome plc, for the illustrations on pp. 138 (lower) and 150 (from *One hundred years of Wellcome*)

Dr John Hume, for the illustration on p. 277 (from his own collection)

ICI plc, for the illustrations on pp. 36, 128 and 129 (from their own collection), and p. 254 (from posters of the early 1960s)

Institute of Petroleum Technologists, for the illustrations on p. 260, top right and lower left (from *Petroleum: twenty-five years retrospect, 1910–1935*)

Ironbridge Gorge Museum Trust, for the illustrations on pp. 279 and 280 (from their publicity material)

Kingston-upon-Hull Museums and Art Galleries, for the illustration on p. 139 (from a photograph of a reconstructed chemist's shop at Wilberforce House, Hull)

John Sherratt & Son Ltd., Manchester, for the illustration on p. 206 (from *A history of calico printing in Great Britain* by G. Turnbull)

Laporte Industries Ltd., for the illustrations on pp. 183 and 343 (from their own collections)

Macmillan Press Ltd., for the illustration on p. 39 (from *The emergence of a scientific society in England* by G. W. Roderick, 1967)

The Metals Society (Institute of Materials), for the illustration on p. 283 (from *The archaeology of industry* by K. Hudson, 1976)

Plastics Historical Society Library, for the illustrations on p. 340 (from their own collection)

Rhône-Poulenc Rorer, for the illustrations on pp. 140 and 149 (from *A history of May & Baker 1834–1984* by J. Slinn, Hobsons, Cambridge, 1984)

Science Museum, for the illustration on p. 242

Shell U.K. Ltd., for the illustrations on p. 261 (from *Shell Haven Refinery*, 1953, and *Refining in Britain by Shell*, 1952) and p. 15 (from *Modern chemicals from oil*)

Shire Publications Ltd. and Mr W. A. Jackson, for the illustration on p. 134 (from *The Victorian chemist and druggist* by W. A. Jackson)

The Society of Chemical Industry, for the illustration on p. 238 (from the *Centennial history of the Liverpool Section of SCI* by D. W. Broad)

Charles Tennant & Co. Ltd., for the illustrations on pp. 61 and 83 (from *Tennant enterprise*, 1937)

Wellcome Library, London, for the illustration on p. 138 (upper)

We have sought to locate owners of all reproduced material not in our own possession. In a few cases we have been completely unsuccessful, but trust we have not inadvertently infringed any copyrights. Should we have done so inadvertently we shall of course take appropriate action for any subsequent editions.

Finally, we must thank the library staff of the Royal Society of Chemistry for location and provision of many portraits; and our special thanks are due to Mr Alan Cubitt of the RSC who has shown untiring patience and courtesy in seeing through the press a book that, by any standards, called for devotion and skill of a high order.

Chapter 1 # *Records of the British Chemical Industry*

C.A. RUSSELL

When Henry Ford announced that 'history is bunk' he was telling us more about his own ill-informed prejudices than about any serious study of the past. If, however, he had proclaimed that 'history is *junk*' he might have been rather nearer the mark. For our knowledge of previous ages depends very much on our use of documents and objects that, in their own day, became of little value to the owners and were often thrown away. But to us they can be invaluable sources of information. This holds good for almost any kind of history: military, social, ecclesiastical or whatever. Yesterday's rubbish may turn out to be today's records. So modern students of history are encouraged to rummage among old letters, bills, receipts, prescriptions, memoranda, photographs, legal agreements, deeds, wills, diaries and the like. Amongst such material may emerge evidence of immense value in reconstructing aspects of a past age. Nor is such evidence limited to the written or printed page (or to computer discs if the date is very recent). One may well have recourse to small articles like buttons, bottles, bullets and bracelets, to furniture and machinery or (on the grand scale) to a cathedral or a factory. They may be in mint condition or in a ruinous state; to the trained observer they can be invaluable. Their study is the preoccupation of the museum curator and the industrial archaeologist. We return to them at the end of the chapter.

## 1 Written Sources

The most obvious kind of object to tell us about the past is, of course, the handwritten or printed page. All such objects, contemporary with the period of interest, are termed in the jargon 'primary sources', in contrast to 'secondary sources' like textbooks that may have been written decades or even centuries after the events they purport to describe. Preoccupation with primary sources is nowadays a mark of historical maturity. It is also a relatively new feature of student training. Gone are the old days of 'mugging it up from a text-book'; instead exposure to contemporary material is more informative and exciting. Textbooks still have their place but, as in science, they need revision as new data accumulate, in this case from new primary sources that are still being discovered.

If that is true of historical training and research in general, it applies with even

1

greater force to history of science and technology, and particularly to that part dealing with the rise and development of the chemical industry. We therefore begin this book by asking where relevant information may be found, and suggesting a number of answers (some of which are not a little surprising). Some important primary sources will be considered first, and later a number of significant secondary sources will be mentioned.

## (a)   Primary sources

What might be called 'the archives of the British chemical industry' were until recently largely unknown territory. A systematic investigation of what might still be available was undertaken a few years ago in a project of the Open University.[1]

Something of the scale and limitations of the enterprise were indicated as follows:

> It is obvious that only a minute proportion of the records that once existed has survived to the present time. Apart from destruction wrought by fire, explosion, air raids or simple modernisation, there are other threats that imperil old documents. They might occupy space needed for other purposes, and are sometimes seen as irritating reminders of an out-of-date image that the current owners of the plant are anxious to shed. However, there were reasons for hope that a good number might have found refuge in local record offices, libraries, and other safe havens, and a few of the largest companies have actually retained them in the care of their own archivists. Such provision seemed likely to be exceptional, however, with the remainder at considerable risk ...
>
> We began our search of materials by calling attention of industrial chemists to the matter, through letters or short notes in the literature.[2] Next a list of probable or known record holders was drawn up and letters sent to each. This revealed a number of sources unknown to us and also disclosed many companies that cheerfully denied having any archives at all ... In parallel with this went a systematic approach to all record offices in the United Kingdom and to all public libraries known (or suspected) to hold appropriate records. Again, there were many unexpected discoveries, but also a disturbingly high number of record offices which reported having no records of the chemical industry whatever ... Following these and other consequent enquiries, an extensive travel programme was embarked upon and a list of holdings gradually compiled *in situ*.

It emerged from this study that, for the chemical industry, the most important kind of source material is sometimes to be found in the archives of individual firms. However, as was also discovered, the vast majority had been destroyed, lost or given to record offices and libraries or (more rarely) museums. In these haunts of local historians and ancestor-hunters may be surprisingly found some of the richest deposits of material from the chemical industry. Often it is not classified as such but might appear under such varied titles as fertilisers, leather, iron, copper, gunpowder, vinegar, photography and so on, or under the names

[1] P.J.T. Morris and C.A. Russell, *Archives of the British chemical industry, 1750–1914: a handlist*, BSHS Monograph, no. 6, Faringdon, 1988.
[2] *E.g.*, C.A. Russell, *Ind. Chem. Bull.*, 1982, **1**, 90.

of individual companies or persons. A little gentle pressure and diligent enquiry may often reveal some astonishing treasures. Occasionally records may still exist in private hands. Identifying where they are, and then gaining access, has proved to be a somewhat specialist task involving an unusual combination of skills. It is not to be undertaken lightly!

Out of the vast mass of paper brought to light in recent years it is clear that archives of the chemical industry (as of almost anything else) are of an astonishing variety. There are the usual minute books, business memoranda, order books, accounts, receipts and bills as one might expect. Hardly less important are maps, plans and diagrams of plant and equipment. There are laboratory analysis books, 'recipes' (sometimes one can only call them that), technical data on reagent and product specifications. Of course one rarely finds all of these for one firm or in one place. Fortunate indeed is the researcher who stumbles upon extensive correspondence or (better still) a letter book. For here will be found discussion on an immense range of social as well as technical matters, on workers as well as work, on schooling as well as sales, and (if litigation was in the air) on pollution as well as profit. Since a location list has already been published,[1] readers are referred to that for further information as to where such gold-mines may be found. Two related projects are represented by the *Scottish brewing archive list*[3] and by a study in progress of the archives of the pharmaceutical industry. But beware: all that has been noted in recent years is but a tiny fraction of what must once have existed. So caution in interpreting the results is just as appropriate as in any area of experimental chemistry where the data are sparse.

How these sources are used depends partly on the skill of the researcher and partly on his or her objective. It may be that one wishes simply to produce a straightforward history (if such a thing exists) of an individual company or factory. Existing accounts, if any, may prove woefully inaccurate in the light of the new documentary evidence, so a 'revisionist' history will be needed. However it may well be that the task in hand is rather more subtle, and involves *asking new questions* of data that may well have been worked over before. Thus, instead of highlighting the business or technical aspects of a firm one might wish to enquire what can be learned about its place in the local community, its attitudes to employee education, how it related academic research to technical growth or what environmental policies were explicit or implicit in its records. That is not to say that more cannot be learned about the kinds of chemistry involved in all departments of the firm, for much recent writing has tended to gloss over chemical issues in the emphasis on economics. Of course an ideal history (which has not been written) will cover all these aspects in one global synthesis.

There are three other places where primary material on the chemical industry may be found. First there are the *collections of local and national newspapers* that exist in many large civic libraries, and supremely at the National Newspaper Library at Colindale in Middlesex. Rarely indexed, local papers are a rich field

[3] *Scottish brewing archive list*, Heriot Watt University, Edinburgh, 1983.

requiring patience, good eyesight (many are now on microfilm) and an eye for the relevant. Their immense advantage is that they give contemporary views of the industry in its immediate context. Amongst national newspapers we are fortunate that *The Times* has been fully indexed and, provided one knows exactly which subject to search for, has much of value. Then, there are detailed listings of many firms at *Companies' House* in Moorgate, London. They tend to be relatively recent. Older material may be found at the *Public Record Office* at Kew, though it is only a fairly small percentage of that which once existed. Particularly relevant is the huge BT31 class, consisting of files for dissolved companies registered in London after 1856. Also important is the PRO's BT41 class, files of companies registered under the 1844 Act.

Most of the source-material so far mentioned is hand-written. Finally it is worth recording that contemporary comment may be found in abundant measure in journals and books. Amongst the journals known for their coverage of industrial chemistry were those mentioned in Table 1. Those with an asterisk (*) are the most generally useful.

**Table 1** *Journals reporting contemporary developments in the chemical industry*

| Journal | Dates | Abbreviation used |
|---------|-------|-------------------|
| Chemical News* | 1859–1932 | Chem. News |
| Journal of the Newcastle Chemical Society | 1865–1882 | J. Newcastle Chem. Soc. |
| Journal of the Society of Chemical Industry* | 1882– | J.S.C.I. |
| Chemistry and Industry* | 1923– | J.S.D.C. |
| Chemical Trades Journal* | 1887–1966 | Chem. Trades J. |
| Journal of the Society of Dyers and Colourists | 1884– | Chem. & Ind. |
| Journal of the Society of Leather Trades Chemists | 1897– | J.S.L.T.C. |
| Journal of the Institute of Petroleum | 1913– | J. Inst. Petr. |
| Journal of the Oil & Colour Chemists' Association | 1918– | J.O.C.C.A. |
| Chemical Age* | 1919– | Chem. Age |

As for books, any short list would have to contain the massive and encyclopaedic work by that remarkable Liverpudlian chemist Sheridan Muspratt: *Chemistry, theoretical, practical and analytical as applied and relating to the arts and manufactures*, William Mackenzie, London, no date but about 1860. Muspratt presented grateful readers with his own portrait as frontispiece, followed at a respectful distance by those of the most eminent chemists of his day. These included Liebig, his own teacher, but not Frankland; with both of these he managed to engage in bitterest controversy. However, his literary labours were not in vain for he has bequeathed to posterity an unrivalled account of the chemical industry in mid-Victorian Britain. Rather earlier, less encyclopaedic, but still valuable was another two-volume work, *Applied chemistry: in manufactures, arts and domestic economy*, by Edward Parnell (Taylor and Walton, London, 1844). On a more limited front were books like John Lomas, *A manual of the alkali trade*, Crosby Lockwood & Co., London, 1880, and the long-lasting works on sulphuric acid, alkali and coal tar by Europe's most learned alkali chemist, Georg Lunge. All these volumes give invaluable insights into the chemical industry a century and more ago. Though available in print they are not to be despised as important primary sources.

## (b)   Secondary sources

For a quick first acquaintance with a subject most people turn to one of the well-established and reputable secondary sources. These fall into many categories, from general textbooks to very specialist monographs.

There are several textbooks available that deal with the history of the industry as a whole, a fact that immediately prompts the question: why yet another one? The short answer is that (1) most of our predecessors did not purport to deal with the industry in its social and environmental context, (2) a number of their books are seriously out of date and (3) most are out-of-print and quite hard to find. Table 2 lists the most important, in order of date.

**Table 2** *Some general histories of the chemical industry*

| | |
|---|---|
| 1906 | L. Fenwick Allen, *Some founders of the chemical industry*, Sherratt & Hughes, London |
| 1931 | S. Miall, *A history of the British chemical industry*, Benn, London |
| 1938 | G.T. Morgan and D.D. Pratt, *British chemical industry: its rise and development*, Arnold, London |
| 1952 | A. and N.L. Clow, *The chemical revolution: a contribution to social technology*, Batchworth, London [reprinted Gordon & Breach, Philadelphia, 1992] |
| 1953 | T.I. Williams, *The chemical industry: past and present*, Penguin, Harmondsworth [reprinted 1972, EP Publishing, Wakefield] |
| 1966 | D.W.F. Hardie and J.D. Pratt, *A history of the modern British chemical industry*, Pergamon Press, Oxford |
| 1969 | L.F. Haber, *The chemical industry during the nineteenth century: a study of the economic aspect of applied chemistry in Europe and North America*, Clarendon Press, Oxford |
| 1971 | L.F. Haber, *The chemical industry 1900–1930: international growth and technological change*, Clarendon Press, Oxford |
| 1971 | W.A. Campbell, *The chemical industry*, Longman, London |
| 1991 | P.J.T. Morris, W.A. Campbell and H.L. Roberts (eds.), *Milestones in 150 years of the chemical industry*. Royal Society of Chemistry, Cambridge |

Of these excellent volumes the one by the Clows is specially memorable, showing how the chemical industry was an integral part of the early industrial revolution. Over half of Williams's book deals with the industry of the early 1950s. As their subtitles indicate, the two books by Haber are not restricted to Britain and focus specially on the economic aspects of the story. The most comprehensive historical views, compactly expressed, are to be found in the books by Campbell and by Hardie and Pratt.

However, not all books attempt comprehensive coverage; some deal with specific parts of the industry only. Table 3 lists a few books of this kind.

Other books look at specific areas of the country where the chemical industry flourished. Area-based studies may also be found in local publications, notably journals devoted to local or regional history. In that connection Tyneside and N.W. England are particularly well-served, as Table 4 illustrates. It is remarkable that London's chemical industry has yet to have a book-length history.

**Table 3** *Histories of specific branches of the chemical industry*

| | |
|---|---|
| Alkalis | K. Warren, *Chemical foundations: the alkali industry in Britain to 1926*, Clarendon Press, Oxford, 1980 |
| Alum | C. Singer, *The earliest chemical industry*, Folio Society, London, 1948 |
| Coal-tar | W.M. Gardner, *The British coal-tar industry: its origin, development, and decline*, Williams & Norgate, London, 1915 [reprinted 1981, Arno, New York] |
| Dyestuffs | M.R. Fox, *Dye-makers of Great Britain, 1856–1976: a history of chemists, companies, products and changes*, ICI, Manchester, 1987 |
| Explosives | E.A.B. Hodgetts (ed.), *The rise and progress of the British explosives industry*, Whittaker, London, 1909 |
| Pharmaceuticals | L.G. Matthews, *History of pharmacy in Britain*, Livingstone, Edinburgh, 1962. |

**Table 4** *Some local studies of the chemical industry*

| | |
|---|---|
| Cheshire | W.F.L. Dick, *A hundred years of alkali in Cheshire*, ICI (Mond Division), 1973 |
| Edinburgh | S.A. Blackden, *A tradition of excellence: a brief history of medicine in Edinburgh*, Duncan, Flockhart & Co., Edinburgh, 1981 |
| St. Helens | T.C. Barker and J.R. Harris, *A Merseyside town in the industrial revolution: St. Helens, 1750–1900*, Cass, London, 1954 [reprinted 1959] |
| Tyneside | W.A. Campbell, *The old Tyneside chemical trade*, University of Newcastle-upon-Tyne, 1964 |
| Tyneside | W.A. Campbell, *A century of chemistry on Tyneside, 1868–1968*, SCI, London, 1968 |
| Widnes | D.W.F. Hardie, *A history of the chemical industry in Widnes*, ICI (Gen. Chem. Div.), 1950 |

Many journal articles cover local aspects of the chemical industry, including its progress in Scotland,[4] London[5] and (of course) Tyneside.[6]

Individual firms may also have their 'histories' written for them, though here a word of warning is appropriate. Commemorative histories commissioned by a company (for example to mark its centenary) are notoriously prone to the danger of high selectivity. No company wants skeletons in its own cupboard displayed for public view, and even if an author is given freedom from formal editorial control, it must be extremely difficult not to paint an unduly favourable picture. This is even more likely in official brochures or local newspaper articles. Nevertheless many company histories are models of sound historical judgement, including the examples quoted below.

Of all the great British chemical companies ICI has probably had most written about it. The literature includes I. Watts, *The first fifty years of Brunner, Mond & Co., 1873–1923* (Brunner & Mond, Northwich, 1923), J.R. Lischa, *Ludwig Mond and the British Alkali Industry* (Garland Publ. Inc., New York and London, 1985) and A.S. Irvine, *A history of the Alkali Division* (ICI Alkali

---

[4] Anon, 'Scotland's chemical industry', *Chem. Age*, 1958, **79**, 813–826; A. Clow, 'Chemistry in Scotland, and its pioneer contributions to textile technology', *J. Textile Inst.*, 1961, **52**, 204–218; A. and N.L. Clow, 'Some aspects of the Chemical Revolution in Scotland', *Aberdeen University Rev.*, 1940 (July), 1–11.

[5] W.A. Parks and E.A. Rudge, 'London's chemical industry', *Chem. Age*, 1946, **54**, 359–362; 1946, **55**, 167–169.

[6] W.A. Campbell, 'A synoptic view of the early chemical industry of the North East', *Cleveland Industrial Archaeologist*, 1975 (4), 1–6.

Division). But it also has perhaps the most admired of all company histories in the chemical history, two magisterial volumes by the distinguished historian W.J. Reader: *Imperial Chemical Industries: a history*, Oxford University Press, London: vol. I, 'The forerunners, 1870–1926', 1970; vol. II 'The first quarter-century, 1926–1952', 1975. Other firms have also been well-served by historians, including those in Table 5.

**Table 5** *Some company histories*

| | |
|---|---|
| Allen & Hanbury | E.C. Cripps, *Plough Court: the story of a notable pharmacy 1715–1927*, Allen & Hanbury's Ltd., London, 1927 |
| Allen & Hanbury | D. Chapman, J. Huston and E.C. Cripps, *Through a city archway: the story of Allen & Hanbury's 1715–1954*, John Murray, London, 1954 |
| Bayer | J. Roddom, *Bayer in Britain, 1870–1999*, privately printed, Newbury, 1999 |
| Castner-Kellner | D.W.F. Hardie, *Fifty years of progress, 1895–1945*, Castner-Kellner Co., Birmingham, 1945 |
| Chance Bros. | J.F. Chance, *A history of the firm of Chance Brothers*, privately printed, London, 1919 |
| Courtaulds | D.C. Coleman, *Courtaulds: an economic and social history*, Clarendon Press, Oxford, 1969–1980 |
| Crosfields | A.E. Musson, *Enterprise in soap and chemicals: Joseph Crosfield & Sons Ltd. 1815–1965*, Manchester University Press, 1965 |
| Ilford | R.J. Hercock and G.A. Jones, *Silver by the ton: the history of Ilford Limited, 1879–1979*, McGraw Hill, London, 1979 |
| May & Baker | J. Slinn, *A history of May & Baker, 1834–1984*, Hobsons, Cambridge, 1984 |
| Reckitts | B.N. Reckitt, *The history of Reckitt & Sons Limited*, A. Brown, London, 1958 |
| Unilever | C. Wilson, *The History of Unilever*, Cassell, London, 1954 |
| Whiffen | R.S. Law, *The End of A Chapter: the story of Whiffen & Sons Limited, Fine Chemical Manufacturers*, Fisons, *ca.* 1973 |

And at this level of specialism one occasionally encounters valuable material in shorter papers and articles in both chemical and historical journals. The growing field of business history also impinges on the chemical industry, and helpful articles are to be found in the journal *Business Archives* and also in the five-volume *Dictionary of Business Biography*.[7] Some of these specialist papers will appear in references throughout the book. There are also academic dissertations that may be consulted, though this is something of a specialist task unless Inter Library Loans can help. Two examples are A.W. Slater, 'Howards, chemical manufacturers 1797–1837: a study in business history' (London University MSc thesis, 1956), and E. Chicken, 'Ultramarine: a case study in the relationships between industry, science and technology, with particular reference to the firm Reckitt and Sons, and their successors' (Open University PhD thesis, 1993).

## 2  Objects

Anyone who tries to understand the history of the British chemical industry from papers alone is rather like a historian of warfare who never inspects a gun or a student of railway history who has never seen a train. Whatever else such persons

[7] D.J. Jeremy (ed.), *Dictionary of business biography*, Butterworths, London, 1984–1986.

may miss, one thing is a *feel* of the subject, a sense of what it must have been like to fire at Waterloo or to travel behind *Flying Scotsman*. In addition to the powerfully evocative effects of real-life encounter with historical artefacts there is their ability to provide hard information. In the case of the chemical industry that is as true of a sample of partially burnt pyrites (which can be analysed) as of a bank of coke-ovens in Co. Durham (which can be visited). From the small to the immense, many objects await detailed analysis that could lead to deeper understanding of the fortunes of this important part of British industry.

## (a)   The use of museums

In 1981 a booklet was published under the auspices of the Federation of European Chemical Societies entitled *A guide of European museums and expositions on chemistry and history of chemistry*.[8] Great Britain had six entries (in London University College, the Royal Institution and the Science Museum, the Cavendish Laboratory at Cambridge, the Museum of the History of Science at Oxford and the Royal Scottish Museum at Edinburgh). Since then it has been possible to identify a number of other museums in the UK with objects of chemical interest. Sadly, very few of these institutions have much on *industrial* chemistry. The one exception is the Science Museum which has many objects relating to the alkali and dyestuffs industries, as well as plastics, metallurgy and biomaterials. It also has some splendid models of 19th century chemical plant. Further material, particularly relating to local industries, may be found in the Museum of Science and Technology at Newcastle-upon-Tyne and the Greater Manchester Museum of Science and Industry. Indeed many local museums own objects that are related to the chemical industry, though they rarely feature in main displays. Good examples include the Birmingham Museum of Science and Industry and the Tullie House Museum in Carlisle. There must be dozens of others.

A place that many might regard as only marginally relevant to the chemical industry is the Salt Museum at Northwich in Cheshire. They would be seriously wrong. Salt was crucially important in the alkali and related industries and this museum has samples, equipment and models relating to the extraction of salt and its use in industrial chemistry. It is particularly rich in illustrations (photographs, paintings, plans) and publishes a number of excellent information leaflets on the chemical industry based on sodium chloride.

Pharmaceutical chemistry is an important part of the Wellcome collection, currently housed at the Science Museum in London. It is also represented at the Oxford Museum of the History of Science. There are several preserved (or reconstructed) pharmacies, of which that at the North of England Open Air Museum at Beamish, Co. Durham, is the most interesting. It comes from Stockton-on-Tees, and includes material from the Stockton druggist John Walker who invented the friction match in 1826.

[8] J.W. van Spronsen, *A guide of European museums and expositions on chemistry and history of chemistry*, Museum for Science and Technology, Budapest, 1981; since then a second, and now a third, edition has appeared (*Chem. Europe*, 1998, **6** (3), 6).

The glass-making industry is featured in the Pilkington Glass Museum at St. Helens, and agricultural chemistry plays a small part in the Institute of Agricultural History and Museum of English Rural Life at Reading.

For a museum devoted entirely to the chemical industry, however, we have to travel northwards from Northwich to the town of Widnes (now part of the Borough of Halton). Here is the Halton Chemical Industry Museum, better (and appropriately) known as Catalyst. Beginning in 1982 in the Old Town Hall of Widnes it has now acquired a four-storey tower of the famous Gossage Works, at the edge of the Spike Island site, once filled with industrial plant of the Victorian chemical industry. An additional storey at the top of the tower contains many samples, pieces of equipment and replicas relating to the alkali industry. The panoramic views today are related to what would have been visible 100 years ago and foundations of some buildings may be traced in the grass below. Catalyst seeks to be a tourist attraction (which is no bad thing) but also caters for serious students of the industry, and is the only museum in the UK devoted to industrial chemistry. The emphasis is mainly, but by no means exclusively, on the manufacture of alkali, chlorine and sulphuric acid, the trio at the heart of the heavy chemicals industry and never more truly so than on Merseyside.

If the chemical industry is conceived to include the burning of lime, gunpowder manufacture, metallurgical processes and gas production then the scope is considerably widened and one enters the world of industrial archaeology.

## (b)　Industrial archaeology

This subject has acquired wide public interest in the last twenty years. As a result objects of industrial archaeological interest may be found in lonely places on remote hillsides or in the bizarre world of the theme park and thus part of the mass entertainment industry. Many sites worth visiting are somewhere between these extremes.

Taking metallurgy first, it is possible to trace the whole process from mining for ores to the purification and fabrication of the extracted metals. A range of early iron-mine tunnels may be explored, for example, at the Clearwell Caves in the Forest of Dean; hardly chemical, perhaps, but an essential preliminary to a chemical process of the first importance. The extraction of iron from its ores may be followed at a number of sites that have been listed elsewhere.[9] The most impressive of these must be the huge complex known as the Ironbridge Gorge Museum which is an imaginative blend of conventional museum, industrial archaeology site and theme park. Exhibits include Abraham Darby's iron-smelting furnace from 1709 and the remains of slightly later blast furnaces, to say nothing of the first cast iron bridge in the world. Next in importance must be the Industrial Hamlet at Abbeydale, Sheffield, with several steel-melting

[9] W.K.V. Gale, *Iron and steel*, Longman, London, 1969, pp. 139–141.

furnaces, tilt-hammers and forges. There are many lesser sites where quite substantial remains of iron-making can be traced.

Other metals were of course extracted, and some of these processes have left discernible traces. Recently several museums devoted to mining have been opened in the North of England, but there is little attention given to the chemical processes of extraction. Lead mining and extraction was common in West Durham and there are many relics to be found in this high and bleak Pennine country. Lead and copper were two metals to be won from the fells of the Lake District and again much can be discovered (though here, particularly, caution is needed on account of hidden shafts and unstable ground conditions). Similarly, Cornwall had its tin mining and smelting industry, with museums at Tolgus and Redruth.

Also on the fringe of the chemical industry properly so-called was the widespread conversion of limestone or chalk into quicklime. Lime kilns are probably the most conspicuous and ubiquitous relics of any of the extractive industries. Excellent examples may be seen from Betchworth in Surrey and Duncton in Sussex, through massive remains on the Western Pennines from Kirkby Stephen to Settle, to the largest surviving range of kilns in the country (14 in all) at Charlestown on the Forth.

The conversion of coal into coke and tar (as well as gas) has left several impressive relics, such as the coke ovens at Hunswick and Rowlands Gill in the North East. The resultant tar was usually distilled in the 19th century, and a few early stills exist, as in a distillation plant at the Scottish Tar Distillers, near Falkirk.

A few examples of early nitration plant survive, by which hydrocarbons derived from tar were converted into intermediates for the explosives industry. A few are at the former government establishment at Powfoot on the Solway Firth. There are some relics of the early days of dynamite manufacture at the ICI plant at Ardeer, in Scotland. But long before organic explosives were introduced gunpowder was being made all over Britain where the right combination of good access, charcoal availability and abundant water power were to be found. Of all the many survivals of that industry perhaps the most impressive are to be found at Faversham, in Kent, and at the government establishment at Waltham Abbey in Essex.

Of the many other branches of the chemical industry a surprising number have survived in East Anglia, possibly because these examples were not in the class of heavy industry, so prone to accident and subject to such pressure for land. They include a manure works at Duxford, a xylonite works at Cattawade, essential oils factories at Hitchin and Long Melford, and the general chemical works of Prentice Bros. in Stowmarket. Their condition and current usage vary widely.

Finally it is sad to record that of the large-scale examples of industrial enterprise, typified by the alkali industry but not restricted to it, very little indeed remains. A few street and pub names indicate something of the past activity on north Tyneside, but all factories have been swept (or burnt) away. Only the immense tips of 'galligu' can sometimes be located; then the still perceptible smell of hydrogen sulphide is all that remains of an industry that was central to Britain's economy.

## (c)    Archive film

It must be apparent that the physical remains of the chemical industry are very vulnerable. Not only accidental fire and explosions, but also deliberate destruction, can ensure that the number of surviving objects and plant decreases each year. There is one consolation from this regrettable fact. It has encouraged a number of film-makers to record while they have the chance processes and plant that are like soon to disappear. Much of this activity has originated with the Open University, as a by-product of its collaboration with the BBC in making TV programmes for certain undergraduate courses in history of science and technology. Table 6 summarises the main processes and locations, together with the Open University course reference number.

**Table 6**  *The chemical industry on archive film*

| Process/objects | Location | OU Course | No. |
|---|---|---|---|
| Beehive coke-ovens | Blaydon-on-Tyne | AST281 | 2 |
| Production of coal-gas in 1840's style horizontal retorts* | Newton Stewart gasworks | AST281 | 2 |
| Distillation of coal-tar | Scottish Tar Distillers, Falkirk | AST281 | 2 |
| Nitration of benzene | Disused factory, Derbyshire | AST281 | 2 |
| Lead chamber process for sulphuric acid* | Leathers Chemicals Ltd., Seaton Carew | AST281 | 3 |
| Alkali waste tip | St Helens | AST281 | 7 |
| Alkali works | ICI Mond, Winnington | AST281 | 7 |
| Leblanc Process: black-ash converter* | [ICI archive film] | A283 | 7 |
| Alkali waste tips | Tyneside | A283 | 6 |
| Gunpowder factory | MOD Waltham Abbey | AST281 | 8 |
| Nobel's explosives factory | ICI Ardeer | AST281 | 8 |
| Lab. scale production of nitroglycerine* | ICI Ardeer | AST281 | 8 |
| TNT factory | Disused factory, Powfoot | AST281 | 8 |
| Ironbridge and Coalbrookdale | Ironbridge | AST281 | 9 |
| Puddling process* | T. Walmsley & Sons, Bolton | AST281 | 9 |
| Bessemer steel-making process* | Workington [Bessemer Steel Corp. film] | AST281 | 9 |
| Forth Bridge: showing construction | Forth Bridge | AST281 | 9 |

* Working processes

## 3   The Wider Context

All that has been said so far, and most of what is to follow in the book, is specifically related to the development of the chemical industry in Britain. Those are our terms of reference and we have tried to keep to them. Nevertheless it would be misleading to imply (by silence) that little has been written on the industry's growth overseas. That is not the case. Books have been written on the history of industrial chemistry in America, France, Germany (a specially well-served country) and elsewhere, and the volumes by Haber (Table 2) have a strong international emphasis. More recently three further books have appeared that give a wider context to the ensuing discussion of British developments. One is a straightforward account of events in a largely chrono-logical fashion.[10] It attempts to mesh in the industrial changes with the growth of chemical understanding in general.

Two other books, more analytical but much less comprehensive, have been supported by the European Science Foundation and naturally emphasise the situations in Europe. Both are published by Kluwer as part of their 'Chemists and chemistry' series. The first[11] has the ambitious agenda of identifying the determinants of the industry from 1900 to 1939. It begins by focusing on one important new technology for the period, the conduct of reactions under high pressure, discusses the impact of the First World War, goes on to examine some case studies of the relations between science and industry, and concludes with economic-politico issues such as fiscal policy, competitiveness and state inter-vention. Particularly relevant to the UK are chapters on industrial organic chemistry in the inter-war years, attempts to provide an intellectual basis to heavy chemical manufacture, and the use of measuring and controlling instrumentation. A second book in the series[12] looks at patterns of industrial-isation over several European countries, addresses questions of pollution in Britain and elsewhere and finishes with a miscellany of themes that includes a study of chemists in the British alkali industry.

Clearly the chemical industry now is an international (not to say a multi-national) phenomenon. Even its origins do not lie in one country alone. In its meteoric rise in the 19th century the lead passed from Britain to Germany, though they were far from the only players in the drama. It is against that cosmopolitan backcloth that we now begin to examine the distinctive history of one immensely important piece of human endeavour: the British chemical industry.

[10] F. Aftalion, *A history of the international chemical industry*, trans. O.T. Benfey, University of Pennsylvania Press, Philadelphia, 1991.
[11] A.S. Travis, H.G. Schröter, E. Homburg and P.J.T. Morris (eds.), *Determinants in the evolution of the European chemical industry, 1900–1939*, Kluwer, Dordrecht, 1998.
[12] E. Homburg, A.S. Travis and H.G. Schröter (eds.), *The chemical industry in Europe, 1850–1914: industrial growth, pollution, and professionalization*, Kluwer, Dordrecht, 1998.

Chapter 2  # *The Shape of the British Chemical Industry*

N.G. COLEY

## 1  Introduction

Since the 18th century, the British chemical industry has developed in areas close to its raw materials and to other industries which use its products. It is an industry of exceptional diversity as chemical products play an important role in the manufacture of most other goods and very few areas of life escape their use. Yet only a relatively small number of chemical products are commonly recognised by the public. These include drugs like aspirin and penicillin, fabrics like rayon or 'artificial silk', nylon and terylene, plastics like formica, melamine and polythene, or fuels such as bottled gas, petrol and diesel oil. All of these add to convenience, comfort, or safety in everyday life, but there are people who do not choose to recognise these benefits and some would even assert that the applications of chemistry and chemical products in everyday life have made the quality of life worse rather than better. The very word 'chemical' has adverse connotations in the public mind, conjuring up images of pollution, poison, and *ersatz*, synthetic materials of doubtful quality rather than the natural materials which are commonly perceived as wholesome.

This book will examine the allegation that chemical industries have often been guilty of polluting the atmosphere and water courses, exerting adverse effects on the environment first in the vicinity of each chemical plant and later on a wider national scale, until in the 20th century its environmental and ecological effects have become world-wide. Certainly the public image of the chemical industry has suffered and its reputation for causing environmental damage has become a political issue. The confidence that earlier surrounded its activities has been undermined as the manufacture of certain products has been banned and financial losses have followed. The industry's reputation fell dramatically in the 1970s and 1980s even though efforts to protect the environment were being made by most chemical manufacturers. The industry as a whole is still failing to persuade the public of its good intentions in this respect, so it is crucially important to set out the facts as plainly as possible.

In this chapter we sketch the position of the current industry and set the scene for an analysis of how the present situation was reached.

# 2  A High-technology Industry

Perhaps the most obvious characteristic of the chemical industry and one which causes problems of presentation to the public stems from its high technology profile. New developments in the industry depend wholly on its scientific and technical foundations; the construction and operation of chemical plants demands advanced engineering skills and modern large-scale plants rely extensively on computers for automatic control. For these and other reasons the modern chemical industry is heavily dependent on capital investment.[1]

To the lay person the complexity of chemical technology is mystifying, yet most of the techniques used throughout the industry are fundamentally simple. Solution, crystallisation, precipitation, filtration and distillation are used in nearly all chemical manufacturing processes. Scaled up to industrial proportions these techniques require special reaction vessels, towers and pipework, giving chemical plants their characteristic external appearance. It is here that the scientific basis of the industry, becoming visible, creates an apparently insuperable barrier to public understanding. Chemical works often seem to present a menacing aspect, heightened by mistrust and media reports of serious accidents which occur from time to time across the world. Consequently, chemical industry tends to be regarded with suspicion based on lack of knowledge, fear of pollution and of the likely consequences of plant failure or malfunction. It is true that similar attitudes also apply to other industries, notably the nuclear industry. However the immense size and diversity of the chemical industry (due to widespread demand for its products) has given rise to deep public concern about its causing damage to health and to the environment. There is little public understanding of the industry and of its importance to the national economy.

A certain secretiveness about the activities and products of the industry has done little to improve its image, although since the 18th century there have been good reasons for inhibiting the publication of technical details. Products and processes are commonly the subjects of patents and in a fiercely competitive industry there is much commercial and industrial information which cannot be divulged. Many specialised compounds have been developed after considerable capital outlay in research and development – solvents, oxidants, stabilisers, accelerators, inhibitors and other industrial chemicals with detailed specifications and functions. Naturally chemical manufacturers wish to maintain a competitive advantage in an industry where minute details of technique can make all the difference between financial success or failure.

By far the greater part of the chemical industry's output is used within the confines of industrial sites where chemicals are used as intermediates in most manufacturing processes. Hence bulk chemicals become apparent to the public only when they are in transit by road or rail. The fear of accidents in which corrosive, inflammable or poisonous chemical spillages might occur is one of the perceived hazards of the industry. By no means all chemical accidents are life-

[1] C.A. Heaton (ed.), *An Introduction to Industrial Chemistry*, Leonard Hill/Blackie, Glasgow 1984, 2nd ed. 1991, p. 9.

*An early pilot plant for a Shell petrochemical plant; the next stage is 'scale-up' (p. 347) to a huge modern installation*

threatening and many chemical products are benign, but media reports seem incapable of mentioning 'chemicals' without the adjective 'dangerous', especially in the context of any kind of leak.

The great diversity of chemical industry means that, rather than a single industry, it must be regarded as a loose collection of related enterprises which, though different, show some similarities to each other. Some of these are very large multi-national conglomerates like Du Pont or ICI, each with many divisions making various chemical products across a wide range, but there are also many small chemical companies specialising in particular types of product. Consequently, chemical plants vary widely in size from large-scale plants producing up to 600,000 tonnes of chemical intermediates per year, usually in computer-controlled continuous production, to small plants with an annual output of no more than 100 tonnes or so. The latter are generally operated on a batch-process system; their products are often specialised chemicals made in small quantities one after another and the plant is operated manually. It will be obvious that the training, qualifications and skills required by chemical operatives in these two extremes of the industry must be very different.

Yet, despite this diversity, there are some broad divisions within the industry. One significant sector is concerned with the manufacture of primary materials, including the so-called 'heavy' chemicals, manufactured in very large quantities from natural raw materials such as common salt, sulphur, limestone, coal, sand, borax, oil and natural gas, air and water. During the 19th century coal-tar provided the fundamental raw material for the organic chemicals industry, but since the 1950s coal-tar has been gradually superseded by crude oil and natural gas. Today the petrochemicals industry, manufacturing polymers, plastics and synthetic rubber, is an important sector of the heavy chemical industry. Another large and growing sector depends on fermentation processes. In the quantity and fundamental importance of its products the heavy chemical industry ranks in the national economy with other basic industries such as fuel and power, mining, construction and transport.

## 3 Primary Source Materials

The history of the manufacture and uses of the most important chemical products will be considered in more detail in later chapters. Here we shall briefly introduce some of them to present a rapid overview of the industry.

### (a) Sulphur, sulphur-bearing ores and sulphuric acid

Sulphur, perhaps the most important raw material in chemical industry, is one of the few chemical elements to occur in its native state. It is found in volcanic regions, and Sicily was an important European source until the mid-19th century. When in 1838 the Sicilian Government licensed a French commercial society to be the sole exporter of sulphur, the price of Sicilian sulphur rose sharply and alternative sources had to be found to maintain sulphuric acid production in Britain. Useful quantities of sulphur dioxide could be obtained from burning iron pyrites ($FeS_2$), copper pyrites ($Cu_2S.Fe_2S_3$) and other sulphur bearing ores such as galena (PbS) and zinc blende (ZnS). 'Spent oxide', formed by removing sulphur compounds from crude coal-gas using moist ferric oxide ($Fe_2O_3$), also contained about 50% of sulphur, and after 1888 sulphur recovered from alkali-waste by the Chance–Claus process became available. For making sulphuric acid all these sources were suitable, although the sulphur dioxide produced by burning them was impure. When the contact process began to supersede the older lead chamber process in the 1880s, the impurities, especially arsenious oxide ($As_2O_3$), could not be tolerated and methods of removing them were needed. By this time however the sulphur beds of Texas and Louisiana had been tapped using the Frasch process which extracts sulphur of 99.5% purity ready for immediate use.

### (b) Alkalis and chlorine

Like sulphuric acid, sodium carbonate ($Na_2CO_3$) or 'soda', has long been an important product of the British chemical industry. The demand for soda began to grow as the textile trades expanded during the industrial revolution. Extracted first from kelp and barilla, the ashes of marine plants, soda began to be made from salt and sodium sulphate in the 18th century as demand outstripped the capacity of the traditional methods. Early attempts to manufacture soda were only partially successful, but the Leblanc process, invented in

*Hugh Lee Pattinson's alkali works at Felling (1843)*

1789 by the Belgian Nicolas Leblanc, was used successfully in Britain from the 1820s (see Chapter 4). The Leblanc process, worked on a very large scale throughout the 19th century, was a major culprit of the chemical industry's poor environmental reputation. It was gradually superseded from the 1880s by the cleaner, more elegant ammonia–soda process. In addition to its use for the manufacture of soda, common salt is also the main source of chlorine and hydrochloric acid, as well as caustic soda (NaOH) and metallic sodium, all of which are obtained by electrolysis.[2]

## (c)  The nitrogen industry

The 'fixation' of atmospheric nitrogen (conversion to reactive nitrogen compounds) forms the basis of another large sector of the heavy inorganic chemicals industry.[3] Over 50 million tonnes per year of fertilisers in the form of ammonium nitrate, ammonium phosphate and urea are used and a further 10 million tonnes per year of fixed nitrogen goes into explosives, dyes and resins. The primary processes of nitrogen fixation are still based on the Haber–Bosch process for ammonia synthesis from hydrogen and nitrogen, introduced in Germany about 1911, but during the past half-century many improvements have been made (Chapter 5). Originally the gases used for ammonia synthesis

[2] R.W. Purcell, 'The chlor-alkali industry', in R. Thompson (ed.), *The Modern Inorganic Chemical Industry*, Royal Society of Chemistry, London, 1977, pp. 106–133; see also Chapter 4.
[3] S.P.S. Andrew, 'Modern processes for the production of ammonia, nitric acid and ammonium nitrate', *ibid.*, pp. 201–231.

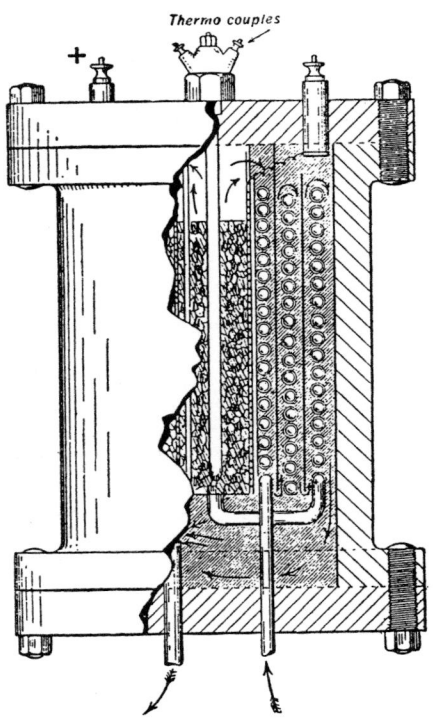

*An early version of Haber's cell in which ammonia is synthesised from nitrogen and hydrogen over a metal catalyst. The vessel is made of thin steel, electrically heated and placed in a large steel pressure container*

were obtained by passing air and steam alternately through a bed of red-hot coke, yielding a mixture of 'producer gas' (carbon monoxide and nitrogen) and 'water gas' (carbon monoxide and hydrogen), but in the modern industry the feed-stock for the process is 'synthesis gas' obtained almost exclusively from natural gas or reformed petroleum fractions.[4]

Nitric acid, produced by the catalytic oxidation of ammonia in stainless steel vessels, is also made in large quantities for use in the manufacture of fertilisers and explosives. Urea ($H_2NCONH_2$) is another important product in this sector, now made largely from ammonium carbamate ($H_2NCOONH_4$). It is manufactured on the same site as synthetic ammonia because the necessary carbon dioxide is available cheaply. Cyanamide ($H_2NCN$) is also produced in large quantities for the manufacture of melamine, its cyclic trimer. Lastly among the important nitrogen compounds, hydrogen cyanide is made by allowing ammonia to react with hydrocarbons (often in the presence of oxygen). From hydrogen cyanide cyanogen (($CN)_2$) and cyanogen halides (*e.g.* ClCN) are prepared – the value of cyanogen chloride lies in its use for making triazine for herbicides and dyestuffs manufacture. Thus, in this branch of the heavy

[4] *Ibid.*, pp. 209–217.

chemicals industry there are close links between inorganic and organic syntheses.[5]

## (d)   Limestone and phosphates

Among the bulk raw materials for the chemical industry won by mining and quarrying, the commonest is limestone (calcium carbonate, $CaCO_3$). This is required for many purposes in the heavy chemical industries including the manufacture of alkali, cement, glass and as a flux in metal smelting. Lime, made from limestone by calcination (lime burning), is used in paper-making, tanning, cement manufacture and agriculture. Phosphates, another important primary product, are quarried in bulk mainly as 'phosphate rock' or apatite. The latter was first mined in Britain early in the 19th century, but there are no longer any commercial deposits in Europe and the largest reserves presently being worked are in Morocco, North America and Russia. There are also large, virtually unexploited, deposits in Australia and South America. The quality of phosphate rock is expressed in terms of its tricalcium phosphate content $[Ca_3(PO_4)_2]$ or 'bone phosphate of lime', BPL. Commercial phosphate rock varies in BPL content from 60 to 83%. Although some elemental phosphorus is extracted, about 90% of the phosphate rock is converted into phosphoric acid ($H_3PO_4$). The largest commercial use of phosphates is as the fertiliser calcium super-phosphate, made by treating phosphate rock (as fluorapatite) with concentrated sulphuric acid:

$$CaF_2 \cdot 3Ca_3(PO_4)_2 + 7H_2SO_4 = 3Ca(H_2PO_4)_2 + 7CaSO_4 + 2HF$$

If phosphoric acid is used instead of sulphuric the product is 'triple super-phosphate' containing about three times as much phosphate available for plant growth. Besides their use in fertilisers, phosphates and polyphosphates are added to detergents and are used as water-softeners. Another large-scale use is in corrosion-resistant paints and oils for metal bearings in machinery and for car-bodies.[6]

## (e)   Silicates

Sodium silicate, one of the oldest chemicals to be made in bulk, has a long history in chemical manufacturing linked to alkali and soap. In Britain, William Gossage at Widnes in 1854 began to make sodium silicate by fusing soda ash and white sand in an open hearth furnace. After cooling, the resulting 'glass' was broken up and powdered before being dissolved in boiling water. The liquor was then concentrated by evaporation. Gossage's technique using sand and soda ash is still the most widely used method for the production of sodium silicate, although in parts of Europe sodium sulphate and carbon sometimes

[5] *Ibid.*, pp. 221–228.
[6] A.F. Childs, 'Phosphorus, phosphoric acid and inorganic phosphates', *ibid.*, pp. 384–395.

replace soda ash. Sodium silicate is the main soluble silicate product. Quantities of sodium silicate amounting to several million tonnes annually place this product among the heavy chemicals. Smaller quantities of potassium silicate are manufactured and limited amounts of lithium and quaternary ammonium silicates are also made.

The soluble silicates are used in detergents for domestic and industrial use. They are widely employed for washing industrial plant in contexts as diverse as dairying and metal processing. Silicates may also be added to cement where they allow thicker slurries to be produced with lower viscosities than would otherwise be the case. In foundry work silicates are used to permit the rapid production of sand cores and moulds. Silicates are also widely used as adhesives, especially in the paper and board industry. They are also used to produce coatings and paints which seal porous surfaces. Much of the sodium silicate is turned into amorphous silica precipitates, gels and silica–alumina compounds used in water softeners or as substrates for catalysts in catalytic crackers.

### (f)  Borax and boron compounds

The chief sources of native borax are in the Mojave desert, California, though supplies also come from Argentina, and Turkey has deposits of calcium borate from which boric acid is obtained.[7] The American ores are mainly tincal (sodium borate decahydrate $Na_2B_4O_7.10H_2O$), kernite (sodium borate tetra-hydrate) and lake brine. The Turkish ore is colemanite, $(Ca_2B_6O_{11}.5H_2O)$. Borax production is based on solution and crystallisation. The crushed ore is dissolved in hot water, the mixture is 'thickened' (*i.e.* concentrated by evaporation) and filtered to remove siliceous matter and the clear solution is then crystallised. A temperature of 100 °C is maintained throughout the process. Crystals of borax are separated from the mother liquor by centrifuging and dried in rotary driers. The weak mother liquor is then used to dissolve more tincal and is recycled through the process. Boric acid is obtained from borax or colemanite by treatment with sulphuric acid. Insoluble calcium sulphate can be filtered off and the clear liquor yields boric acid crystals on cooling. The boric acid obtained needs purification for the higher grades, although the initial product can be sufficiently purified by simple washing. A similar method is used with borax as the starting material; the first crop of crystals obtained on cooling is boric acid. Above 33 °C anhydrous sodium sulphate (salt cake) can be obtained. The main advantage of using borax is that minor impurities have already been removed.

For use in many chemical processes dehydrated borates are required. Calcination at fairly low temperatures removes most of the water of crystallisation from borax and fusion to 850 °C eliminates the remaining traces of water. Boric acid is readily dehydrated to form boric oxide ($B_2O_3$) at relatively low temperatures. The main uses of boron compounds are in glass manufacture

[7] R. Thompson, 'Production and uses of inorganic boron compounds', *ibid.*, pp. 303–313.

where they act as solvents in the fused state for many metal oxides. Borax is added to detergents for its mild alkalinity and buffering qualities, but it is also converted into sodium perborate which is used as a laundry bleaching agent. While boron compounds are non-toxic to mammalian life, they are poisonous to insects such as ants and cockroaches, and they are also useful in fungicides and herbicides. Boron compounds are essential for the health of many plants including apple, pear and citrus fruits, sugar beet, peanuts, alfalfa and most fodder and oil crops. Hence water soluble sodium borate is widely used as an additive to chemical fertilisers. The quantities of borates produced annually (of the order of 2 million tonnes) make this a significant sector of the heavy chemical industry.

## (g) Fluorides, bromides, iodides

The three most important ores of fluorine are cryolite ($3NaF.AlF_3$), fluorspar ($CaF_2$) and fluorapatite [$CaF_2.3Ca_3(PO_4)_2$]. The first two contain about 50% of fluorine but the latter only around 4%. However, the vast tonnages of apatite mined annually and processed as phosphate rock by the fertiliser industry suggest that there is here a large potential source of fluorine which is not at present being tapped.[8] Very large quantities of fluorosilicic acid produced in the processing of phosphate rock are discarded each year. Fluorspar, a glossy mineral varying in colour from white through amber, green and blue to purple, occurs widely all over the world, including deposits in Derbyshire. It is always associated with other minerals, especially quartzite. It has been known since the 16th century that fluorspar lowers the melting point of other minerals and the major use of fluorspar is as a fluxing agent in the metallurgical and related industries. It is also used in the glass and ceramics industry and in the chemical industry for the manufacture of hydrofluoric acid. In steel making fluorspar is still used as a flux to remove impurities and improve the separation of the metal from slag during melting. The shift from the basic open hearth process to the basic oxygen method for steel in recent years has seen a large increase in the use of fluorspar in the steel industry which accounts for just under half the world tonnage of fluorspar produced. Just over half forms the feed-stock for hydrofluoric acid production. This has been used in ever-increasing quantities since the 1930s in the rapidly expanding aluminium industry to make synthetic cryolite which acts as an electrolyte in aluminium production. Hydrofluoric acid is also used for the production of chlorofluorocarbons (CFCs) as refrigerants and aerosol propellants. These uses of the latter are now banned owing to the damage caused by CFCs to the ozone layer. Small amounts of fluorine are also produced for use in the separation of uranium isotopes by diffusion as uranium hexafluorides.

Magnesium, calcium and alkali metal bromides occur in sea-water, the waters of the Dead Sea being an important source. They are also found in the Stassfurt

[8] H.C. Fielding and B.E. Lee, 'Hydrofluoric acid, inorganic fluorides and fluorine', *ibid.*, pp. 149–167.

salt deposits. Average sea-water contains 0.065% of bromine, the Dead Sea contains 4.8% and the Utah Salt Lake contains much more. In many modern processes elemental bromine is extracted using a mixture of chlorine with steam and most of the bromine produced since 1928 has been used to manufacture ethylene bromide as an additive for motor fuel, to remove lead deposits in the engine as the relatively volatile bromide. This use is rapidly declining in the UK (though not in some developing countries) with the dwindling numbers of cars using leaded petrol. Other uses of bromine include production of methyl bromide, a fumigant used to control insects in the food industries, various dyes (*e.g.* the bromo-indigo dyes) and methylene chlorobromide in fire extinguishers.

Iodine occurs mainly as iodides and iodates in sea water, in certain magnesium limestones, calcium phosphate and rock salt. Salt brine from certain oil wells contains iodides and the mother liquor remaining after the crystallisation of sodium nitrate from caliche (crude Chile saltpetre) contains sodium iodate; in both cases the quantities available warrant commercial extraction. For a long time in Britain and France iodine has been extracted on a commercial scale from seaweed, a process which has more recently developed in Japan. Soluble iodides find various uses in medicine, both for treating diseases and in antiseptic compounds; silver iodide is used on a large scale as the basis for many photographic materials.

## (h)  Coal and its products

In the 19th century and until the 1950s coal was the main fuel used for domestic and industrial heating, for raising steam, making electricity and by the railways. It was also used in large quantities for the production of town gas. Coal tar, a major by-product of the gas industry was the basic raw material for the organic chemical industry until, with the advent of petrochemicals from the oil industry and of natural gas, coal rapidly declined in importance as a raw material for chemical industry. Its main use now is as coke in the extraction of metals from their ores. Coke ovens produce large quantities of coal gas which is used as an industrial fuel to heat the coke ovens and other furnaces. Coal is also used by some electricity generating stations, although the cost of eliminating atmospheric pollution now renders the use of raw coal as a fuel uneconomic and it is replaced by oil, natural gas and other cleaner fuels.

## (i)  The petroleum industry

The petrochemicals industry, using chemicals produced from natural gas and the distillation products from crude oil, is a relatively recent sector of the heavy chemical industry (Chapter 9). As a naturally occurring material crude oil varies widely in composition and consistency depending on its origins, although it is always principally a mixture of hydrocarbons. The crude oil is refined by fractionation and the proportion of lighter hydrocarbons is increased by breaking down the heavier fractions in a catalytic cracker; aromatic compounds

*Pipes, towers and chimneys at Shell Haven (1950s)*

are obtained by 'reforming' and de-alkylation processes, followed by extraction in aqueous diethylene glycol and distillation to purify the product and recover the solvent. The many constituents of crude oil include alkanes, alkenes, benzenoid aromatics, naphthalene derivatives, polycyclic hydrocarbons, sulphur compounds such as hydrogen sulphide and thiophenes, alcohols, phenols, organic acids, pyridines, quinolines, indoles, pyrroles and inorganic compounds such as water, salt and sand. Crude oil also contains dissolved gases which escape when the oil reaches the surface. A large proportion of the gaseous low molecular-weight hydrocarbons also separate out in the ground and together with the oil there is often a reservoir of natural gas, consisting mainly of light alkanes like propane ($C_3H_8$) and butane ($C_4H_{10}$). Natural gas is also found separately, often in very large quantities.

The petrochemicals industry uses oil fractions to produce reactive compounds such as acetylene ($CH{\equiv}CH$), ethylene ($CH_2{=}CH_2$) and propylene ($CH_3CH{=}CH_2$) which form the primary materials for the production of polymers. Aromatics like benzene ($C_6H_6$), toluene ($CH_3C_6H_5$), xylenes [$(CH_3)_2C_6H_4$] and naphthalene ($C_{10}H_{10}$) form the feed-stocks needed to produce dyes, drugs, other chemicals and consumer goods. Up to the 1950s acetylene was made from calcium carbide and the aromatics came largely from coal-tar distillation, but with the rise of the petrochemicals industry, oil and natural gas have taken over almost entirely and the variety of primary organic chemicals from this source is now very great. Nevertheless, fermentation processes have long been used to provide certain unsaturated aliphatics such as alcohols, aldehydes and ketones and this remains a very important and growing source of these primary compounds in this sector.

*A reforming plant of the 1950s, treating 4.5 m litres of oil per day*

## (j)  Oils, fats and waxes

Oils, fats and waxes are obtained from animal and plant sources in considerable variety and a large sector of the chemical industry is concerned with extracting and purifying them. Plant oils such as sunflower and rapeseed oils are used in the manufacture of margarine, thus linking the chemical industry to agriculture and food technology. But the major use of oils and fats is in the manufacture of soaps for industrial and domestic purposes. Since soap forms scum (insoluble calcium and magnesium soaps) with hard water, other surfactants are also used for industrial and domestic purposes. These are products of the petrochemicals industry and are introduced to the domestic market to increase the quantities of detergents available and to supply a suitable domestic detergent for use in washing machines and dish-washers. The commonest synthetic detergents of this kind are alkylbenzene sulphonates, but fatty alcohol sulphates and other compounds are also used. Similar surfactants are employed for degreasing in many industrial situations. However, some synthetic detergents have been found to cause pollution problems because they are non-biodegradable and pass unchanged through sewage treatment works where they prevent bacterial action and so reduce the quality of the effluent. They reduce the solubility of oxygen in water and in rivers they have been found to kill plants and fish; they also cause lathering. These problems have contributed to adverse views about the chemical industry.

# 4 Chemicals and the Domestic Consumer

Although many of the domestic uses of chemical products will be encountered again in this book it may be useful to offer a brief and broad summary of the situation as it is today.

## (a) Dyestuffs

The manufacture of dyes is a very ancient craft. Early dyes were extracted almost entirely from plant sources; the range of colours was limited and the natural dyes were often not fast to light (Chapter 3). Nevertheless, dyes such as madder, indigo and saffron remained important throughout the 19th century, despite the fact that they began to be displaced by chemically manufactured dyes after William Henry Perkin discovered the aniline dye mauveine in 1856. With the introduction of new fibres and the demand for an improved range of fast dyes the industry developed rapidly from the late 19th century. It has become an important sector of the modern chemical industry and there are now many hundreds of different dyes to provide both the wide range of colours and the variety of chemical properties needed by the different natural and synthetic fibres now in use (Chapter 8).

The chemistry of dyestuffs and the dyeing industry is complex, but a brief survey will serve to indicate the highly technical nature of this sector of the chemical industry. Cotton and rayon, both composed of cellulose, can be dyed by adsorption or reaction of the dye with the cellulose molecule. It is also possible to use a mordant such as alum to bind the dye to the fibres. Wool and silk, on the other hand, are protein fibres the molecules of which contain both acidic and basic groups. They can therefore be dyed directly by acid and basic dyes, both of which form ionic bonds with the fibres. Nylon is also a polyamide with properties similar to those of wool and silk, though it is more hydrophobic and has fewer acidic and basic groups. Nylon can be dyed in carefully controlled acid conditions and by dyes which bond to amino groups. Cellulose acetate fibres are even more hydrophobic than nylon and must be treated with dyes which have low solubility in water. These dyes become dissolved by the fibres while others are formed *in situ* occluded within the fibres. Polyesters like terylene, which are not like any known natural fibres, also have to be dyed by dissolving or occluding the dye in the fibres. The acrylics are also hydrophobic and do not resemble any natural fibres. They have to be modified, for example with sulphonic acid groups, to allow combination with basic dyes. Polyolefins are even more difficult as they are chemically inert. They are most easily dyed by introducing the colouring agent into the polymer before the fibres are spun. It must also be remembered that dyes themselves are chemicals which may alter the properties of the fibres, while the chemical conditions under which dyeing is carried out can degrade fabrics, reducing their quality and value unless carefully controlled.

### (b)  Pigments and fillers

The manufacture of pigments and fillers forms a very large sector of the heavy chemicals industry. Pigments are usually linked with paints and very large quantities of coloured materials are mined or manufactured for this purpose. When it is considered that protective coatings are applied to buildings both inside and outside, the girders and cables of bridges, motor cars, railway coaches and wagons, and the superstructures and hulls of ships, the extent of this industry becomes apparent, yet almost as much pigmenting material is used as fillers for rubber, linoleum, plastics and artificial leather, while large quantities are also used in enamels and ceramics.

The main pigments used are metallic oxides, sulphides, sulphates and basic carbonates. Prussian Blue (ferric ferrocyanide), Chrome Yellow (lead chromate) and some other compounds are also used. Among white pigments titanium dioxide is the most often used, though lithopone (30% ZnS and 70% $BaSO_4$) is also made in considerable quantities. White lead (basic lead carbonate) is still made for certain purposes as is zinc sulphide. The oxides of lead (litharge and red lead) are also used for protection against rust as well as their colouring power. Carbon black is used as a filler for rubber. All these substances are in common use in paints, plastics, rubber goods and ceramics.

### (c)  Polymers

The man-made fibres and fabrics, including various kinds of rayon and synthetic fibres like the polyamides (nylon) and polyesters (terylene) are well-known to the public. Demand for these materials by the fashion industry has resulted in considerable expansion of the textile trades and new synthetic fibres, or improvements to established ones are constantly being introduced. Many of these new fibres, like the acrylics, vinyls or polyolefins result from chemical research. The production of polymers using products of the petrochemicals industry has expanded rapidly in the present century (Chapter 9).

Plastics form a large sector of the chemical industry and, along with the development of synthetic rubber, this has offered the organic chemist a rich field for the discovery and invention of products with pre-determined chemical and physical properties. Plastics have revolutionised many aspects of everyday life in the 20th century; they have made available affordable, attractive and serviceable materials as substitutes for the more expensive wooden, leather, ivory, metal and glass products of earlier years. Celluloid was used for many small domestic items, but its most significant social effects were in amateur photography and the film industry which developed rapidly in the 1930s. Bakelite, a common household material invented by Leo Hendrik Baekeland about 1905, was used in electrical fittings and the cases of 1930s radio sets. After the war new plastics such as polythene, perspex and PVC were introduced with exaggerated claims for their properties. Some of these failed to fulfil expectations; they became discoloured and the surface of plastic work-tops for kitchens in hospitals, restaurants and the home became crazed with hairline cracks which could

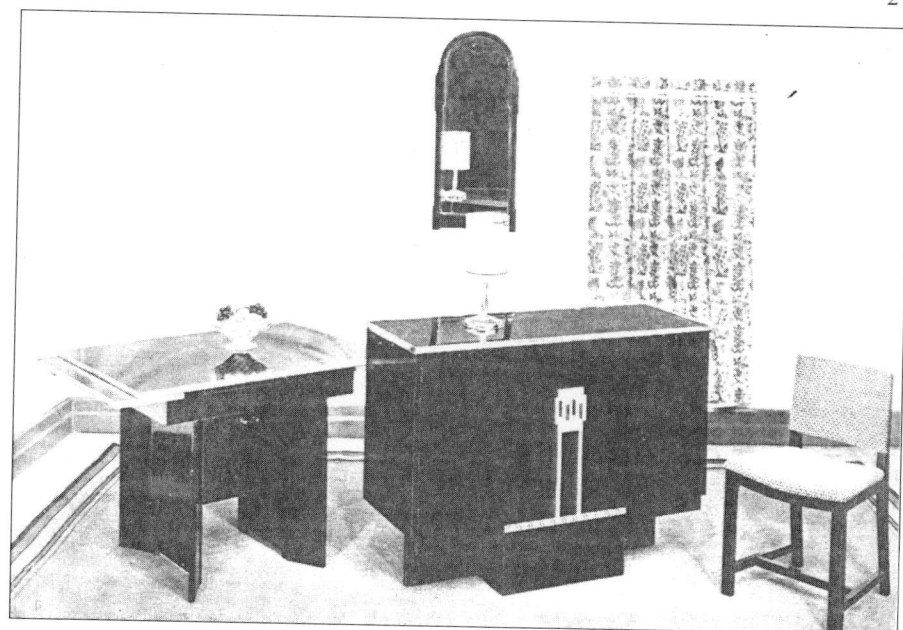

*Bakelite furniture, 1930s*

harbour germs. Others became brittle at low temperature and nearly all the new plastics of that era appeared to be poor substitutes for the natural materials they were intended to replace. Generally recognised as products of the chemical industry this damaged the public perception of the industry's products and, although plastics brightened up many working class homes, they were often thought to be 'cheap and nasty'. A great deal of research to improve plastic materials, together with persuasion and clever advertising, was required to overcome these problems and there is little doubt that the poor performance of these early plastics reflected badly on the public image of the chemical industry as a whole.

## (d)  Soap and detergents

The manufacture of soap and detergents is another very large industry founded on the applications of chemical principles and methods. Soap, a mixture of the sodium and potassium salts of fatty acids of high molecular weight, is made by saponifying animal and plant fats and oils by boiling with caustic alkalis. It has been known since early times, but in Britain demand for soap increased with the expansion of the textile trades during the 18th century. The disadvantage of soap as a detergent in normal urban conditions results from the calcium and magnesium salts which make most public water supplies 'hard'. Soap forms a scum of insoluble calcium and magnesium compounds in such water and with the introduction of the electric washing machine from the 1880s this problem became more acute. The development of the petrochemicals industry leading to the introduction of synthetic detergents offered a solution to the problem.

Used first in industry with trade names like 'Teepol' and 'Lissapol', synthetic detergents based on the sulphonates of alcohols of high molecular weight began to replace soap for some purposes in industry and the home from the 1940s; they were especially valuable for the removal of grease. As the new detergents were developed and successfully improved for the domestic laundry, soap advertising grew. The industry has had a powerful social impact in the present century and has indulged in such extensive advertising, especially by television, that it gave rise to the phrase 'soap opera' to describe any television drama serial in which the episodes are interspersed with commercial breaks filled with advertisements for soap products. The emphasis was on progress through improved convenience, domestic and personal hygiene and beauty. However, there has been little or no attempt to use these advertisements as a means of improving the public image of the chemical industry.

## (e)    Perfumes and pharmaceuticals

The manufacture of perfumes from natural plant and animal sources is a large-scale industry which has retained much of its traditional character. Although distillation techniques are widely used and externally the plant bears resemblance to other organic chemical plants, there is still a large element of craft skill involved. However, most of the perfumes in wide use are now manufactured by organic synthesis and perfumes form a notable part of the output of the organic fine chemicals industry. The products are used in a wide variety of contexts from cosmetics and soaps to aerosol sprays and washing powders.

Like perfumes, pharmaceuticals were originally extracted from plant and animal sources. During the 19th century with the rise of the German organic chemicals industry, new drugs and medicinal chemicals began to be manufactured from coal-tar distillates and other sources and it was only from the 1920s that the large-scale manufacture of pharmaceutical chemicals began to develop in Britain, America and other countries. Even then most of the early products were copies of German drugs, supplies of which had been cut off owing to the war. This situation changed about 1943 when penicillin was introduced. The benefits were so great that after the war the pharmaceutical industry set out to discover other such wonder-drugs. Their endeavours extended the activities and techniques of the chemical industry well beyond chemistry alone. Biologists, botanists, microbiologists, pharmacologists, pharmacists, clinicians and physicians co-ordinated their research efforts with chemists in an integrated effort to discover new drugs for specific diseases.[9] The pharmaceuticals industry has become a very large and important, if highly specialised, sector of the modern chemical industry (Chapter 6).

[9] R.B. Smith, *The Development of a Medicine*, Macmillan, London, 1985.

### (f)   Agrochemicals

The manufacture of chemicals for use in agriculture and horticulture is another large-scale operation. Chemical fertilisers are widely used to increase crop yields; pesticides include herbicides to control weeds, insecticides to kill insects which damage crops and fungicides to protect growing crops from plant diseases. Stored grain, fruit and vegetables are also treated to minimise damage and losses and to control ripening and storage qualities. Despite all the efforts made to reduce losses, billions of pounds worth of growing crops are destroyed annually by pests and it is clear that the use of these chemicals is necessary, despite the fact that they are all highly toxic. In the late 19th century there were hardly any controls on the sale or use of such poisons as copper or lead arsenates, pastes containing red phosphorus, strychnine, thallium and mercury compounds, sodium fluoride and cyanide for use in the garden or on the farm. In 1919 some cases of arsenic poisoning occurred in England due to eating American apples which had been treated with arsenious oxide. This led to the establishment of legal levels of tolerance, first for arsenic and later for other poisons used as pesticides. From the late 1920s regulations controlling the sale and use of pesticides were progressively introduced, until they are now amongst the most thoroughly supervised of all chemicals and drugs.

Insecticides and fungicides are frequently inorganic compounds, such as copper and lead arsenates, sodium fluoride and the fluosilicates and aluminates (Chapter 7). Sulphur is also used both as the element and in various compounds such as dithiocarbamates. Organic pesticides from oil include herbicides, and solvents or carriers for insecticides such as pyrethrins. There are also plant derivatives such as nicotine, pyrethrum and rotenone from the derris plant, but the greater number of pesticides consists of synthetic organic compounds. Among these DDT [1,1,1-trichloro-2,2-bis-(*p*-chlorophenyl)ethane] has become the most notorious. When first introduced in 1943 to destroy lice it was so successful that after the war farmers began to use it liberally as a universal pesticide. In the 1960s it was remarked that with the destruction of insects the birds which live on them had also been decimated. When DDT was identified in the meat and milk of dairy cattle and other food animals its use was first strictly controlled and later banned.

In addition to these core industries there is also a very large manufacturing sector which uses chemical products and methods in the manufacture of *consumer* goods. This sector consists of secondary chemical industries which exert a very important influence on the public perception of the chemical industry as a whole because their products are in daily use and are consequently much better known.

## 2.5   Defining the Chemical Industry

Since the chemical industry consists of so many related but distinct parts, it is not easy to establish criteria which will enable us to recognise with any degree of certainty what does and what does not belong under its umbrella. To take just one example, the differences between polymers and agrochemicals are so marked that they must be considered as quite separate industries. The question is whether they both belong in the same degree to the chemical industry? Clearly this is debatable and the difficulty is to decide where to draw the line between 'true' chemical industry and peripheral industries which, though they use chemical methods and products in manufacturing consumer goods, should not be regarded as part of the chemical industry's core. In practice it seems there are no generally accepted rules. Thus, most would accept that the manufacture of dyestuffs is part of chemical industry and many would include the paint industry, but relatively few would consider the manufacture of synthetic fibres, soaps and detergents or the extraction of metals. What valid definition, however, could include heavy chemicals, fertilisers and explosives while excluding polymers and fine chemicals?

### (a)   Classification by product-types

Given this complex picture let us consider how chemical industry might perhaps be defined in terms of its products. The American Bureau of Census has introduced the description 'chemical and allied products'; it includes nine categories:[10]

1. Industrial inorganic chemicals
2. Industrial organic chemicals
3. Drugs and medicines
4. Soaps and related products
5. Paints and allied products
6. Fertilisers
7. Gum and wood chemicals
8. Vegetable and animal oils
9. Miscellaneous chemical products

To be even more comprehensive we could include industries which use chemicals and chemical processes in the course of their manufacturing activities under the heading 'chemical process industries'.[11] This would include the separation and purification of chemicals from natural raw materials. These processes also employ engineering techniques, mechanical or electromagnetic methods of separation, machines for crushing rocks and ores, mechanical agitators for dissolving or lixiviating soluble material from natural mixtures, filter presses and other mechanical devices. Sublimation and distillation are in common use; electrolytic methods are also quite widespread. In petroleum refining, besides fractionation of the crude oil, catalytic

---

[10] J.A. Kent (ed.), *Riegel's Industrial Chemistry*, Van Nostrand/Reinhold, New York, 1962, p. 1.
[11] R.M. Stephenson, *Introduction to the Chemical Process Industries*, Reinhold Pub. Corp., New York, 1966.

cracking is used to increase the proportion of low boiling fractions in the product.

The manufacture of chemicals for use in other processes includes the synthesis of materials from these purified raw materials. The products, which generally have specific properties required by industrial customers, are no more than intermediates to be used by other large industries whose end products may be consumer goods. The inclusion of all the chemical process industries would also extend the bounds of the chemical industry to embrace paper and glass manufacture, metal extraction and refining, fabric treatment and finishing, tanning, brewing and a host of other industries which are dependent upon, but peripheral to the chemical industry. The demands made by these chemical process industries are important to the economic health of the chemical industry itself. They support manufacturing activity and by their demands for new products they stimulate research and development. The picture of the chemical industry which begins to emerge, therefore, is one of a hard core in which the products are primary chemical compounds, linked to zones of manufacturing which depend on the chemical industry, but have decreasing claims to be considered part of it as they are further removed from the manufacture of chemical products and instead are engaged in using them. According to this model the chemical industry is a multi-layered system with central and peripheral components as in Figure 1. The demarcation lines are blurred and there are many 'grey' areas

## (b)   Classification by techniques

If this appears to present an ill-defined identity for the chemical industry, other criteria might be considered in a bid to achieve better definition. For example, we might try to build a definition of the chemical industry on the basis of *chemical* techniques which link its different branches. These are procedures involving the chemical transformations of matter, beginning with the extraction and purification of raw materials from natural sources. The extraction of oxygen, nitrogen, argon and other gases from the atmosphere is an obvious example, and so are the processes by which common salt is mined and purified, sulphur is extracted from the earth, organic chemical feedstocks are obtained by distilling crude oil or from natural gas, while salts may be extracted from sea-water. Useful primary chemicals are manufactured in bulk from these raw materials; alkalis, hydrochloric acid and chlorine from salt, sulphuric acid from sulphur, industrial gases from coke, organic chemicals from petroleum distillates and natural gas, or wood, and sulphur obtained from crude oil or from the gases produced in smelting sulphur bearing ores. These heavy chemicals would be at the heart of the chemical industry since chemical methods are used to make them. So also would be the extraction of most metals from their ores, though these are generally excluded (probably on the basis that metallurgy is a large enough subject on its own!).

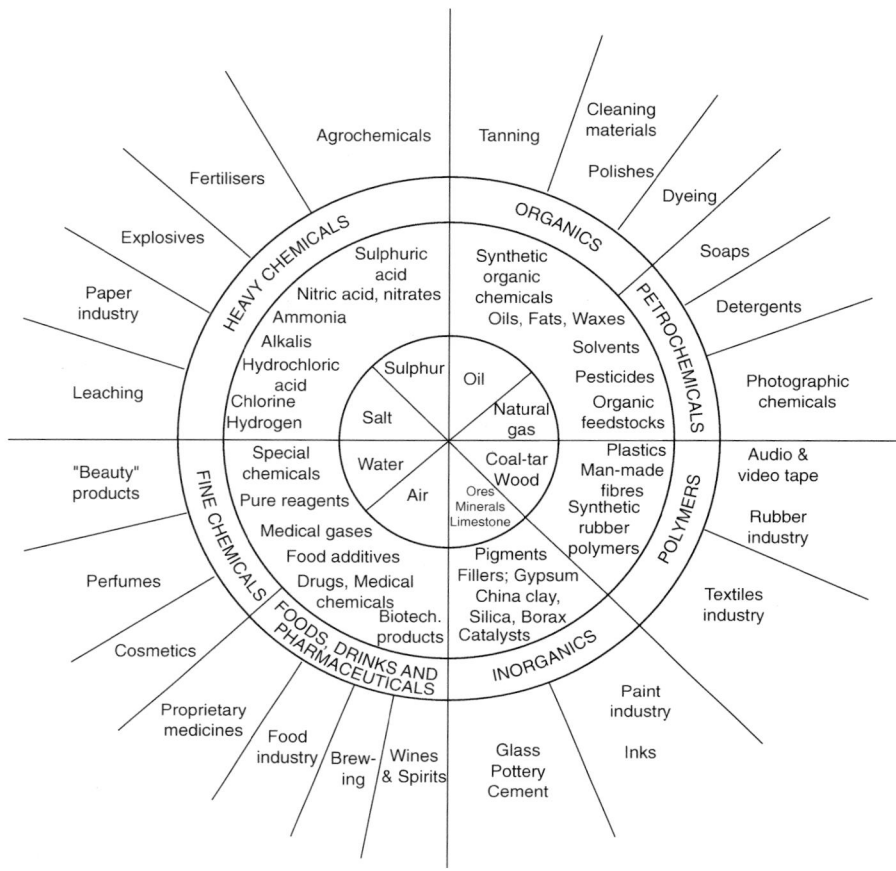

**Figure 1**  *Relationships in the chemical industries*

The manufacture of fine chemicals and pharmaceuticals, involving the synthesis of complex substances from simpler materials in a high state of purity, clearly occupies an important place in any definition of the chemical industry and there are also those industries in which chemical products and processes are employed to manufacture consumer goods such as polymers, fibres, paper and glass. All of these use chemical techniques and must be covered in a wider definition of the chemical industry.

Biochemical methods such as fermentation are increasingly used in chemical manufacturing, but here again we come into conflict with tradition. Fermentation is essential in other industrial processes like wine and vinegar manufacture, brewing and bread-making, none of which belong to the chemical industry. In modern industry there is an increasing use of biochemical and biological methods. Micro-organisms drive fermentation processes which produce a number of chemical substances including not only ethyl alcohol and acetic acid, but also food additives and a range of therapeutics and pharmaceutical chemicals such as antibiotics and vitamins. These biochemical processes produce many compounds which are not obtainable in commercial quantities by purely chemical methods. The first commercial fermentation process to require aseptic conditions was the butanol–acetone process developed in the first world war to produce acetone for the manufacture of cordite.

Thus, although it might appear at first sight that a definition of the chemical industry based on the use of chemical techniques could be established, in practice there will always be room for debate about which category to choose for many industries. Decisions are arbitrary and influenced by traditional attitudes. It therefore seems unlikely that a clear-cut definition of the chemical industry which could be universally accepted can be derived from these criteria either and we find ourselves forced to the conclusion that no classification is wholly satisfactory and capable of yielding a completely unambiguous definition.

# 6   Chemical Industry and Economic Viability

## (a)   The economic feasibility of chemical waste recovery

Chemical industry requires heavy capital investment in land, plant and machinery, raw materials, research and development and labour. To ensure profitability a complex web of factors must be considered. These include physical and chemical constraints, plant maintenance and labour costs, the amounts to be spent on research and development, the cost of raw materials and the price the market will bear for the product, along with other economic and market forces such as taxation and undercutting by the competition.

On the technical side all chemical processes have features which must be taken into account in the design and operation of the plant. In accordance with the laws of conservation of matter and energy, *all* the raw materials and energy supplied to any process must reappear in the products. In an ideal situation *all* the initial reactants would be transformed into saleable products and there would be no waste. In practice this is rarely if ever achieved. However nearly a process approaches the ideal, there is usually some residual waste, but by-products 'lock up' valuable raw materials and it is economically desirable either to find markets for the by-products or to convert them into more useful substances.

Chemical processes always involve energy changes, usually, though not always, in the form of heat. In exothermic processes the heat evolved is as much a by-product as the material ones. Exothermic processes can sometimes be made self-sustaining by using the heat evolved, but in any case the heat must be conserved if the economic viability of the process is to be maintained. Energy must be paid for and wasted energy represents financial loss just as wasted materials do. Thus, what began as a process to obtain a particular product, quickly grows into a complex of inter-related processes which seek to maximise the use of the raw materials in saleable products whilst minimising waste and conserving energy. If the chemical manufacturer is to maintain competitiveness in world markets it is essential to ensure the most efficient and economical use of raw materials, energy and intermediate products.[12]

However desirable, the recovery of usable materials from waste products is sometimes uneconomic and providing the waste does not cause environmental damage it may be cheaper simply to discard it. The manufacture of titanium dioxide pigment used in plastics, rubber and paints[13] is an example of this. The common ore of titanium is ilmenite (ferrous titanate, $FeTiO_3$), which usually contains no more than 20% of titanium dioxide. There are several methods of extraction in use but the commonest involves converting the ore into a mixture of titanium and ferrous sulphates. The waste products of this process amount to about seven times the yield of titanium dioxide by weight and since the quantities of $TiO_2$ produced annually are of the order of 2–3 million tonnes the total quantity of waste is very great. In theory it would be possible to extract the sulphur or convert the waste into saleable products, but the processes so far available are uneconomic. Titanium dioxide plants have instead been sited on fast-flowing rivers or near the coast so that the waste can be discharged directly into the river or dumped from barges into the sea. The results have been closely monitored and so far there seems to have been little measurable effect on the environment. Nevertheless, the ethics of the practice seem questionable and it will probably not continue to be acceptable as European Community controls on environmental pollution are tightened.

Sometimes a dangerous waste product for which there is no obvious use is produced in large quantities as an inevitable by-product of an industry. For example, in the extraction of non-ferrous metals such as copper and zinc, arsenious oxide is a by-product. Very large quantities accumulate since there is little use for this substance. It is virtually impossible to render this poisonous material environmentally 'safe', or to convert it to useful purposes. Instead it must be held in secure conditions indefinitely until a solution to the problem of disposal is found. In the meantime the costs of maintaining secure storage fall on the non-ferrous metals industry.[14] Even more difficult problems arise over the treatment and disposal of nuclear waste. Compared with the quantities of waste produced by other industries, the amounts of nuclear waste are quite small, but the technical problems of collection, transport, treatment and disposal are very considerable. The long-term effects of radiation from nuclear waste are still uncertain and the environmental issues involved are very sensitive.

Re-processing spent nuclear fuel involves the separation of unwanted fission products by treating the fuel with nitric acid. The actinide nitrates are then extracted from aqueous solution using an organic solvent such as tri-n-butyl phosphate in liquid hydrocarbons. A large re-processing plant can treat 7 tonnes of irradiated fuel per day and after extracting the uranium and plutonium the remaining solution is still highly radioactive. This is partly evaporated and

[12] Kent (ref. 10), Chapter 1.
[13] R.S. Darby and J. Leighton, 'Titanium dioxide pigments' in Thompson (ref. 2), pp. 354–374.
[14] See Chapter 10.

stored in double walled stainless steel tanks shielded by several feet of concrete. The contents of this waste include elements such as aluminium, iron, chromium, nickel, manganese and zinc which come from the cladding of the fuel rods and the processing vessels. The waste emits appreciable quantities of heat and requires storage for many years with constant surveillance. For radioactive waste with very long half-lives a more permanent method of disposal is needed. The only viable method which seems certain to hold the waste intact for very long periods is vitrification in a glass or ceramic matrix followed by storage in water-filled ponds for several years and ultimate disposal in the deep ocean, under the ocean floor or in some geological formation which is likely to remain untouched by human contact for the foreseeable future. Such secure containment for several hundred thousand years is considered necessary.

In view of the difficulties of solving the problems of nuclear waste disposal, public concern about nuclear fuel re-processing is understandable, yet it may be useful to reflect that nuclear waste is not all of long half-life, its lethal potential will ultimately decline and the precautions taken to avoid leakage into the environment are extreme, whereas the arsenious oxide produced by the non-ferrous metals industry can only increase in quantity and will never lose its potency. In all such areas the chemical industry faces difficult ethical and moral issues (including those of the environment) which must be balanced against complex technical and economic concerns.

Chemical industry employs ingenious methods for using the cheapest available raw materials and linking processes so that the waste products of one process form the starting point for another. Thus a chemical works becomes a complex of buildings, plant, machinery, energy sources and raw materials all of which require capital investment. Financial gains are made by increasing the efficiency of the plant, improving the product and reducing the human input as far as possible. Throughout the 19th century the chemical industry was labour-intensive, but the automation of chemical plants and more recently computer control of continuous production processes has reduced the need for unskilled manual labour. As the industry has developed a highly trained workforce, efficient management and a sophisticated marketing structure has become essential.

## (b)    Research and development

In common with all high-technology industries the chemical industry invests heavily in research and development (R & D).[15] At any one time in a chemical works of moderate size a number of projects is being examined or developed. Some of these are short-term efforts to improve the efficiency of a process or the quality of a product. These are usually undertaken within the company itself. For example, it may be necessary to develop a manufacturing process for a new product introduced to meet the specific needs of a customer. Others involve a

[15] F.R. Bradbury and B.G. Dutton, *Chemical Industry: Social and Economic Aspects*, Butterworths, London, 1972, Chapter 7.

*An early industrial chemical laboratory, established by UAC at Widnes in 1891 for F. Hurter*

longer-term exploratory programme, such as the search for a new catalyst or additive to speed up a process or improve the yield. Such long-term research, although directly linked to the requirements of industrial success, is closer to the 'pure' research usually associated with university science and it may be sponsored, especially by a large organisation, in a suitable university chemistry department. But the company may be at risk when industrial secrets are involved.

Chemical R & D needs a considerable degree of freedom and in some respects it may almost be considered a business within a business. It absorbs significant financial resources and each project must be evaluated before deciding to allocate funds to enable expensive development work to proceed. Successful R & D is the fundamental source of the firm's prosperity and very close links must be maintained with other parts of the chemical works, including not only the actual operation of the processes, but also the marketing strategies of the company so that any research carried out can be geared to future production and marketing plans. The final decision to pursue any research project is always a financial one; unless it is likely to result in a promising new product or a cheaper method of manufacturing an existing one, the expenditure of resources may be wasted.

Balancing long- and short-term projects is also difficult. Long-term projects are more uncertain than short-term ones since it often cannot be known whether or not there will be a future market for some new product. Changes in costs and future customer demands are extremely difficult to estimate and cost–benefit analysis is much easier to carry out on short-term projects. Nevertheless, R & D is an essential commercial activity without which a chemical company would quickly find itself at a competitive disadvantage.

The amounts spent on R & D vary widely. In the manufacture of heavy chemicals changes to the processes are slow and relatively little research is required, but in the manufacture of fine chemicals, especially pharmaceuticals, where innovations are rapid and one product is quickly superseded by another, the need for intensive R & D is much greater and expenditure upon it is correspondingly higher. Thus, while Shell Chemicals may spend only around 2.5% of its income from sales and the proportion for ICI is around 4%, Glaxo by contrast spends nearer 11–12%.[16] (Research in the pharmaceutical industry is a special case discussed more fully in Chapter 6.) Small firms may buy in research effort from larger ones, though there is always the danger of losing confidentiality and thus gaining a bigger and more powerful competitor.

## 7  The Chemical Industry and its Work Force

In the mid-18th century when an embryonic British chemical industry was required to meet the demands of a rapidly growing textile industry, the chemical product in greatest demand was soda. Even with the advent of the Leblanc process and during the 19th century alkali manufacture remained dirty, inefficient, wasteful and labour intensive. A similar situation prevailed in the manufacture of sulphuric acid by the old lead chamber process. The production of these chemicals together with alum, copperas, red and white lead, vitriols, lunar caustic and a few others[17] were crafts operated by artisans and the special skills for each process were passed down from one generation to the next through apprenticeship, just as they were in other trades. Workers in the British chemical industry in the 18th and much of the 19th century needed physical strength and works experience rather than knowledge or intelligence. Innovations were introduced as a result of trial and error. The invention of bleaching powder is a good example. In 1810 Charles Tennant in Glasgow observed that slaked lime would absorb chlorine to form 'chloride of lime' which did not give off much chlorine and could be used as a safe bleaching agent instead of chlorine gas itself. The discovery of bleaching powder was purely empirical; it did not require a knowledge of theoretical chemistry and could have been made by any observant artisan.

From the first quarter of the 19th century onwards, however, chemistry and with it chemical industry began to take on a different aspect. Demands for purer, better quality products in larger quantities led to the need for something more than craft skills alone. A grounding in chemistry was becoming necessary for the new generation of chemical operatives. The mechanics institutes which had been established in the 19th century to educate artisans in science had failed – what was needed was an organised system of education beyond school with an

[16] Heaton (ref. 1), p. 90.
[17] See Chapter 3.

examination structure to provide incentives and a method of measuring standards of attainment. For ordinary chemical operatives in the mid-19th century, attendance at night school in the new technical colleges was becoming essential. This was especially true in the new field of organic chemistry which entered the chemical industry with the introduction of the distillation products of coal-tar. In the British chemical industry this amounted to no more than the purification of products like naphtha, paraffin wax and phenol, but in Germany the beginnings of synthetic organic chemistry were already evident before the middle of the century. The need for research into the chemical properties and reactions of these organic products was becoming apparent and in Germany the chemical industry began to seek university trained chemists.

Before 1826 there was very little opportunity of obtaining a chemical education as such in Britain. The Scottish universities had long included chemistry as an auxiliary subject in their medical curricula, but even this was largely concerned with the preparation of medicines. When University College London was established in 1826 chemistry and physics appeared in the curriculum and courses in practical chemistry for medical students were offered; this was emulated by King's College soon afterwards. Other courses in practical chemistry could also be found in the 1830s and 1840s scattered across various Institutions like the Putney College for Civil Engineers, the Royal Polytechnic Institution and the Pharmaceutical Society, but the first attempt to provide a thorough chemical education along German lines came in 1845 with the establishment of the Royal College of Chemistry. The first professor August Wilhelm Hofmann had been trained as a research chemist by Justus von Liebig in the small German university of Giessen. Liebig's students had already become established as teachers of chemistry in many other German universities and foreign students, including some from Britain, went to study with Liebig to learn his analytical and research techniques in organic chemistry. With the arrival of Hofmann at the Royal College of Chemistry, Liebig's methods were introduced into Britain some twenty years after they had begun in Germany.[18] This delay put British chemical industry at a disadvantage in comparison to Germany, particularly in the organic field. Thus, when W.H. Perkin, a pupil of Hofmann, discovered the aniline dye mauveine in 1856 and began to manufacture it, the search for other synthetic organic dyes was on, but it was ultimately taken up in Germany not in Britain.

In 1841 London chemists established the Chemical Society as a forum in which to meet and discuss chemical issues. Papers were to be published, lectures given and the Society was to be open to all who were interested in chemistry, whether or not their livelihood depended upon it. In fact the majority of members of the Chemical Society were engaged in university or other professional work. Most chemists employed in industry were educated to lower standards and were unable to benefit from the Chemical Society's academic papers and discussions. But, as more chemists were employed in chemical

[18] G.K. Roberts, 'Chemical training before 1877' in C.A. Russell, N.G. Coley and G.K. Roberts, *Chemists by profession*, Royal Institute of Chemistry/Open University Press, Milton Keynes, 1977, pp. 75–77.

*The Royal College of Chemistry, established in 1845*

industry it became apparent that an organisation was needed to protect their interests and in particular to establish qualified chemists as members of a profession with recognised standards of education and competence. There was also the problem of distinguishing the industrial and professional chemist from the 'chemist and druggist' and from the pharmacist who also called himself a chemist. These latter groups had arisen from the long established profession of the apothecary, who often also had some medical competence. With these ideas in mind, some members of the Chemical Society proposed the establishment of an institute which would offer chemical education and establish professional standards of attainment and qualifications by examination. In 1877 after a long and involved arguments and counter-proposals the Institute of Chemistry was finally established.[19]

[19] *Ibid.*, pp. 135–157.

While the early aims of the Institute were concerned with analysis, especially water analysis, pressures from chemists in industry ensured that the qualifications of the Institute would be recognised as of a high standard. The Institute received its Royal Charter in 1885 and its aims and activities became more academic as it strove to establish the Associateship as the equivalent of a good honours degree in chemistry, though with a strong practical bias. The RIC set out to establish the profession of the chemist in industry and to protect the interests of its members. On the one hand links with chemistry departments in British universities were forged, on the other the RIC established the credibility of its qualifications with chemical industry. In these ways the professional status of the chemist in industry, whether in R & D, analytical work, as a works chemist or in marketing was assured. While individual chemists benefited from the support and activities of the RIC, chemical manufacturers also felt the need for mutual support which the RIC was not able to offer. This need, first expressed by members of the Newcastle Chemical Society, led in 1881 to the establishment of the Society of Chemical Industry (SCI). As chemical matters began to impinge more often upon political and social issues, public bodies, including the Government, increasingly required technical advice. Unfortunately, with their different emphases the Chemical Society, the RIC and the SCI did not speak with one voice and it was not always clear which of these bodies to approach for the most reliable advice. Moves were therefore made from the late 1960s to amalgamate the Chemical Society, the Royal Institute of Chemistry and one or two other bodies. After long and difficult negotiations the Royal Society of Chemistry was established in 1980. The Society of Chemical Industry remained aloof and continues as an independent organisation.

The diversity of the chemical industry is reflected in the variety among its workers. These range from highly qualified and experienced research directors to unqualified manual workers. In the modern industry staff trained in operating computer and other automatic control systems are employed, as well as mathematicians, physicists, biochemists and others. At the highest level the Board of Directors, as in other industries, will usually be dominated by accountants and economists, but it is also essential to have well-qualified chemists on the Board since only they know what is feasible on the practical level. The works management and leaders of process teams will be experienced industrial chemists, but manual workers trained to a lower standard are still needed for many day-to-day operations.

All who work directly with chemicals must obey essential safety precautions, both for their own protection and that of their colleagues. Chemical products include many poisons and corrosive substances which require workers to use protective clothing to prevent contact with the skin. Masks are often needed to prevent respiratory damage. Hot liquids and gases, often under pressure, circulate in the plant, and fire hazards and other potential dangers must be recognised and treated with respect. Apart from the dangers inherent within chemical works, there are also the problems of the effects these works have on the immediate surroundings. There are dangers from chemical poisons not only to the workforce but also to the public whether or not they live near chemical plants. It is extremely difficult to avoid entirely the escape of gases and vapours into the atmosphere. This means that in the vicinity of chemical works there will often be smells, acid vapours or fumes in the air. Escape of effluents is easier to control, but it too is difficult to avoid altogether and streams or rivers near chemical works are usually more or less polluted. Besides these disadvantages there is always the risk of human error or plant failure resulting in fires, explosions, massive escapes of noxious gases or vapours which may well be lethal. In recent times the chemical industry has been subject to some serious accidents, several of which have reached disaster proportions and have further tarnished its environmental reputation (Chapter 11).

More recently there have been determined efforts to overcome these problems, either to comply with Government or European Union legislation, out of economic necessity, or from a genuine desire to improve its public image.[20] Modern reaction vessels are usually sealed and the continuous operation of large-scale chemical processes is automatically controlled by electronic methods, which leave the factory floor almost clinically clean. Regulations now demand that the escape of acid vapours into the atmosphere from sulphuric acid plants must be less than 100 parts per million and the aim is to reduce that still further.[21] Quite apart from the necessity to avoid polluting the environment, as we have seen it makes economic sense to use the initial reactants and energy supplied in any chemical process to the fullest possible extent, avoiding waste as far as possible.[22] Although waste, accidents and other disasters cannot be denied, they must be seen in the context of the vast size and complexity of the industry and the large number of industrial processes producing toxic chemicals which are carried on daily without mishap and with the minimum of pollution.

[20] Heaton (ref. 1), Chapter 8; M.E. Trowbridge, in Thompson (ref. 2), pp. 30–45.
[21] A. Phillips, 'The modern sulphuric acid process', in Thompson (ref. 2), pp. 183–200.
[22] N.G. Coley, 'Removing mountains: industrial waste and the environment', in M. Fetizon and W.J. Thomas (eds.), *The Role of Oxygen in Improving Chemical Processes*, 6th BOC Priestley Conference, Royal Society of Chemistry, Cambridge, 1993, pp. 123–130.

Chapter 3 ## *Origins of the British Chemical Industry*

N.G. COLEY

## 1 Pre-history

The origins of chemical industry lie in pre-historic times; evidence of chemical activity is found in archaeological discoveries, ancient manuscripts, art, and artefacts. Using these sources we may speculate about the uses of chemistry in antiquity. Heat and fire were used in preparing food, making charcoal or pot-ashes from wood, firing pottery, making glass, working metals like bronze or iron and other purposes. Fermentation appears to have been among the earliest of chemical processes in common use; brewing was known by nearly all ancient peoples, just as today most primitive tribes have methods of making intoxicating liquors. It has always been necessary to combat disease and relieve suffering, and though there was a strong element of magic and religious belief involved in attempts to effect cures, medicines were prepared from certain plants from very early times. Cosmetics were also known to Egyptian, Greek and Roman civilisations. Tanning, the treatment of animal skins with lime, dung, oak galls and bark to make leather for clothing, footwear, animal harness, tents and other domestic items was also practised from ancient times. Colouring matter for dyeing cloth was extracted from the flowers, stems and roots of certain plants at least as early as 2000 BC, while coloured earths and minerals were used as pigments in paints, pottery, coloured glass and cosmetics, by the ancient Chinese, Sumerians and Egyptians. Many of these processes are depicted in ancient Egyptian tomb-paintings dating from about 4000 BC and there is archaeological evidence of even earlier technologies such as brick and tile-making, but secrecy and the belief that the everyday work of slaves was not worthy of notice has ensured few written records, and details of the methods used in these ancient crafts are scanty. Yet there is no doubt that chemistry and chemical techniques have played a significant role in human affairs from earliest times.

Alchemists, in their futile search for the so-called 'philosophers' stone', which would turn base metals into gold, practised a crude metallurgical science using furnaces, hearths and cupels. Various forms of distillation and sublimation for the extraction of 'spirits' were used and alchemical techniques also contained the germs of later methods used to assay ores and precious metals.[1] Sometimes

[1] E.J. Holmyard, *Alchemy*, Penguin, Harmondsworth, 1957, pp. 41–57.

*Distillation: important from alchemical times to the present day* (from Biringuccio's *Pyrotechnica* of 1540)

*The alchemist, by David Teniers, 17th century*

called 'puffers' from their use of bellows to drive their furnaces and stills, the alchemists polluted the air in the vicinity of their laboratories with smoke and noxious vapours. For them, though not necessarily their neighbours, the desirability of their ultimate goal outweighed the disadvantages of their methods.

Despite the secrecy and mysticism of their craft, alchemists were nevertheless engaged in a practical enterprise and their experiments revealed useful chemical compounds such as *lunar caustic* (silver nitrate), *sugar of lead* (lead acetate), *butter of antimony* (antimony chloride), *sal ammoniac* (ammonium chloride), *aqua fortis* (nitric acid), *aqua regia* (a mixture of nitric and hydrochloric acids) and the vitriols. They discovered the chief properties of these and other chemical substances; they also devised useful furnaces and reaction vessels. Sublimation and distillation figured largely in their work and this led to the invention of special apparatus like the retort and several varieties of stills, including primitive fractionating columns for separating the components of liquid mixtures. Through these activities the alchemists made valuable contributions to early chemistry and many of the alchemical names for chemical substances together with some of their ideas about reactions and chemical combination formed the basis for early theories and practices in chemistry. Thus some of the origins of chemical industry can be found in the jumble of alchemical writings from which, like the metals themselves, they were later extracted and refined.

## 2  Early British Chemical Industry

In Britain vestiges of chemical industry date from the Roman occupation (55 BC–*ca.* 450 AD) and even earlier.[2] Urban historians and industrial archaeologists have discovered traces of chemical industry relating to the preparation of dyes like woad, saffron or madder, glass making and the extraction of metals such as lead, copper and iron. There is also evidence that quite powerful drugs and medicines were being prepared from plants at this time. A few chemical substances like alum and copperas were made in sizeable quantities and in the industrial parts of many ancient British towns remains of tanneries, dye-works, glass-works, soap boilers and other related trades have been found. These were often clustered together and were mutually dependent on one another. Other proto-chemical industries were established near supplies of their raw materials. For example, kelpers were found in the Scottish Highlands and in parts of Ireland, iron founders in the Forest of Dean and the Sussex Weald. It was on the strength of such scattered 'cottage industries' and ancient crafts that the British chemical industry grew up from the middle of the 18th century when the foundations of the modern industry began to be laid. Expansion was stimulated by the growing demands of industrialisation. The iron industry required ever-increasing quantities of charcoal and limestone, the burgeoning textile trades required soap for scouring raw wool and dilute sulphuric acid for bleaching and dyes. Many of these products had to be extracted, or manufactured, from plant or mineral sources and increasing demand led to the rapid expansion of chemical industries. Indeed, it may be said with some justification that the British chemical industry was one of the main pillars of the 18th century Industrial Revolution, along with improvements in agriculture, the iron industry and the textile trades.

In discussing the evolution of technology in society Lewis Mumford identified a sequence of successive stages which he labelled eotechnic, paleotechnic and

[2] T.I. Williams, *The Chemical Industry*, Penguin, Harmondsworth, 1953 [reprinted EP Publishing Ltd., Wakefield, 1972], pp. 11ff.

neotechnic.[3] The first of these was characterised by the use of 'wood and water', the second by 'coal and iron' and the third by 'electricity and alloys'; in this chapter the first two phases concern us. Mumford's classification arose from an attempt to integrate the history of technology with social development and this is a significant concern of the present study, though we are also interested in the environmental effects of industrial developments. While Mumford's categories do not precisely fit the rise of the chemical industry, they were used by A. and N.L. Clow[4] in their classic study of early chemical industry in Scotland. Until the 19th century chemical industry, like other technologies, was operated by artisans and remained largely devoid of theoretical considerations; it belonged to the period of 'eotechnology'.

The siting of heavy chemical industries was determined by the proximity of raw materials such as limestone, coal, salt, sand and metal ores. Consequently, the industry has often been identified as the major cause of environmental damage resulting from mining and quarrying for these mineral resources. However, it should not be forgotten that many of these products were used for purposes other than chemical manufacture, or that environmental damage also arose from sources other than the chemical industry. For example, burning fossil fuels pollutes the atmosphere, though much of this was caused by domestic fires. Leaching, as in the alum and copperas industries and by tanneries, often polluted water-courses, but at least as much pollution resulted from the effluents from abattoirs. Thus, the chemical industry has never been the sole source of environmental pollution, nor arguably even the main one.

## (a)   Wood

Until the mid-18th century the most versatile material for a wide variety of purposes, both domestic and industrial, was wood. Large quantities of wood were used in the construction of buildings, in mining and shipbuilding. Wood in the form of charcoal was needed for smelting metals and as a constituent of gunpowder. As a fuel wood was used in lime-burning for mortar and in metal-working, but all processes which depended on heating, boiling liquids or firing furnaces depended on wood as the fuel. Wood products such as oak galls and bark were also used in tanning animal hides to make leather, but in addition to all these uses the destructive distillation of wood yielded pitch and tar for shipbuilding and the by-product of this process was 'pyroligneous acid', a kind of crude vinegar. Produced during the distillation of most kinds of non-resinous woods, the acetic acid in this liquid became very important towards the end of the 18th century when aluminium acetate began to replace alum and sugar of lead (lead acetate) as a mordant in dyeing. It was also used to make ferrous acetate for reaction with tannic acid to produce printers' ink. Lastly, the ashes left after burning wood and other vegetable matter contained *potash* essential for the manufacture of glass, soap, alum and saltpetre. With all these applica-

[3] L. Mumford, *Technics and Civilisation*, Harcourt, Brace & Co., New York, 1934.
[4] A. and N.L. Clow, *The Chemical Revolution*, Batchworth, London, 1952 [reprinted Gordon & Breach, Philadelphia, 1992], pp. xiii–xiv, and *passim*.

tions it is obvious that industry in general, and chemical industries in particular, were heavily dependent on wood and by the 18th century deforestation had become a major problem. With increasing demand supplies of wood and its products, especially charcoal and potash, became so scarce that it was necessary to seek imports.

## (b) Coal and coke

The British iron industry, which had been centred on the Sussex Weald, has a long history going back at least to the 15th century. By the mid-17th century the industry was in a decline until the growing demands for iron in the 18th century revived it. The Industrial Revolution brought about an increase in British iron production which outstripped the available supplies of charcoal and this encouraged the transition to coke. Already used in the brewing industry for drying malt and hops and in the production of brass, coke was introduced for iron smelting at Coalbrookdale by the Quaker iron master Abraham Darby in about 1709.[5] The iron industry developed rapidly in 18th-century Britain with a concomitant increase in demand for coking coal which was in abundant supply (Chapter 10). Coke production gave rise to by-products, most notably coal-tar, for which there was at first little or no use, and led to an increase in atmospheric pollution for which the chemical industry was to become notorious. During the 19th century coal displaced wood as the most important raw material and became the foundation on which British industry was built. It was not only a valuable fuel for raising steam, but was also essential for the production of iron and steel. With the increasing demand for coke, its by-product, coal-gas, began to be used for lighting and heating from the 1820s. Ammoniacal liquor, another by-product of coal-gas production was a source of nitrogen, useful as a fertiliser. Coal-tar, however, remained a noxious waste product until, with the discovery and isolation of its many components by fractional distillation, it became an essential source of primary materials for the developing organic chemical industry of the mid-19th century (Chapter 8).

## (c) Limestone (calcium carbonate)

An abundant native raw material, used in large quantities for building, agriculture and the metal and glass industries from the middle ages onwards, limestone has long been quarried in large quantities to be burnt with wood or coal in lime-kilns for conversion into quicklime:

$$CaCO_3 = CaO + CO_2$$

The main use of quicklime in ancient times was for lime-mortar, but from the 17th century it began to be used on the land in the form of slaked lime

---

[5] A. Raistrick, *Quakers in Science and Industry*, David & Charles, Newton Abbot, 1968, pp. 94, 158–159.

[Ca(OH)$_2$] which was found to improve fertility. In Scotland during the 17th and 18th centuries, lime-burning for agricultural purposes reached very considerable proportions. Limestone was also used as a flux in the extraction of metals from their ores. The charge in iron furnaces, even in Roman times, consisted of iron ore, limestone and charcoal. Glass manufacture used limestone together with soda and sand, fused together at high temperatures in a glass furnace. As these were all growing rapidly with industrial expansion in the 18th century, limestone was always an important raw material in the chemical and related industries.

## (d)  Barilla and kelp

Soda (sodium carbonate) and potash (potassium carbonate), while not themselves found native in Britain, were nevertheless required in increasing quantities from the 18th century onwards (Chapter 4). The traditional source of alkalis was the ashes produced by burning wood, kelp (seaweed) and other vegetable matter. As demand rapidly increased and suitable supplies of wood for making potash became scarce, it was necessary to look for other sources including imports. In Spain, salsola, a kind of seaweed, was cultivated for burning to make *barilla*, a crude alkali containing soda. Native soda, *trona* (Na$_2$CO$_3$.NaHCO$_3$.2H$_2$O), occurs as a mineral deposit in Egypt, and in Northern Europe large conifer forests provided wood from which potash could be obtained. All three sources were tapped by 18th century chemical manufacturers, yet there was still a shortage of alkali and further supplies had to be sought.

In Scotland William Cullen investigated the possibilities of burning seaweed to form *kelp*. He also tried bracken, nettles, thistles and other vegetable matter, such as ferns and the waste from felled timber, as possible sources of kelp. As wood ashes became scarcer and more expensive, kelping developed into an important industry in the Scottish Highlands and islands.[6] Kelp was found to be an excellent fertiliser and vast quantities were used in agriculture. In Ireland it was the source of alkali used in bleaching linen. Kelp is a mixture of salts, whose proportions differ according to the source, and each chemical manufacturer preferred to purchase a particular kind of kelp best suited to his needs. Samples containing as much as 25% of common salt were preferred by soap boilers; others contained a larger proportion of magnesium sulphate or potassium chloride and were used for glass making or in the alum trade. Scottish kelp was sold in large quantities to chemical manufacturers on Tyneside and Merseyside and in this way an extensive kelping industry, sited in Northern Scotland, was connected with the rise of the heavy chemical industry before the introduction of chemical methods of making soda from salt.

## (e)  Salt

Long used as a means of flavouring and preserving food, salt was considered

[6] Clows (ref. 4), pp. 70–74.

*Kelping on the island of St. Kilda (upper) and burning sea-weed to produce kelp on the Brittany coast (lower)*

such an important dietary component that it became a mark of hospitality and a symbol of prosperity. From very early times salt was an item of commerce. In hot countries impure salt can be obtained directly from sea-water by trapping it in shallow lakes (salt-pans) at high tide and allowing it to evaporate in the heat of the sun. The salt crystallises on short poles set in the base of the salt-pan. In colder climates sea-water may be evaporated by heating, though this is expensive in terms of fuel. Such activity took place on the coast of North Eastern England at least from the 13th century, the smoke from the 'pan coal' being visible in the early 18th century to Daniel Defoe from the top of the Cheviot, 40 miles away.

In many places there are salt deposits deep in the earth, the dried-up beds of ancient seas and salt lakes.[7] These deposits may produce salt-springs, or they can be mined as in Cheshire. The presence of the salt beds there, together with the Lancashire coal-field and the port of Liverpool, made that part of Britain ideal for the development of glass manufacture and other chemical industries, especially the soap-boilers who used animal fats and vegetable oils which were imported as the industry grew in size and diversity. Salt attracted fiscal attention and was heavily taxed. In domestic use this was an inconvenience rather than a real problem, but as salt came to be used in industry for making alkali, the salt-tax became a crippling burden for the alkali manufacturer.

## (f)   Sulphur

As industry expanded during the 18th century the demand for sulphuric acid rose. It was used for 'pickling' metals, bleaching cloth, preparing other chemicals like nitric and hydrochloric acids and from the end of the 18th century in growing amounts by the alkali industry. Small quantities of sulphuric acid had long been made by distilling ferrous sulphate crystals (copperas), a material made by exposing iron pyrites to the oxidising action of air and water (see below). When sulphuric acid was required in larger quantities it was necessary to use elementary sulphur, a material which is not indigenous to Britain and had to be imported. Found in volcanic regions, the main source of elementary sulphur in Europe was in Sicily. Although sulphuric acid is the most important compound used in chemical manufacturing at all levels, it is itself a manufactured product and in a country like Britain, lacking supplies of native sulphur, yet requiring ever growing quantities of sulphuric acid to maintain its industries, the production of this acid was linked to other industrial processes such as the extraction of metals from sulphide ores. As the manufacture of sulphuric acid on the large scale began with sulphur dioxide, it was possible to use the gases given off when sulphide ores like galena ($PbS$), zinc blende ($ZnS$), copper pyrites and iron pyrites were roasted.

[7] R.P. Multhauf, *Neptune's Gift*, Johns Hopkins University Press, Baltimore & London, 1978.

## (g) **Metals**

These sulphide ores were important raw materials for the extraction of lead, copper, zinc and iron (Chapter 10). Although the smelting of metals from their ores is not generally regarded as part of the chemical industry, the sulphur dioxide formed during roasting is used for sulphuric acid manufacture. Smelting also involves reduction with charcoal or coke and for these reasons the metal industries have a claim for inclusion as part of the history of the British chemical industry. The extraction and uses of lead, copper and iron all have a long history and are fundamental to the development of heavy industry in Britain.[8]

Lead mining in the Mendips, begun in Roman times, flourished during the 11th century at the height of the cathedral-building era, when lead was used extensively for church roofs. The lead ore was broken up with hammers or mechanical stampers and smelted in a crude furnace or 'bole' shaped like a lime-kiln and driven by a natural wind-draught. Later furnaces, like the blacksmith's forge, with a stone hearth 12–18 inches deep and hand-operated bellows to provide the air-blast were used. Brushwood was used to fuel boles, while charcoal or peat was used in the later hearths. Various modifications of both forms were known and similar furnaces appeared in many parts of Europe. In 1556 Georgius Agricola provided descriptions of them and the methods of working them.[9] A large bole could smelt up to 6 tons of lead in 2 days. A 'turn-hearth' furnace as used in the Mendips, could be turned to face the wind. These techniques continued in use up to the mid-18th century after which simple reverberatory furnaces and small blast-furnaces were used.

Lead ores occur in conjunction with many other metals, notably silver which, if present in sufficient quantity, it was desirable to recover owing to its value. This was done by cupellation, a process in which the lead was oxidised on a flat bone-ash dish or 'cupel'. All the lead was oxidised to litharge in order to separate the precious metal from it and the lead oxide then had to be re-smelted to recover the lead. The cost of this process was usually recovered in the value of silver extracted from the lead during refining. Quaker mining companies operating in Derbyshire, Alston and Wales during the 17th century accumulated so much high quality silver that they were able to supply all the needs of the Royal Mint for silver coinage from 1705 to 1737 and long afterwards the Mint still purchased silver from the Quaker Lead Company.[10] In the 19th century Hugh Lee Pattinson introduced a method of extracting the silver by fractional crystallisation from molten lead. As the liquid lead cools crystals of pure lead are formed which can be removed from the mixture leaving the silver dissolved in the remaining lead. The enriched silver-bearing lead can then be cupelled.

From the 16th to the 18th century copper mined in Cornwall was shipped to South Wales where copper smelters were established first at Neath in 1582 and

---

[8] See Chapter 10
[9] Georgius Agricola, *De Re Metallica*, translated by H.C. Hoover and L.H. Hoover, Dover edn., New York, 1950, Bk. IX, pp. 353–437.
[10] Raistrick (ref. 5), pp. 180–182.

*Cupellation to extract silver from lead in the 16th century* (from Biringuccio's *Pyrotechnica* of 1540)

from 1717 at Swansea which then remained the centre of the industry. Both have easy access by sea and there were good supplies of coal to permit rapid smelting. The method used was similar to that described by Agricola, but the large scale of operations led to the discharge of vast quantities of sulphur dioxide into the atmosphere. The chief volatile impurity, arsenic, also sublimed as arsenious oxide. The resulting pollution destroyed vegetation over a wide area around the works. Efforts were made to disperse the acid vapours but it was not until well into the 19th century that steps were taken to absorb the sulphur dioxide in water. Sulphide ores were roasted to burn out much of the sulphur as sulphur dioxide which could then be used in the lead chamber process for sulphuric acid production. The resulting copper was impure and had to be refined in a reverberatory furnace to remove the iron, but much of the copper was made into brass by the traditional method of fusing it with calamine.

The history of the British iron industry certainly extends to 1200–1300 BC and probably earlier. The commonest ores used in iron manufacture were magnetite, haematite and some native iron carbonates and the first stages began with roasting to remove moisture and carbon dioxide. The prepared ores were then smelted in crude furnaces using charcoal as the fuel and a silicate rock such as lava to form slag. The furnaces used were at first merely holes dug in the hillside and lined with refractory clay. As there was no way to drive the air-blast apart from the wind, the process was unreliable and temperatures reached were only moderate so that the iron produced formed a spongy mass or 'bloom' which contained slag in its pores. The bloom had to be removed from the furnace, repeatedly heated in a forge and hammered to squeeze out the slag. Improvements to the ventilation of the furnace were made during the 17th century, but the processes of iron-making continued to rely on these primitive methods. Thus the bloomeries found in the Forest of Dean and the Sussex and Kentish Weald were still using these methods while 18th-century iron masters like Abraham Darby were developing larger and more powerful furnaces, with an air-blast driven by large water wheels, for the production of iron in bulk. However, when Darby began to use coke, the cast iron produced was found to be brittle and could not be used for many of the purposes for which the wrought iron of the bloomeries was ideal. It could not be welded and was useless for nails, horse-shoes and iron rails, in fact in any situation where it would need to withstand shocks. Coke-smelted cast iron was, in effect, a new product and new uses had to be found for it while the demand for wrought iron remained steady and it was produced in small quantities by bloomeries until they were super-seded from 1783 by the 'puddling process' devised by Henry Cort (Chapter 10).

## 3   Some Products of Early Chemical Manufacturing

It was the expansion of the textile trades during the 18th century which, more than any other aspect of the industrial revolution, stimulated the rise of the British chemical industry. As the textile trades grew from a small-scale cottage industry to a factory system, powered first by water-wheels and later by the steam engine, ever larger volumes of water together with various chemical products were required for washing, bleaching and dyeing. Wool needed large quantities of soap for scouring to remove its natural fats before spinning could take place. The manufacture of soap required alkalis and the demand rapidly outstripped the capacity of traditional methods for making them from kelp or wood ashes.

### (a)   Alkalis

Already in the 18th century the search for a commercial method of converting salt into soda on a large scale had begun in earnest. The alkali industry became an important driving force, along with coal and iron, for all the other industries which arose during the industrial revolution. The manufacture of alkalis stimulated the mining of salt and coal, lime-burning and, from the end of the 18th century with the introduction of the Leblanc process, the manufacture of sulphuric acid. When John Roebuck opened his sulphuric acid works at Prestonpans in 1749 he also began to explore ways to manufacture other chemical products. One of these was a method of making soda from sea-salt and lime. Roebuck became a close associate of James Watt, a chemist himself and a friend of Joseph Black, professor of chemistry at Edinburgh. It may have been through this connection that Roebuck was led to try the manufacture of synthetic soda. The minimal reaction between common salt and lime meant that the method could never have proved economically viable, especially in view of the high level of duty then levied on salt (£30 per ton). In the 1770s there were other attempts to make soda from sea-salt (which contains a small proportion of sodium sulphate that would react with the lime), or from sodium sulphate alone. Perhaps the most successful of these attempts was that by James Keir, an 18th-century chemical manufacturer from Tipton in the Black Country near Birmingham.

*Early sulphuric acid manufacture*

In 1771 Keir protested against the granting of patents for the manufacture of soda from common salt. Later, when Alexander Fordyce petitioned Parliament for the return of the tax on salt used in making soda, a Parliamentary Committee was appointed to investigate the manufacture of mineral alkali from salt and in 1781 a Bill was passed reducing the duty on *foul* (*i.e.* unrefined) salt used for this purpose. Several patents were then filed by various chemical manufacturers, including Keir. As a student at Edinburgh Keir had met Erasmus Darwin, a fellow medical student. Later, when Keir moved to Birmingham, Darwin introduced him to a group of men interested in science, the Lunar Society, where he met Matthew Boulton and James Watt. Through them Keir was introduced to the manufacturing scene in the Birmingham area, first as a glass manufacturer at Stourbridge, where he also made red and white lead for his own use and for sale to other glass houses. About 1770 he began to manufacture soda, soap and other chemicals. Keir's alkali process involved passing a *dilute* solution of sodium sulphate *slowly* through a thick bed of lime:

$$Na_2SO_4 + Ca(OH)_2 = CaSO_4 + 2NaOH$$

Keir thought that a large excess of lime relative to the quantity of sodium sulphate was vital for success and he also believed that it was essential to give the

*James Keir*

chemicals time to react. He seems to have had some success and his chemical and soap works became a tourist attraction equivalent to Boulton and Watt's Soho Foundry.

In the north of England and in Scotland other alkali works were established in which common salt or sodium sulphate were treated with lime. When the salt-tax was repealed in 1823 there were already several alkali works on Tyneside and new ones began to spring up on Merseyside where James Muspratt began to manufacture soda and other chemicals, close to the Cheshire salt fields. The success of the Leblanc process marks a turning point in the fortunes of the large kelping industry. It also marks the true beginning of the heavy chemical industry in Britain which is usually dated from the establishment of Muspratt's alkali works in Liverpool. The main customers for alkalis were soap-boilers whom Muspratt persuaded to use his soda instead of the more usual potash. He soon found the demand greater than he could meet and in 1828 he formed a partnership with Josias Gamble which lasted only two years. Muspratt then built a new works at Newton on the St. Helens canal where he worked the Leblanc process (see Chapter 4).

## (b)  Soap

Soap manufacturers were the largest users of mineral alkalis (soda and potash) and 18th century soap-works were often found attached to the alkali factories. Most of the product was used for scouring and washing fibres, yarns and fabrics in the burgeoning textile trades. As the demand for soap increased alkali works turned to soap-boiling as a more lucrative trade. Keir's chemical works at Tipton became better known as a soap works and Gossage at Widnes and Warrington also used his alkali for soap manufacture. Gossage's soap works became especially famous for the mottled soap which was produced there for more than a century. In soap making fuel was needed to raise steam for heating the vats and salt was added after the boil to separate the soap from the alkaline lye. Consequently proximity to a coal-field and to a source of salt was valuable to the soap manufacturer. The vegetable oils and animal fat were imported from Africa, America and other distant parts of the world so that access to a major seaport was also desirable. For all of these reasons soap manufacture prospered on Merseyside.

## (c)    Glass

During the 18th century glass-making increased both in quantity and variety. The glass industry required not only alkali and limestone, but also large quantities of sand (silica). Apart from washing, no special preparation of the sand was required, but different types of sand produced different grades of glass. For ordinary crown glass sand, soda and limestone were fused together in a kiln fired by coal or coke. There were small glass-works in many parts of England and Scotland, supplying glass for windows or glass bottles for a local industry as in the case of the West Country cider and mineral water trades. Glasshouses were often found close to lime burners and soap boilers since the glass manufacturer could use soap boilers' waste which contained both lime and alkali. As with other industries the fuel commonly used in glass-making was wood and the proximity of suitable supplies of this commodity also helped to determine the siting of glass works. Besides ordinary soda glass, flint glass was also made by fusing flint (pure silicon dioxide) with litharge to give a dense, highly refractive glass containing 66% lead silicate which can be cut and polished. The manufacture of flint glass requires lead oxides and the larger glass-works often chose to produce this ingredient on site rather than rely on external suppliers. Lead glazes were also used in pottery, where mixtures of various kinds containing metal salts and oxides ('frits') were also needed. These too could be produced at the glass-works for sale to pottery manufacturers. Thus a glass-works might become the nucleus for manufacturing a limited range of chemical substances.

## (d)    Copperas

Copperas (green vitriol crystals, $FeSO_4.7H_2O$) was manufactured at various sites in England and Scotland. There was a thriving copperas industry in Britain from the 16th century based on the atmospheric oxidation of iron pyrites ($FeS_2$) of which there were plentiful supplies. The process was very slow and the yield was low. At Queenborough on the Isle of Sheppey the dilute liquors draining away from the iron pyrites were 'ripened' in wooden troughs before being led into large boilers holding up to 12 tons. Iron dust was added and the liquid was heated for three weeks over a coal fire, fresh solution and iron being added as required. The resulting liquid was then allowed to cool and the crystalline product was collected.[11] A similar process was carried out at Deptford in London in the 17th century. In 1748 copperas manufacture was started at Hartley near Newcastle, and a copperas works opened at Walker-on-Tyne in 1789 was still in operation a century later, producing 2000 tons per year by a method similar to that originally operated at Queenborough.[12] Copperas was also made in Scotland, as a sideline to alum manufacture at Hurlet and Campsie where the coal contained considerable quantities of iron pyrites. Copperas became a major commercial product initially for sulphuric acid manufacture and later as a mordant in dyeing.

### (e)  Sulphuric acid

There has been a sulphuric acid industry in Britain since at least the 16th century, but by the 18th the acid was in use in more processes than any other chemical and its manufacture became a matter of great economic significance. Until the industrial revolution sulphuric acid was made by strongly heating copperas, distilling off and condensing the acid vapours formed. Decomposition of the copperas yielded sulphur trioxide and water which combined to form 'oil of vitriol', an acid of very high concentration. As carried out at Nordhausen in Germany this process yielded 'fuming sulphuric acid', a solution of sulphur trioxide in concentrated sulphuric acid, later called oleum. The by-product of the process is ferric oxide.

In the early 18th century demand for ordinary sulphuric acid was already growing as it was being used for bleaching linen cloth and for many other industrial purposes. To supply the necessary quantities of sulphuric acid an alternative to copperas was required and elementary sulphur seemed to provide the obvious answer. It had been found that although sulphur burned readily enough to form sulphur dioxide, its conversion into sulphuric acid was far from easy. German chemists had observed that potassium nitrate added to the burning sulphur increased the yield of sulphuric acid and this method may have been introduced into England by Cornelius Drebbel some time in the 17th century. It was already in operation in France and Holland and was described with an illustration in Nicholas Lefèvre's *Compleat Body of Chemistry* published in 1670.

The first sulphuric acid works in England about which there is any detailed information was set up by Joshua Ward, a quack doctor, at Twickenham in 1736. Ward began to manufacture sulphuric acid 'by the bell', a phrase he used to distinguish his product from acid made by distilling copperas. Ward's acid was made by burning the sulphur in glass vessels and dissolving the resulting

[11] J.U. Nef, *The rise of the British coal industry*, 2 vols., Routledge, London, 1932, vol. 1, p. 184.
[12] W.A. Campbell, 'The copperas trade, 1750–1850', *Pharm. Historian*, 1980, **10** (3), 9–11.

*Joshua Ward*

gases in water.[13] The method was very inefficient but was greatly improved when he began to burn charges of sulphur mixed with nitre. The acid was then concentrated by evaporation. In 1740 Ward moved his operations to Richmond in Surrey. The glass vessels used in this process had a capacity of 40–60 gallons, a fact which suggests that there was considerable skill among 18th-century English glass-blowers. In 1749 Ward took out a patent for his method of increasing the production of the acid by adding nitre to the burning sulphur. Ward's innovation dramatically increased the yield of acid and he was able to reduce the price by as much as 95%.

One important use of sulphuric acid was in 'pickling' metals to clean the surface of any oxides, ready for coating with a thin film of a precious metal like gold or silver. In a reverse process, mixed with potassium nitrate the acid was used for stripping these precious metals from base metals. Both processes were commonly used in the Birmingham jewellery trade, hence the extensive demand for sulphuric acid in that region. As transport was costly it is not surprising to find sulphuric acid manufacture springing up in and around Birmingham. John Roebuck, a Sheffield physician, had settled in Birmingham where he met Samuel Garbett. Roebuck and Garbett worked as consulting chemists whose main business was in recovering and refining precious metals. Requiring large quantities of sulphuric acid they began to manufacture their own at their works in Steelhouse Lane, Birmingham in 1746. The glass vessels used by Ward and his partner John White were expensive, fragile and limited in size, so Roebuck and Garbett substituted very much larger vessels made of sheet lead which the 17th century German chemist, Johann Rudolph Glauber, had found

[13] F. Sherwood Taylor, *A History of Industrial Chemistry*, Heinemann, London, 1957, p. 96.

to be unaffected by the acid. This increased the scale of the manufacturing operation and brought down the price of the acid still further. The Steelhouse Lane factory continued to operate for over a century, changing hands several times, but in 1749 Roebuck and Garbett established a second vitriol works at Prestonpans on the Firth of Forth near Edinburgh. In this case, although a small proportion of the acid produced was used to make Glauber's salt (sodium sulphate $Na_2SO_4.10H_2O$), it is likely that its main use was in bleaching Scottish, and perhaps Irish, linen.

## (f) Chemicals for bleaching

Various chemical products beside soap are used in the textile trades. Wool, linen and cotton all need thorough washing with soap, but this was never enough to produce whiteness as there were usually substances left in the yarn or fabric which would produce a dingy look in wear. To prevent this bleaching was necessary. The traditional method of bleaching had been by laying out the material on the ground, exposed to sunlight and air in 'bleachfields' where it was treated with water and sour milk. The process took anything up to a year to complete; it tied up good agricultural land and caused long delays in the manufacture of textiles such as linen. There were extensive bleachfields in Scotland and Ireland, but up to the middle of the 18th century the centre of the bleaching industry was in Holland where sour buttermilk was plentiful and the bleaching process using it had been improved by using an alkaline lye before the buttermilk.

In the second half of the 18th century several improvements to the bleaching process were made in Britain, all related to developments in the chemical industry. The first improvement depended on the use of a dilute solution of potash in which the yarn was boiled for three hours before being washed. Then it was repeatedly soaked in dilute alkali, wrung out and watered for some days until ready to be woven into cloth. The latter was then again bleached but this time by souring with buttermilk for forty-eight hours. In other processes acid solutions were used, such as the juice of crab apples or lemons. After 'souring' the cloth was spread out on the bleachfield and watered to wash out the acid. It was then rinsed and washed with soap. The whole process was repeated six or seven times until the required degree of whiteness had been attained. This represented some improvement on the traditional process as far as the speed of the operation was concerned but it was still very slow and inefficient. Many of the improvements in bleaching methods were introduced by Scottish linen manufacturers who appear to have been willing to try out new methods in the search for faster, more effective processes.

One such improvement came with the use of dilute sulphuric acid in bleaching which was investigated by Francis Home in the 1750s, and with the production of cheap sulphuric acid by Roebuck bleachers began to take up the new method soon afterwards. Large firms set up their own sulphuric acid plants, as the Lancashire firm of Richard Bealey & Co., Ltd did about 1791, or Charles

*'Grassing': After boiling in an open vessel (upper) the cotton fabric was exposed to sunlight for many days (lower). Good agricultural land was often used for these bleachfields*

Tennant & Co. in the early years of the 19th century.[14] Lime was also sometimes used in the chemical bleaching process, but unless thoroughly washed out of the fabric this caused undue wear and its use along with sulphuric acid was inadvisable. If hydrochloric acid was used instead of sulphuric, the lime would be converted into soluble calcium chloride instead of the relatively insoluble calcium sulphate and this was done by some bleachers.

In 1789 C.L. Berthollet in France, who had already investigated the bleaching properties of dilute sulphuric acid, showed that chlorine gas could be used to bleach cloth. Chlorine had been discovered in 1774 by C.W. Scheele in Sweden

---

[14] C.A. Russell and P.J.T. Morris, *Archives of the British Chemical Industry 1750–1914*, British Society for the History of Science, Monograph no. 6, Faringdon, 1988, pp. 14–19, 181–183.

*Two views of the St. Rollox chemical works in the 19th century: upper 1844; lower 1890*

and later he, as well as Humphry Davy, investigated its bleaching properties. At first impure chlorine was dissolved directly in water and the cloth to be bleached was steeped in the solution, but the fumes given off began to affect the workers and although the process was used up to about 1830, the need for a safer method of bleaching with chlorine was recognised. For a time a solution of chlorine in dilute alkali was tried, but it was Charles Tennant who solved the problem most effectively with his invention of bleaching powder, 'chloride of lime'. Tennant moved to the St. Rollox works near Glasgow in 1799 where he began to manufacture bleaching powder by absorbing chlorine in slaked lime. The firm prospered as the demand for bleaching powder increased. Later a process invented by Henry Deacon yielded chlorine mixed with air, but the product was quite satisfactory for making bleaching powder. Tennant was so successful that he was soon able to gain financial control of the St. Rollox works (Chapter 4).

## (h)    Nitric acid

Both Lazarus Erker and Georgius Agricola described the manufacture of nitric acid in the 16th century by a process in which saltpetre (potassium nitrate $KNO_3$) was heated with ferrous sulphate from which most of the water of crystallisation had been driven off. The apparatus used was made of glass, pottery or preferably iron.[15] The ferrous sulphate yielded sulphuric acid which liberated nitric acid from the saltpetre. Alum was sometimes added and if *aqua regia* – a mixture of concentrated nitric and hydrochloric acids – was required, salt was also added. Beginning with gentle heat the temperature was gradually raised to red heat. The fumes of nitrogen tetroxide ($N_2O_4$) and nitric acid evolved were led into a receiver containing a little water. The flask in which the heating was carried out might be of glass, though that was apt to break under the high temperatures used and pottery or iron vessels were preferred since nitric acid merely renders iron passive. The acid was often purified by adding a little silver to precipitate any chlorides present. The chief use of nitric acid (*aqua fortis*) free from chlorides was for separating silver from its alloys with gold. However, if the alloy contained more than 25% of gold it resisted attack by nitric acid. In that case *aqua regia* was used. This dissolved the gold as its chloride while precipitating the silver. The solution of gold chloride was then decanted and distilled when the acid volatilised leaving gold chloride which was readily converted into pure gold by heating on a cupel. The silver could also be recovered from the silver chloride precipitate.

The manufacture of nitric acid from nitre or saltpetre (potassium nitrate, $KNO_3$) continued throughout the 18th and 19th centuries. Other uses for the nitrates included fertilisers and gunpowder. In the 18th century nitre was often extracted together with saltpetre from farmyard manure by lixiviation and strenuous efforts were made to improve the quality of the product. For gunpowder saltpetre was detrimental as it is deliquescent and pure nitre was needed. In any case the supply was precarious and it was realised during the Napoleonic wars that a more reliable and copious source was needed. The problem was solved when the nitrate beds in Chile were discovered in the 1870s (Chapter 5).

## (i)    Alum

If soda and sulphuric acid were the most important large-scale chemical products in the 18th century, alum is almost certainly the oldest.[16] The main use of alum since ancient times was as a mordant in dyeing cloth. The process was known to the Greeks and Romans; indeed the mordanting of madder with alum dates back to about 2000 BC. However, as the names 'alumen' or 'stypteria' were applied to *any* astringent salt which acted as a mordant it is not always possible to identify the precise substance referred to in Latin and

[15] Sherwood Taylor (ref. 13), p. 92.
[16] C. Singer, *The earliest chemical industry*, Folio Society, London, 1948.

*Peter Spence (1806–1883). A pioneer of the alum industry*

Greek texts. The alum was made on the islands of Stromboli and Melos by roasting alum-rock, weathering and then boiling successive samples of it with the same quantity of water until a solution strong enough to crystallise on cooling had been formed. In the 15th century the papal states developed a monopoly in alum production when it was found that the salt could be extracted by treating the volcanic alunite rocks around Civitavecchia on the Italian coast.

Other countries had no similar supplies of alunite, but it was found that certain shales when roasted produced aluminium sulphate which could be converted into an alum by treatment with an alkali such as kelp or stale urine containing ammonia. In England shale quarried in East Yorkshire and calcined by burning with wood was used from the 17th century onwards. Iron pyrites contained in the shale was converted into ferrous sulphate and at the high temperatures achieved during calcination this decomposed aluminium silicate into aluminium sulphate which could be extracted from solution in water. Kelp, wood ashes or stale urine were then added to convert the aluminium sulphate into alum after which the solution was boiled down and crystallised. Such methods involving the use of alkalis were adopted by alum manufacturers throughout Europe during the 18th century. The ferrous sulphate produced was also used as a mordant.

In 1796 Charles Macintosh realised that it would be possible to make alum from aluminous shales found in coal-pit wastes at Hurlet near Paisley in

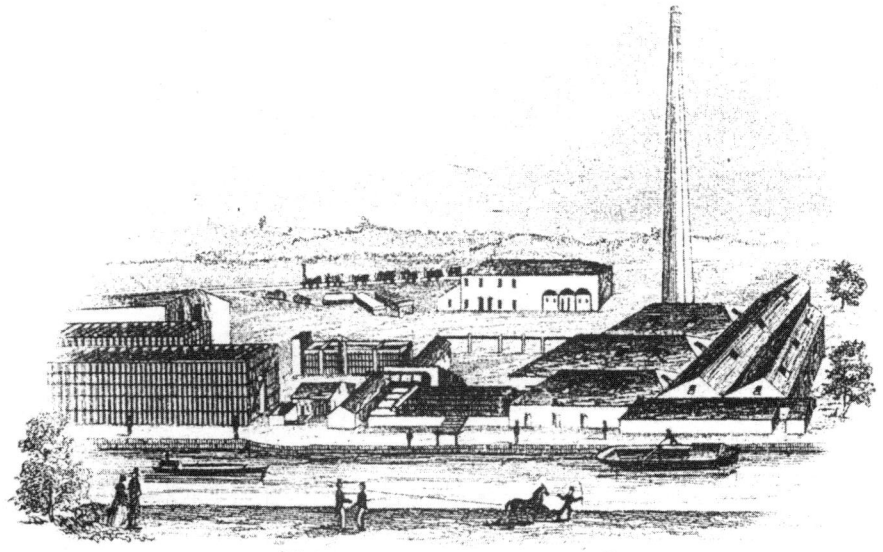

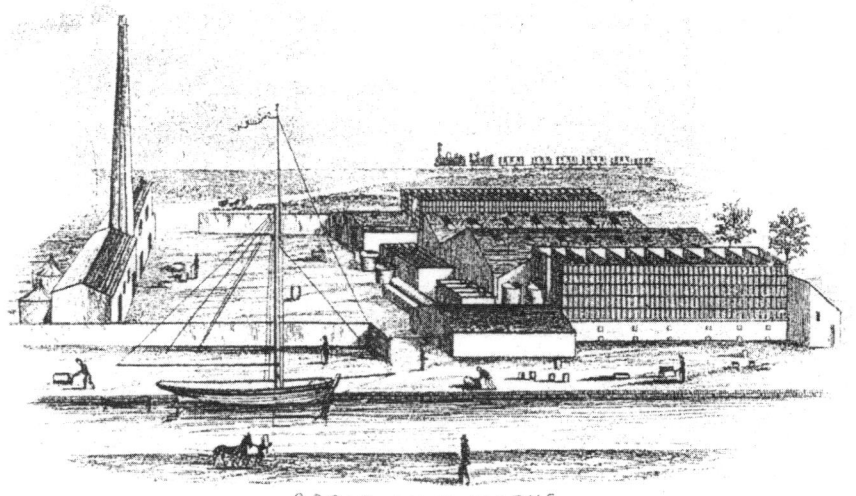

*Peter Spence's alum works at Pendleton (1846) and Goole (1854)*

Scotland.[17] Some of this material had been lying exposed to the atmosphere for two hundred years and the surface layers were completely acidified. Production of alum was started in 1797 and the product could be sold more cheaply than the Yorkshire alum. The Hurlet works was so successful that in 1808 a similar factory was opened at Campsie in Stirlingshire. By 1835 they were making 2000

[17] Clows (ref. 4), pp. 235–255; Morris and Russell (ref. 14), pp. 107–111.

tons of alum per annum, and production continued during the 19th century to about 1880 when the supply of shale was becoming exhausted. Another important 19th-century alum manufacturer was Peter Spence with works at Pendleton near Manchester, Birmingham and Goole.[18] Spence took out patents which revolutionised the manufacture of alum and gave him such success that he became the largest alum manufacturer in the world. He also made other chemicals by processes which utilised the waste products of other industries. These included copper hydroxide, carbonate and chloride, ammonium sulphate, potassium sulphate, and aluminium sulphate

## 4 Dyes and Dyeing

The origins of dyes and dyeing lie far back in the ancient past in Egypt, India and Turkey. The processes were kept secret but the use of metallic salts (mordants) to fix the dye to the yarn or fabric was known in Europe by the 17th century. Cornelius Drebbel discovered a method of dyeing cloth bright scarlet using cochineal and a mordant of tin chloride. Most early dyestuffs were extracted from plants, insects or certain shellfish and the cultivation of certain crops or the harvesting of shellfish for the dyers became important in some parts of the world. The common blue dye was obtained from a genus of plants known as the *Indigofera*, grown in tropical countries. Solid indigo was imported from India where the cultivation of indigo plants was a major industry. In colder European climates the blue dye was obtained from plants of the *Isatis* genus, woad, which also contain indigo, though of an inferior quality. The process involved fermenting the insoluble indigotin in the dye-bath with lime. Alternatively indigotin was boiled with honey and lime. In either case the blue colour was lost by reduction and the dye rendered soluble. Fabrics were then steeped in the colourless solution and hung in the air for the dye to oxidise on and in the fibres so regaining its blue colour. In mediaeval Europe woad was common, but by the 15th century Indian indigo had supplanted European woad.

Yellow dyes were of several kinds. The oldest known is the safflower, found on ancient Egyptian mummy wrappings of about 2000 BC. This is not the same as saffron which comes from the stigmas of the saffron crocus and is the dye used in dyeing the yellow robes of Buddhist monks. Weld, or dyers' weed, a plant related to mignonette, was the commonest source of yellow in the Middle Ages; this gave a clear yellow with alum as a mordant. Fustic, a Mediterranean shrub, gave a brown-yellow and in the 16th century turmeric was introduced from India. Later, towards the end of the 18th century, quercitron, made from the wood of the American oak, began to be used.

Madder, the oldest red dye, was derived from various species of *rubia*, plants which grow wild in the Mediterranean region, known in Egypt from about 1500 BC. With alum as a mordant it gave a fine bluish red. From the Graeco-Roman period crimson was produced by dyeing with insect parasites of the species *Coccus* (kermes). Cochineal, related to these insects, gives the strongest colour,

[18] *Ibid.*, pp. 176–179.

but this comes from Mexico and only reached Europe in the 16th century. Before that time the bodies of the female kermes insects, parasitic on holm oak, were collected and crushed. The product was used for dyeing with an alum mordant. The name 'crimson' comes from the Arab word kermes, but the same insects were also called vermiculi (little worms) and this gave rise to 'vermilion'. A brilliant red dye discovered by Drebbel was called 'scarlet' to distinguish it from crimson. Other red dyes were obtained from brazil wood which gave a reddish brown with alum as mordant, and 'lac', a resin made from ivy or some other creeper. There was also archil, an ancient red dye made from lichens which grow near the sea. None of the natural red dyes is resistant to fading in sunlight.

Ancient purple, the most expensive dye of all, was obtained from various species of shellfish of the genera *Murex, Purpura* and *Helix*. They possess a gland filled with a creamy liquid which on exposure to air is oxidised to produce the well-known Tyrian purple. This dye is insoluble in water and very fast to light and washing. The earliest known centre of this part of the dye trade was in Crete, but in Roman times it centred around the Phoenician ports of Tyre and Sidon. Owing to its scarcity purple was regarded as a precious commodity comparable with gold and silver in ancient times. Dyeing with purple continued at Byzantium until its capture by the Turks in 1453, but the details of the art became lost.

In the history of dyestuffs manufacture one of the most interesting and successful products was *cudbear*. It was discovered by George Gordon, a coppersmith who when repairing a boiler in a London dye house noticed that one of the methods used there was similar to that used in the Highlands for making 'crottel' from a lichen (*Lichen tartareus*). He found that he could obtain a purple dye from this which he called 'cudbear'. For industrial purposes the lichen was obtained by scraping it off rocks and the method of obtaining the dye involved first cleaning the lichen with water, drying it and pounding it with 'spirit of soot' (ammonia solution). Lime was then added and the mixture was left for fourteen days. The purity of the ammonia was important for producing good colours and for the industrial preparation of cudbear ammonia was obtained by distilling urine. Increased demand for cudbear led to the use of as much as 250 tons of lichen annually and 2000 gallons of urine were collected daily and tested with Twaddell's hydrometer to ensure that the natural product had not been diluted with water. When Scottish supplies of the lichen became exhausted it was imported from Sweden, Norway, the Canary Islands and Malta; other varieties of the lichen were also used. The manufacture of cudbear was surrounded with great secrecy, but in 1793 the details were disclosed by a disloyal employee and a rival company producing the dye was established in London. However, this was not a success and it seems that the purity of the materials was an essential part of successful operation. The colours produced by cudbear varied from bright pinks and reds through purple to bright blues. It dyed silk and wool (protein fibres) but not cotton and was often used to improve the colours obtained from other dyes like indigo and madder.

Black was usually obtained by reacting tannic acid extracted from oak galls with iron. This compound was also used in so-called 'indian ink'. There was no satisfactory green dye from natural sources, though buckthorn and sloe berries with alum as a mordant gave a green pigment. Cloth boiled with verdigris (basic copper carbonate) and alum took on a copper green shade, but none of these colours was fast and it is probable that greens were obtained by dyeing with mixtures of one of the yellows with indigo. Most of the natural dyes were more or less unsatisfactory; few were resistant to fading and most were only loosely attached to the fibres so that with exposure to sunlight and repeated washing the colours faded. Before the mid-19th century however there was no hint of more satisfactory dyes and when Perkin's synthetic purple was introduced in 1856 it caused a sensation (Chapter 8).

# 5   Chemicals in Agriculture and Allied Industries

## (a)   Fertilisers

In farming, traditional methods of cultivation handed down from the Middle Ages had shown the benefits of spreading marl (a mixture of lime, clay and sand) or lime alone on the fields, rotating crops and periodically allowing the soil to lie fallow. In the 17th century the Royal Society collected information about agricultural methods from various parts of Britain. It was found that in addition to lime and marl, dung, and other materials such as woollen rags and sea sand, were in common use to improve the fertility and tilth of the land. A controversy began which lasted well into the 19th century over doubts about the relative value of different manuring methods. It was generally held, on the strength of experimental results by J.B. van Helmont and others in the 17th century, that the main constituent for plant growth taken up from the soil was water. But Glauber in Germany showed that *saltpetre* caused a marked increased in plant growth and concluded from this that saltpetre was an essential principle for vegetation. In 1672 Nehemiah Grew at the Royal Society in London confirmed that plants absorb salts from the soil, but much more knowledge of plant physiology would be required before the processes of absorption were fully understood. A further step in plant physiology was taken by Stephen Hales in his *Vegetable Staticks* (1727) when he showed that air as well as water and salts is necessary for plant growth. He also discovered a continuous flow of liquids through the plant and measured the hydrostatic pressure of sap rising from the roots. Later, in the 18th century, Joseph Priestley, Jan Ingenhouz and others observed the essential role of atmospheric exchanges through the leaves of plants involving carbon dioxide, water vapour and oxygen.

In the 19th century Humphry Davy at the Royal Institution was required by arrangement with the Board of Agriculture to lecture on agricultural chemistry. For ten years from 1803 he delivered six lectures annually. These formed the basis for his influential *Elements of Agricultural Chemistry* (1813) in which he described a method of soil analysis, assumed that certain plants were able to fix atmospheric nitrogen and showed that most plant-ashes contained large amounts of earthy phosphates. Davy recommended the use of bones as a form of manure. Shortly afterwards the gas industry began to provide a useful source of nitrogen fertiliser in the form of sulphate of ammonia, a by-product of town-gas manufacture.

The first compound manufactured purely as a fertiliser was calcium super-phosphate. In 1837 Sir John Bennet Lawes, experimenting on the elements of plant-food, found that spent animal charcoal from sugar refineries was a good fertiliser. This substance is a mixture of calcium phosphate and charcoal and Lawes discovered that its value was increased after treatment with sulphuric acid. This led him to treat other phosphatic materials with sulphuric acid and he was soon manufacturing the artificial manure 'superphosphate of lime' on a large scale at Deptford. From 1843 Lawes carried out field experiments with Sir Joseph Henry Gilbert at Rothamsted which showed that non-leguminous crops needed a supply of nitrogen compounds while legumes like beans and clover did not. They also found that the essential requirements of crops were combined nitrogen, phosphates and potash, suggesting that sulphate of ammonia or saltpetre, with superphosphate of lime or powdered bones and wood-ashes, were capable of supplying the needs of most crops.

## (b)   Sugar, starch and fermentation

While common salt was the first mineral substance produced in quantity as a condiment, sugar from the sugar cane was the first foodstuff to be extracted from a plant source and obtained in a pure crystalline state in large quantities. The sugar industry is among the earliest branches of the organic chemical industry along with starch products and fermentation processes. The sugar cane, a large perennial grass growing to a height of 10 ft (3 m) or more, was originally found in New Guinea and is thought to have spread westward through Indonesia, the Philippines, the Malay Peninsula, Indochina and the area around the Bay of Bengal. From there it was transported to many other parts of the world. Columbus is said to have carried sugar cane plants to San Domingo on his second voyage in 1493 and within a relatively short time it spread to the West Indies, Central America and the Northern parts of South America. Already by 1600 the production of raw sugar from cane grown in tropical America was the largest industry in the world. There was a thriving sugar trade between Barbados, Antigua, Jamaica and Britain from about 1625. The raw sugar imported into Britain was refined in London where the Venetian method of making sugar loaves had been introduced as early as 1544.[19] The use of sugar in Britain developed during the 18th century and may have been linked to the spread of tea-drinking; later in the century other culinary uses of sugar were added. These superseded the use of honey which in earlier times had been the chief saccharine agent.

Starch, like sugar, is of plant origin and its manufacture was an important industry, prominent in France. Some starch was also converted into dextrin, used for dressing textiles and paper, and in the preparation of colours for calico-

[19] Clows (ref. 4), p. 520.

printing. The methods of separating starch from plant sources differ from sugar production. In the case of roots like the potato, the first stage is to grind or rasp the material to a fine pulp. This breaks up and exposes the starch-containing cells, water is added and the starch is separated from the liquid, originally by settling, but later in a centrifuge. Further purification stages include treatment with sulphur dioxide and sieving through silk screens. Originally most of the starch produced was used in the textile trades, but in the 19th century it was also converted into glucose for beer-making and the production of alcohol. When starch was converted into dextrose it was necessary first to hydrolyse it by heating gently with dilute hydrochloric or sulphuric acid. All these commercial processes were well-established by the mid-19th century; they include some origins of the organic chemicals industry and are linked to the food, fine chemicals and drugs industries as well as the brewing trade.

## (c)   Brewing

The brewing of ale and beer is as old as civilisation and certainly goes back to the Egyptians and Sumerians (3000 BC). The drink then brewed was ale, made from barley and flavoured with various plants, spices or honey, but not with hops which were introduced by the Germans in the 8th century, though not used in England until the 16th century. Already by the 15th century brewing had become a craft protected by powerful Guilds, especially in England and Germany. It was recognised as a source of revenue by feudal lords who set up a system to license the brewers. In the 16th century John Caius described the brewing of ale using malt made from barley in a method reminiscent of more modern times. Taxation of beer and of the materials used to make it appeared in the 17th century. Although brewing undoubtedly relies upon chemical processes, it cannot be regarded as part of the chemical industry until the late 19th century when these changes began to be understood in chemical terms. The same is not true, however, of the production of spirits from fermented liquors. The distillation of spirits was an invention of Greek alchemists about the first century AD and the later development of large-scale distillation apparatus occurred in response to demands for spirits. It was thought that the distilled spirits might be capable of transmuting base metals into gold. It is not until the 12th century that references to the distillation of wine to produce *aqua vitae* are found. This was then used as a medicine which not only carried the qualities of the fifth element or *quintessence*, but could extract the virtues of plants to form healing tinctures. This led to a vogue for alcoholic extracts of plants in sweetened drinks similar to modern liqueurs and as these became popular the demand for alcohol increased during the 16th and 17th centuries. The distillers introduced simple forms of fractionation and in the 17th century the manufacture of spirits became a large-scale industry. As the products contained some water as well as the essential oils and esters which gave them their special aroma and taste, it was not necessary to seek an efficient fractionating column which would separate the alcohol alone, but it was out of this industry that the manufacture of alcohol free from water and contaminants grew up. The purification of alcohol involved repeated distillation to produce so-called rectified spirits or spirits of wine and ultimately pure alcohol.

### (d)   Tanning

Another ancient industry linked to agriculture by its dependence on natural products is the tanning of animal hides to produce leather. The hides are first hung in running water to remove blood and dirt and are then treated with lime and decomposing dung, the sulphides in which loosen the hair and epidermis and make the texture more open. The hides are then scraped with blunt knives to remove the epidermis and hair; they are then ready for tanning. The traditional process used tan bark obtained principally from oak trees. The hides were placed in layers in pits containing tan liquor and were moved from one pit to the next, each containing liquors of increasing concentration over a period of six to eight weeks. They were then placed in 'layers', each hide being dusted with dried, powdered tan bark laid one on the other, the whole being covered with strong tan liquor which was drawn off and replaced several times during a period of three to six months according to the thickness of the hides. In a faster process, developed in the 19th century extracts from oak and chestnut wood and various wood barks were made by treatment with hot water followed by concentration. Using these the time required to complete the tanning process could be reduced to a few weeks. In still other processes alum and common salt were used, as were basic iron and especially chromium salts. There remains some doubt about the mechanism of the process and whether it is in fact a chemical reaction or merely the mechanical deposition of tannin in the pores of the hide, but it seems likely that the changes involved follow the laws of colloid chemistry (Chapter 5).

## 6   Drugs and Pharmaceuticals

The pharmaceutical industries can be traced back to a variety of ancient techniques related to methods of extracting dyes, drugs, perfumes and other substances from plant and animal sources. From ancient times, many curative substances capable of relieving or preventing diseases were extracted from the roots, stems, leaves, flowers and fruits of plants. Methods of extraction involved simple processes like lixiviation, solution, filtering, evaporating, crystallisation and especially distillation. In the Middle Ages and during the 16th century, distillation was regarded as the most important of all chemical techniques. Like most of these processes, it originated with the alchemists' desire for purification. The art of the apothecary, pharmacy and later, the manufacture of proprietary medicines, grew up from these early applications of alchemical methods in the preparation of the so-called 'galenicals', medicines from plant sources.

In the 16th century Paracelsus insisted that the true objectives of alchemy should be to discover and prepare new medicines from mineral sources. Coming from a metal-mining district of central Europe where the practical operations of chemistry using charcoal furnaces, cupels, and stills were well-known, Paracelsus was familiar with the properties of metals and many metallic compounds. In his hands the use of mercury preparations for syphilis reached new heights. Some alchemists, known as 'iatrochemists', followed Paracelsus in using 'chymical' medicines, including powerful poisons like compounds of mercury, antimony and arsenic.

In the same era physiologically active compounds of antimony were described,[20] while Glauber, who has been considered the first chemical manufacturer, prepared many previously unknown inorganic compounds for sale, including especially his *sal mirabile* which was in fact sodium sulphate. The use of such substances as remedies, at first eschewed by the physicians trained in Galenic principles, steadily gained ground throughout the 16th and 17th centuries until, by the early 18th century they had taken their place in pharmacy among the most favoured remedies.[21]

This growing interest in chemical remedies was one important stimulus to the early fine chemicals industry. To satisfy increasing demand by apothecaries and farriers many inorganic chemicals had to be manufactured in a high state of purity. At first they were made by each individual druggist, following the instructions for their preparation given in pharmacopaeias and dispensatories. The Society of Apothecaries of London was established in 1617 under a charter which separated them from the spicers, pepperers and grocers with whom they had previously been associated. One year later, in 1618, the first London *Pharmacopaeia*, enforceable on all apothecaries in England, was issued by the Royal College of Physicians. It contained mainly medicines made from plant sources and very few 'chemical medicines', the chief of which were mercury compounds used to treat syphilis. There were, however, some new discoveries like quinine, digitalis and opium. By 1700 pharmacy stood on the brink of a new era. In England during the early years of the 18th century a number of firms of pharmacists preparing pure chemicals and drugs for sale to apothecaries and physicians were established. These firms form the origins of the pharmaceutical and fine chemicals industry in Britain, though local druggists continued to prepare their own medicines well into the 20th century. Compounded medicines sold to the public were not regarded as *materia medica* – these were the raw materials, natural or manufactured, from which the medicines were made. It was the manufacture of *materia medica* which helped to establish the fine chemicals industry.

One difficulty facing the early 18th century pharmacist was the great variety of substances which, although virtually the same in chemical composition, had distinct names, often derived from the method of preparation, or from the appearance of the product. For example, there were many 'salts' of plants, each made by burning the stems or roots to an ash, but it was only slowly realised that these were all identical with ordinary 'pearl ash' (potassium carbonate),

[20] B. Valentine, *The Triumphal Chariot of Antimony*, 1604.
[21] R.I. McCallum, *Antimony in Medical History*, Pentland Press, Edinburgh, 1999.

*Allen & Hanbury's works at Bethnal Green in the late 19th century*

made by lixiviating wood ashes and evaporating the solution until it would crystallise on cooling. It was also found that pearl ash and 'salt of tartar', made by calcining tartar (potassium tartrate), were the same substance, although the latter was purer than pearl ash and was found to be deliquescent. By allowing salt of tartar to absorb water from the air a very strong solution of potassium carbonate could be obtained called 'oil of tartar'. Heated with lime, oil of tartar was converted into a powerful caustic solution which would solidify on cooling. The improved understanding of the chemical relationships between these and other groups of compounds which began to develop during the 18th century led to some simplification of the pharmacist's work. Nevertheless, there were still plenty of traditional names for chemical products from sodium and potassium salts which, though made in different ways, were chemically similar or even identical.

The pharmaceutical industry was founded by firms like Allen and Hanburys,[22] John Bell & Co., Corbyn,[23] Stacey & Co. and Howards, with Boots, Wellcome and Glaxo[24] (Chapter 6). From the 18th century onwards firms like these manufactured items of *materia medica*, and pure inorganic chemicals used in the preparation of medicines. Each firm also marketed its own proprietary brands. By the first quarter of the 20th century there were many pharmaceutical and fine chemical manufacturers in Britain.[25] Besides medicinal products these firms manufactured compounds which form the origins of the British fine

[22] G. Tweedale, *At the sign of the plough: 275 years of Allen & Hanburys and the British pharmaceutical industry, 1715–1990*, Murray, London, 1990.

[23] R. and D. Porter, 'The rise of the English drug industry: The role of Thomas Corbyn', *Med. Hist.*, 1989, **33**, 277–295.

[24] R.P.T. Davenport-Hines, *Glaxo: A history to 1962*, Cambridge University Press, Cambridge, 1992.

[25] S. Miall, *A History of British Chemical Industry, 1634–1928*, Benn, London, 1931, pp. 128–148.

chemicals industry, but since they were predominantly inorganic compounds, while so many fine chemicals result from organic syntheses, it is now generally considered that the modern fine chemicals industry originated in Germany in the late 19th century. By 1914 Germany had a virtual monopoly in this field, including, besides the synthetic drug industry, the manufacture of synthetic flavours and perfumes as well as organic fine chemicals.[26] During the first world war, however, other countries were forced to manufacture such organic chemicals as could no longer be obtained from Germany and it is from that period that the modern British organic chemicals industry began after what had proved to be a false start in the mid-19th century (Chapter 8).

[26] C. Friedrich, 'The influence of pharmacists on the development of pharmacy and chemistry as academic disciplines', *Pharmazie*, 1992, **47**, 541–546.

Chapter 4 *The Alkali Industry*

W.A. CAMPBELL

## 1 Alkali in the Industrial Revolution

In 1736 James Dunbar of Edinburgh published his *Smegmatologia, or the Art of making Potashes, Soap and bleaching Linen*. If the title had been extended to include glass, it would have epitomised the relationship between alkali and the technology of the time. The ancient crafts of soap-boiling, glass-making and bleaching all required some form of alkali. Soap was made by boiling oil or fat with alkali, and glass by fusing sand and limestone with alkali, while cloth was bleached by soaking alternately in dilute acid and alkali. Makers of alum for mordanting dyes, and of saltpetre for gunpowder both needed alkali.

What then was alkali? For commercial purposes there were three sorts: mineral alkali or soda (sodium carbonate, $Na_2CO_3$), vegetable alkali or potash (potassium carbonate, $K_2CO_3$) and volatile alkali or ammonia solution ($NH_3$ aq.). For most chemical purposes soda and potash are interchangeable, but soap made with potash is too soft for making into bars, and glass made with potash melts at a higher temperature than soda glass and is therefore harder to work. These trades called for soda, but for making alum and saltpetre only potash would serve. Ammonia solution was used chiefly by bleachers and manufacturers of such lichen-based dyes as cudbear.

Soda in the form of sodium sesquicarbonate ('bath crystals' ) had long been known as an encrustation around certain saline lakes in lower Egypt, but this was both scarce and expensive in Europe. Later discoveries of native soda in Mexico and Central Africa did little to increase supplies as transport was expensive and uncertain. As the Industrial Revolution gathered speed, alkali users were forced to depend on vegetable sources, of which three were important.

### (a) Sources of alkali

Wood ashes contained some 80% potash, sufficient for most purposes, but extremely wasteful of wood which yielded only a few percent of ash. If a purer alkali was needed, the ash was crystallised from water to form 'pearl ashes'. Ashing was more efficient if performed as part of a larger wood-based technology, making tar for timber treatment, wood spirit for the varnish trade, or charcoal for gunpowder in addition to ashes. Wood ashes were imported into Britain from Russia, the Baltic states, and from America; by the end of the 18th century Russian ashes cost £40 per ton.

Certain seashore plants such as *Salsola soda* yielded ashes containing about 25% soda. These flourished mainly in Southern Spain. The plants were harvested in autumn, dried, and burned to ash in pits dug in the ground. Under the name of barilla the ash fetched £45 per ton including £10 import duty. Though it furnished soda instead of potash, and was therefore more acceptable to glass-makers, it came from a politically sensitive area and supplies were liable to interruption in time of war, as when France (1778) and Spain (1779) declared war on Britain over American independence.

The third source was seaweed collected at low water, dried in the sun, and burned in pits similar to those for barilla. The ash was called kelp, a dark blue vitreous material containing only a few percent of soda, often adulterated, but subject neither to import duty nor foreign interference. That used in Britain came largely from the coasts and islands of Scotland where it formed a major part of the economy, providing winter employment for crofters. An equivalent product from Normandy was called *varec*. These seaweed ashes contained a large quantity of salt which made them attractive to soap-boilers for 'salting out' the soap.[1]

Depending on the location, there were other minor sources of alkali for hard-pressed manufacturers. *Sandiver*, the scum floating on the surface of molten glass, was a useful source of mixed salts as were the waste salts from soap kettles, containing common salt and soda; these were exported from Leith to England at £8 per ton around 1800.

## (b)    Early soda manufacture

During the 18th century Britain suffered an acute alkali shortage for two reasons. It was here that the great expansion of the textile trade had first taken place, bringing in its train a new demand for soap and bleach, and the American War of Independence (1775–1783) had interrupted supplies of American wood ashes. France suffered for different reasons. The wars in which that country was engaged in the last quarter of the century soon devoured the stocks of gunpowder, so all available potash had to be reserved for making saltpetre. At the same time the war with Spain cut off supplies of barilla on which France depended for most of its alkali. It is not surprising therefore that most of the research into synthetic alkali took place in France and Britain.

Duhamel de Monceau had established in 1736 that the base in common salt was identical with that in soda. This ingenious piece of pure chemistry contained the seeds of future attempts to convert salt into soda on an industrial scale.

British experiments began in the 1770s when James Keir, a Stourbridge glass-maker, converted salt into sodium sulphate which he treated with slaked lime to produce caustic soda. The government saw the importance of such experiments and reduced the duty on salt to be used to make hydrochloric acid or soda; this move resulted in a spate of patents.

[1] A. and N.L. Clow, 'The natural and economic history of kelp', *Ann. Sci.*, 1947, **5**, 297.

Brian Higgins proposed heating sodium sulphate with charcoal and iron, and the physicians Alexander and George Fordyce substituted iron oxide for metallic iron. For a time they worked this process at South Shields. In a similar way Shannon's patent of 1779 prescribed sodium sulphate, charcoal and lime. For making the sodium sulphate these all depended on sulphuric acid. Most of this was supplied by the copperas trade in which iron pyrites was weathered to iron sulphate which was roasted to yield fuming sulphuric acid; the residue of red iron oxide was sold as a pigment, Venetian Red, or for polishing plate glass. Once more the importance of a saleable by-product must be stressed.[2]

Two different approaches were made on Tyneside. Archibald Cochrane, ninth Earl of Dundonald,[3] an impoverished Scottish nobleman with sound scientific instincts but little business sense, effected a simple transformation:

$$\text{common salt} + \text{potash} \rightarrow \text{soda} + \text{potassium chloride}$$
$$2NaCl + K_2CO_3 \rightarrow Na_2CO_3 + 2KCl$$

He worked this process with the brothers John and William Losh at Walker, and it was also worked by the South Shields glass-maker James King, but it was viable only if the less soluble potassium chloride could be sold to the alum manufacturers.

Losh and Dundonald also worked a process based on an observation by Scheele in Uppsala. Salt and litharge (the yellow oxide of lead) were made into a paste with water and left for some days when caustic soda and lead oxychloride were formed. This too could succeed only if the lead oxychloride was recycled as litharge, smelted to lead, or sold as Turner's Yellow, a pigment brighter than yellow ochre and cheaper than arsenic sulphide.[4]

Experiments began in France at about the same time. In 1777 Father Malherbe, a Benedictine abbé from the monastery at St. Germain des Prés, tried heating sodium sulphate with charcoal and iron. At a works at Javelle near Paris he used a reverberatory furnace, setting a trend for future experimenters. In the same year P.-L. Athénas, an apothecary's assistant, decomposed salt with iron sulphate instead of sulphuric acid, but he and others who operated directly on salt without going through the sodium sulphate stage soon aborted their attempts.

## 2 The Leblanc Process

In 1789 there appeared a paper which was to prove seminal. De la Métherie, editor of the *Journal de Physique*, published an account of a process involving sodium sulphate and coal which he mistakenly believed would yield soda. The

[2] L. Gittings, 'The manufacture of alkali in Britain 1779–1789', *Ann. Sci.*, 1966, **22**, 175.
[3] A. and N.L. Clow, 'Lord Dundonald', *Econ. Hist. Rev.*, 1942, **12**, 47; C.A. Russell, 'Aristocracy and alkali', *Chem. Brit.*, 1999, **35**(12), 30.
[4] R.C. Clapham, 'An account of the commencement of soda manufacture on the Tyne', *Trans. Newcastle Chem. Soc.*, 1868, **1**, 162.

paper was read by Nicolas Leblanc, physician to the Duke of Orleans, who had been experimenting since 1784 with the reaction between sodium sulphate and limestone. Although the premise was wrong, de la Methérie's communication suggested to Leblanc that coal might be incorporated into the reaction mixture, and this now became the core of the Leblanc process.

A factory was set up at St. Denis on the outskirts of Paris, largely financed by the Duke of Orleans. The partners in the enterprise, apart from Leblanc and the Duke, were J.J. Dizé, assistant to Jean Darcet at the Collège de France, and Henri Shée, an administrator with military experience. The chemistry of the process, as far as it was understood, was validated by Darcet. From this small beginning (the main factory building measured only 10 × 15 metres) sprang the process which was to dominate the heavy chemical industry of Europe for a century to come.[5]

On paper the Leblanc process seems very simple:

(a)  Salt + sulphuric acid → saltcake + hydrogen chloride

$$2\,NaCl + H_2SO_4 = Na_2SO_4 + 2\,HCl$$

(b)  Saltcake + coal + limestone → black ash

$$Na_2SO_4 + 4C + CaCO_3 = Na_2CO_3 + CaS + 4CO$$

(c)  Black ash extracted with water → soda crystals + alkali waste

In practice it was wasteful and tricky. The first stage, it seemed, consisted merely in pouring sulphuric acid on to salt and heating the mixture to complete the conversion into sodium sulphate or saltcake. Yet this stage gave enormous trouble, depending as it did on stirring the pasty mass without overheating on the one hand or incomplete reaction on the other. The iron rakes lasted only two weeks, so corrosive was the reaction mixture. But unless a good quality saltcake was obtained, the rest of the process could not be successfully worked. It is not the main purpose of this book to dwell on technicalities, but the commercial success of the industry, with all its social consequences – good and bad – depended on getting the details right; it is necessary to dispel the notion that soda manufacture was straightforward. We will therefore examine the second stage, the preparation of black ash, in some detail.

## (a)  The raw materials

First, the selection of raw materials was crucial. Saltcake was inspected for colour, tested for acidity, and crushed with a spade to assess texture. Batches which did not meet the necessary criteria were diverted to other uses such as glass-making. Depending on what was available in the district (to avoid

---

[5] C.C. Gillispie, 'The discovery of the Leblanc process', *Isis*, 1951, **48** (2), 152.

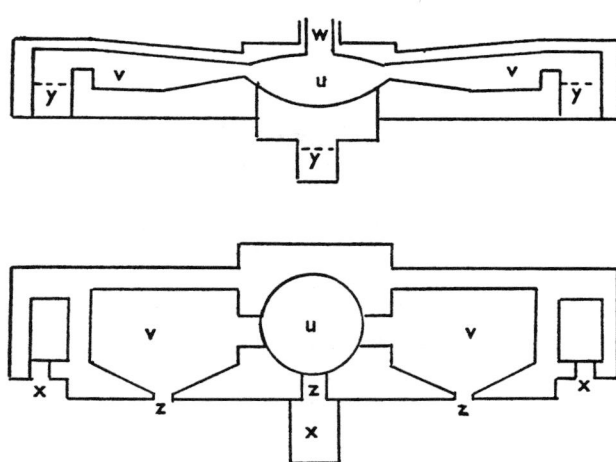

Section and Plan of Saltcake Furance

u. Saltcake pot
v. Roaster bed
w. Flue to acid condenser
x. Fire doors
y. Fire grates
z. Working doors

*Section and plan of saltcake furnace*

transport costs), different factories used different kinds of coal. Coal dust (duff), small coal (slack), and qualities ranging from bituminous to hard anthracite were employed, but each called for a particular mode of working and it was not easy to switch from one to another. Ash and nitrogen content were particularly important, affecting the proportions in the mix, and the purity of the product. In general, manufacturers did not depart too far from Leblanc's prescription of equal weights of saltcake and limestone, with half the weight of coal.

Limestone and chalk are both calcium carbonate, $CaCO_3$, but their physical properties differ in ways which affected the soda makers. Only the purest chalk, free from sand, clay or iron, was suitable, but the texture was also important as very hard minerals were expensive to crush and grind. Tyneside alkali works used 'London chalk' brought back as ballast in the sailing collier ships which took coal to the Thames. The alternative was 'cliff' from the neighbourhood of Rouen, also brought in as ballast, but this was usually too difficult to grind.

Limestone was used when chalk was not available. Most highly prized was Buxton limestone which was delivered to the chemical works already crushed. The presence of magnesium was not acceptable as soluble magnesium sulphide would get into the soda. For this reason the dolomites of County Durham, though so close to the Tyneside industry, could not be used for soda making, though they found a ready application in the manufacture of Epsom salts for the dyeing industry.

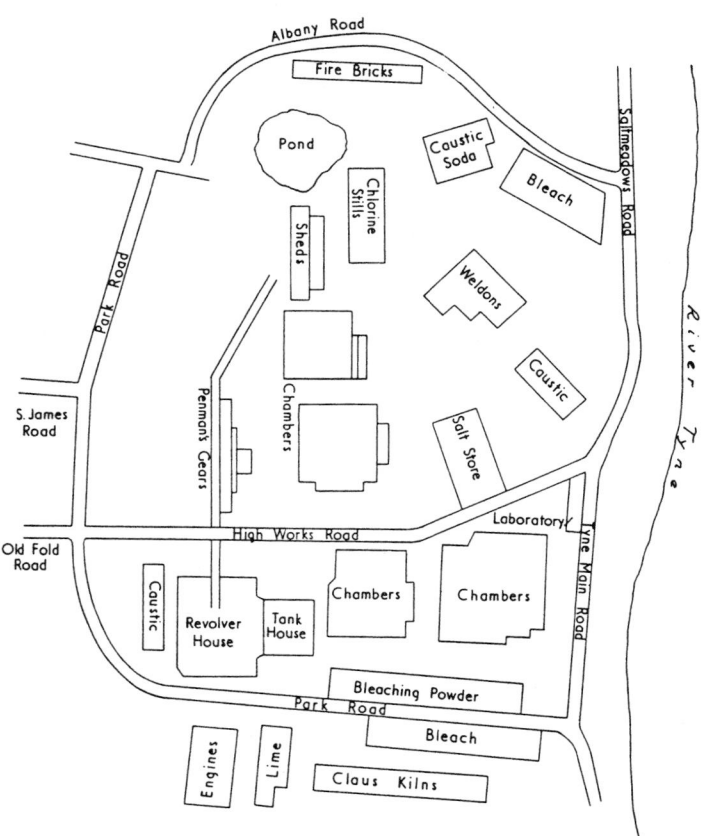

*(Upper) Allhusen's works at Gateshead (1910). This was the last British Leblanc works to survive, ceasing production in 1926. (Lower) Plan of the works, showing the haphazard nature of the layout*

### (b)  Furnace design

Black ash furnaces ranged widely in design. Hand worked furnaces with two or three beds, either long and narrow with doors in the sides, as used on Tyneside, or short and wide with doors in the front as were common in Lancashire; small enough for a charge of eight hundredweight, or large enough to hold a ton of mix – all these existed and made the movement of workers around the industry more difficult that it might have been.[6]

In 1848 W.W. Pattinson introduced at Walker a cylindrical furnace with rakes attached to revolving arms, but the mechanism could not withstand the corrosive environment. Five years later Elliott and Russell of Middlesbrough patented a cylindrical furnace which mixed the reactants by revolving along its axis, but it failed to compete with manual labour since it made an inferior black ash.

The most satisfactory form of 'revolver' was designed by Williamson and Stevenson at the Jarrow Chemical Works. It was cheap to run and made an excellent black ash. Whereas the workman could only stir the mix for a few moments for fear of lowering the temperature by keeping the door open, the revolver's continuous mixing made for a uniform product and also saved fuel.

This catalogue of wide variations in the manner of conducting just one phase of the Leblanc process indicates how difficult it was to manage the whole, and perhaps explains why Leblanc alkali did not immediately displace that from older sources. At Losh's alkali works at Walker the Leblanc process operated alongside double decomposition of potash and salt, and Scheele's litharge process. In France, Leblanc – who in many ways was his own worst enemy – failed either to run his own factory satisfactorily or to make friends with the revolutionary establishment. It is not surprising therefore that his process did not triumph immediately.

### (c)  Fiscal considerations

Taxes on glass, soap and especially on salt bore heavily on all aspects of the early chemical industry. However, the inhibiting effect of the salt tax has been exaggerated. When Losh and Dundonald opened their pioneering works at Walker they were fortunate in possessing a brine spring in a coal mine on their land. It is often stated that Lord Dundonald's influence with the government enabled the partners to avoid tax on salt from their own spring, thus giving them an advantage over the rest of the country. In fact, government had since 1783 permitted bleachers and glass-makers using synthetic alkali to recover most or all of their salt tax, and between 1782 and 1786 all attempts at converting salt into soda were freed from duty. It seems clear that the concession given to the Walker works could easily have been acquired by the others. In spite of the daily presence of an exciseman on the premises, Losh's workmen succeeded in

[6] T. Richardson and E. Ronald, *Chemistry applied to arts and manufactures*, Ballière, London, 1848, vol. 1, p. 86.

*Revolving furnace for soda ash manufacture*

stealing salt by piercing holes in the wooden trunking which conveyed the brine, and filling their water bottles with brine to take home.

The statement has often been made that it was the abolition of the salt tax in 1823 which enabled James Muspratt to found his large alkali works at Vauxhall Road in Liverpool. However, he had already gained experience of alkali-making in a small way in Dublin, and so was able to take the risk associated with a larger enterprise. His most serious obstacle was the soapers' resistance to synthetic alkali. Muspratt slowly wore this down, often by personally supervising the saponification process until the soap-boiler was familiar with the new material. As confidence grew, alkali manufacture spread in three main centres, Tyneside, Merseyside and Clydeside. This last is typified by the St. Rollox works founded by Charles Tennant in 1799 to make sulphuric acid for bleaching. This firm, through its manager James Mactear, greatly influenced the technical development and spread of the chamber process for making sulphuric acid. In 1863 they opened a works at Hebburn on Tyne, managed by the polymathic genius John Theodore Merz, chemist, philosopher, educationist, administrator and pioneer of large-scale electric power generation.

## 3  The Alkali Workers

An alkali works would employ three types of worker: process men who carried out the large-scale chemical reactions, yard men who were general labourers often drawn from former process men whose health had deteriorated, and

*James Muspratt (1793–1886), founder of the Merseyside alkali industry*

*Charles Tennant (1768–1838) first manufactured bleaching powder at his St. Rollox works near Glasgow*

tradesmen such as bricklayers, joiners, plumbers, engine men, wagon drivers, crane men, lightermen and coopers. Census returns are often confused, as a joiner might say that he 'worked at the chemicals' and would be entered by the enumerator as a chemical worker.

Some 80% of the process and yard men were Irish. Some had come into the country as 'navvies' to work on building canals or railways while others had been driven by poverty or famine to seek work in the daunting conditions of the alkali works. They were not popular in the towns where they settled. A Liverpool magistrate wrote to the Home Secretary to complain that seven hundred 'half naked and starving Irish' were arriving in the town each day.[7] Thomas Carlyle in his essay on *Chartism* (1840) described the Irish labourers as the most serious threat to industrial towns, performing as they did such work as called only for strength of arm, dressing in rags, and living in pig hutches on a diet of potatoes and salt. A.P. Laurie wrote 'the trade is one which to a large extent employs the lowest class of labour, and requires little skill or special knowledge'. He further claimed that employers preferred this grade of labour since the less the men understood about the processes, the less they could communicate to any rival manufacturer who might tempt them away.[8]

[7] T.C. Barker and J.H. Harris, *A Merseyside town in the industrial revolution: St Helens 1750–1900*, Frank Cass, Liverpool, 1954, p. 280; see also K. Warren, *Chemical foundations, the alkali industry in Britain to 1926*, Clarendon Press, Oxford, 1980, pp. 73ff.
[8] T. Oliver (ed.), *Dangerous trades*, Methuen, London, 1902, p. 593.

Widespread as these views were, they do not present a true picture of the chemical worker. In the absence of laboratory control of the process, the quality and yield of each product depended upon the men at the work-face. At every stage of the Leblanc trade, the judgement of the workman and the foreman made the difference between success and failure. These men were doing chemistry on a large scale, mixing, heating, dissolving and crystallising, and such work could never be described as unskilled.

It is true that the alkali workers, like the glassmen, were heavy drinkers. John Morrison, a Newcastle chemical manufacturer, whose view of workmen was somewhat jaundiced, described the men as:

> The caustic fishings of potato land . . . The men both physically and morally possess an individuality of their own. An experienced person could pick them out anywhere. During two or three days of the week the bulk of them would furnish happy studies for an artist of the Cruikshank school – but I should not like to be the artist . . . The beginning of the week is of course the worst. On Mondays the foremen are chiefly employed in keeping their men in an upright position, and not until Tuesday or Wednesday is the maximum angle of stability generally secured . . . The major portion of them profess an acquaintance with the principal operations of a chemical works and are prepared to accept almost any description of work . . . To say that they would vex the patience of a saint is an assertion of far too mild a character.[9]

## (a)  The saltcake workers

Of the potmen who made saltcake from salt and sulphuric acid, John Lomas said 'The rough nature of the work and the unpleasant surroundings tend to harden and demoralise the men'. Laurie wrote that work in an alkali factory unfitted a man for any other kind of work. When we examine this work in more detail, a somewhat different picture emerges.

The work was indeed arduous. The rakes and slices with which the batch was stirred were of wrought iron, 12–15 feet long with handles of one inch diameter. The tools were partly supported on hooks attached to chains suspended over the working doors. The unpleasantness of the environment is indicated by the short working life of the tools before they corroded away. Yet the men, wielding these giant tools and exposed to great heat and acid fumes, had to ensure that the semi-solid mass in the furnace was thoroughly mixed and regularly turned, and also to judge when the reaction was complete by the appearance and stiffness of the mass. If they allowed the mix to liquefy, it would set to a glass on cooling and would be useless for the second stage of the process. So essential was the experience of the maligned potman that Lomas recommended that two old hands should be employed elsewhere in the works so that they could be available at short notice 'in case of defection or disorder'.[10]

The potmen were suspected of Luddite activity when their jobs were threatened by machinery. In 1878, goaded by the fact that the men regularly

[9] J. Morrison, 'Gossip from Widnes', *Proc. Tyne Chem. Soc.*, 21 November 1875, 16.
[10] J. Lomas, *A manual of the alkali trade*, Crosby Lockwood, London, 1880, p. 88. See also p. 338 for an employer's view of workmen.

came in drunk and broke the pots, Jones and Walsh introduced a mechanical saltcake furnace at Middlesbrough with predictable results. 'By some mysterious means which they had not the means of ascertaining, the furnace perpetually went wrong at night; a cog broke or the crown wheel gave way, or something or other went wrong.' The trouble ceased only when all the potmen were discharged and replaced by common labourers. The work of three potmen was done by one man, a former cab driver, who was paid one penny per hundredweight of salt cake from the new furnace.[11]

### (b)   Black ash men and finishers

Similar skills were required of the black ash men. A hand-worked black ash furnace did not differ much from that for saltcake. The charge was allowed to react on the back bed and was raked on to the hotter front bed to complete the reaction. Since the chemistry of this stage was less clear, there was more opportunity for individual judgement. Times and temperatures were extremely important, demanding subtle variations from one charge to another. If the working door was opened too frequently, the temperature fell; if it was opened less often there was a danger of incomplete mixing and turning. Each furnaceman had one heavy and two light paddles or slices, three rakes, a poker, a shovel and a hook for removing lumps of clinker. When the mass was ready to be drawn, pointed flames ('soda candles' or 'pipes') appeared on the surface of the melt which became stiffer at this point; choosing exactly the right moment to draw was critical.

Where mechanical revolving furnaces were employed the mixing and turning was carried out mechanically; the only control exercised by the worker was the speed of revolution and the temperature. Again, recognising the exact moment to pour the black ash into the waiting bogies was critical, and the manager was dependent on the skilled worker. Black ash work was piece work, as was the extraction of soda from the cuboid lumps of black ash perversely known as 'balls'. Speed was important, and Lomas observed that 'it is astonishing what amount of work can be got through by an old hand on piece work'.

The curious practice of carbonating the dried extracted soda by calcining it with sawdust once more called for experienced judgement on the part of the furnaceman, for the temperature had to be high enough to burn off all the charred sawdust, but not before it had carbonated all the semi-molten soda. The tools were similar to those of the black ash worker.

The work of producing soda crystals, valued on account of their purity, calls for no special mention, although cases of drunken vat-men falling into crystallising vats were not unknown. One such delinquent, whose clothes were so stiff with crystals that his trousers stood up by themselves, was dismissed by the magistrates as having been sufficiently punished already.

[11] R.C. Clapham, 'On a new decomposing furnace patented by Messrs Jones & Walsh', *Trans. Newcastle Chem. Soc.*, 1876, **3**, 195.

### (c)    Bleaching powder workers

Of an entirely different nature was the work in the bleaching powder section. These men, the bleach packers and the lime dressers, worked with two unpleasant substances, finely powdered lime and chlorine gas. Their periods of exposure were short, as were their hours of attendance at the factory. They wore minimal protection in the form of a roll of flannel wrapped round the mouth to give seven layers. The men breathed in through the mouth and out through the nose. Sometimes leather goggles were worn, and brown paper was wrapped round the trouser legs. The lime dresser spread the lime evenly on the floor or shelves, depending on the design of chamber in use. The bleach packer raked the finished bleaching powder, reeking with chlorine, through a trapdoor into the barrels below. Each operation took about half an hour, but the team would look after several chambers. Between their bouts of labour, they would superintend the admission of chlorine, monitor its absorption by the lime by observing the colour of the atmosphere in the chamber, and sweep out most of the chlorine with a current of air when the reaction was complete. In factories where a series of chambers was operated, this residual chlorine was recycled; in other cases it might be used to make hypochlorite solution.

### (d)    Wages and conditions

Chemical work was not poorly paid as Carlyle had hinted. Table 1 shows hours and wages in different factories. Bleaching powder work was especially well paid, reflecting the risks which these workers took.

**Table 1**   *Average wages in two alkali districts, 1891*

|  | Tyneside | | Widnes | |
| --- | --- | --- | --- | --- |
|  | *Wage (£)* | *Hours* | *Wage (£)* | *Hours* |
| Pyrites burners | 1.60 | 56 | 1.70 | 84 |
| Saltcake potmen | 1.65 | 70 | 2.25 | 72 |
| Revolver men | 2.00 | 70 | 2.00 | 72 |
| Bleach packers | 3.00 | 40 | 3.00 | 36 |

Most firms worked two twelve-hour shifts, though the night shift was sometimes lengthened and the day shift shortened (Table 2).

Attempts at introducing three eight-hour shifts often met with opposition from two directions; the men saw a decrease in their wages, and their wives (backed by the foremen) feared the results of more free time for drinking. Nevertheless, the change from a twelve-hour to an eight-hour shift at Brunner Mond reduced the number of men attended by the works doctor to one half its previous level.[12] In Lancashire the ganging system prevailed. A gaffer would

[12] A.S. Irvine, *History of the Alkali Division*, ICI, Birmingham, 1958, p. 3.

**Table 2** *Hours of work in Merseyside alkali works, 1889*

| | Day shift | | | Night shift | | |
|---|---|---|---|---|---|---|
| | *Start (a.m.)* | *Finish (p.m.)* | *Total* | *Start (p.m.)* | *Finish (a.m.)* | *Total* |
| Monday | 7 | 4 | 9 | 4 | 6 | 14 |
| Tuesday | 6 | 4 | 10 | 4 | 6 | 14 |
| Wednesday | 6 | 4 | 10 | 4 | 6 | 14 |
| Thursday | 6 | 4 | 10 | 4 | 6 | 14 |
| Friday | 6 | 4 | 10 | 4 | 5 | 13 |
| Saturday | 5 | 12 noon | 7 | 12 noon | 7 | 19 |
| Sunday | –* | – | – | 7 | 7 | 24 |
| Total | | | 56 | | | 112 |

* On Merseyside there was no Sunday work beyond maintenance in alkali works.

negotiate a price with the firm and would provide and pay his own men, thus acting as a sub-contractor; the system was far less common on Tyneside.

Moves toward unionisation, though on a strictly local basis, were made in Newcastle in 1875 and Widnes in 1878 but both were short lived. In 1889 P.J. King of St. Helens founded his Chemical and Copper Workers' Union with branches in most chemical districts. One of the Union's proposals deserves special mention. It was suggested that in each alkali area one factory should be owned and operated by the Union; if an employer objected that a certain demand could not be met for economic reasons, the idea could be tested and its feasibility demonstrated in the Union's factory. No other union took up this potentially valuable proposal. In the early 20th century the alkali workers were taken into the Transport and General Workers' Union. The chemical workers' conditions of labour were investigated in 1893–1894 by the Chemical Works Committee of Inquiry.

## (e)   Children in the alkali trade

Very few women were employed in the alkali industry, but the situation with regard to children is more obscure. In response to the 1843 Commission on the Employment of Children, alkali managers maintained that children were employed only in helping the craftsmen, and never in connection with the chemical processes. The evidence of the children, especially that given to J.R. Leifchild (commissioner for the North-East) tells a different story. Many of the boys started when they were nine, and by the age of eleven were engaged in such tasks as mixing the raw materials for the saltcake and black ash sections, and wheeling the spent oxide from the vitriol furnaces. They worked in twelve-hour shifts, starting at five in the morning on the day shift. If his relief did not come on at five in the evening, the boy was asked to work a double shift. The wage was from four to six shillings a week. Some could read easy words and sign their names, but very few could read or write easily. Most were ignorant of common flowers and birds, and were unable to answer such questions as 'Who is the queen?'. One boy when shown a picture of a cow being milked said it was a lion. Many of the boys had fathers engaged in the same works. Indeed, the manager of the Jarrow Chemical Works blamed the parents for demanding work for their children at the earliest age, and felt that a law to restrict the age at which children might be employed in factories would strengthen his hand in dealing with importunate parents.

What prospects existed for the cleverer kind of boy? James Cooper, aged twelve, worked in the rudimentary laboratory from six in the morning to seven or eight in the evening for six shillings a week; thus although the work was skilled and responsible, he got no more than the boy who wielded brush and shovel on the mixing floor. It was also an unhealthy environment. 'When he is trying the copper ore in the crucibles, sulphur comes off and makes him sick. Once every day for 3 or 4 hours, he is busy with this work. It injures his health a good deal; was quite healthy before working here; has been here a year and six months. Father is one of the foremen in the warehouse.' (The commissioner comments that the boy is 'not healthy in appearance').[13]

## (f)  The health of the workers

The health of the alkali workers is a complex question. Some medical officers held that the alkali districts were less prone to visitation by epidemics of infectious diseases. On the other hand, mortality tables show the death rate of chemical workers to be exceeded only by a few trades, including brewers, cutlers and lead workers (Table 3). The figures for deaths from industrial accidents place chemical workers at the top of the list, though these accidents were often due to events of a non-chemical nature such as boilers bursting, walls and chimneys collapsing, and fires which were common in all industries in which furnaces figured widely.

**Table 3**  *Deaths per 1000, males aged 54–55*

| | |
|---|---|
| Earthenware | 43.0 |
| File making | 40.0 |
| Lead-works | 37.6 |
| Cutlery making | 35.6 |
| Glass-works | 32.1 |
| Brewing | 30.8 |
| Chemicals | 30.3 |
| (All males | 21.4) |

The workers themselves usually denied that their trade was unhealthy. When a public outcry arose against Cookson's works at South Shields on the ground that the firm's chimney emitted poisonous fumes, the workers held public

[13] *Report of the enquiry into employment of children*, (c. 431), P.P. 1843, p. L32, no. 111.

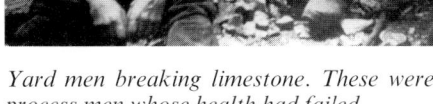

Yard men breaking limestone. These were     Bleach packers wearing their flannel masks
process men whose health had failed

meetings at which one of their number, Roby by name, allowed himself to be exhibited with his immense stature and ruddy complexion as living proof that the works was a healthy place.[14]

It was commonly stated that the saltcake section of a Leblanc works could be identified by the piles of bread crusts which the potmen discarded, being unable to chew. While some of this evidence is anecdotal, Leifchild, ostensibly engaged in examining the children in the works, did succeed in looking at the teeth of the potmen. His comments include 'his teeth were worn down', 'his teeth were nearly gone' and 'his teeth were completely worn down in front'.

Respiratory diseases affected the men in middle and old age, but this may have been due to exposure to heat and draughts – common to most furnace workers – rather than to chemical causes. Tales of regular and repeated gassing were recorded by R.H. Sherard in his *White Slaves of England* (1897), but his information was collected largely in public houses frequented by the men and not on the factory floor. The ramshackle nature of many of the alkali works with their shanty town construction made such accidents almost inevitable.

Process men whose health had broken down were sometimes employed in the yard, especially as 'stone knappers'; these men sat on the ground in front of heaps of pyrites or limestone, breaking the large lumps with hammers to a size more suitable for the crushing machinery. One man said the work was the last

[14] W.A. Campbell, *The old Tyneside chemical trade*, University of Newcastle, Newcastle, 1964, p. 43.

stage before the workhouse, while another claimed it was worse than being on the road except that it was warmer in winter.

# 4   Environmental Aspects

Pollution of the atmosphere and the waterways by the Leblanc factories was a major concern in those centres where the industry was largely concentrated. The scale of the problem can be judged by considering two pollutants, hydrochloric acid gas which was discharged into the air where it combined with atmospheric moisture to fall in droplets over a wide area, and calcium sulphide (alkali waste, tank waste, or galigoo) which was stacked in artificial mountains or dumped at sea.

Every ten tons of salt decomposed yielded six tons of acid gas or nine and a quarter tons of concentrated hydrochloric acid together with six tons of dry calcium sulphide or ten tons of a pasty thixotropic mass. When the acid rain descended on the waste heaps, poisonous and evil-smelling hydrogen sulphide fumes were evolved; moreover, waste heaps underwent slow oxidation, sometimes becoming red-hot and emitting sulphur dioxide which tarnished all the brassware in the neighbourhood. The heaps were physically unstable, behaving like quicksands when disturbed. On Tyneside alone, 100,000 tons of salt were decomposed every year resulting in 92,500 tons of acid and 100,000 tons of alkali waste.

The alkali manufacturer is often portrayed as an ogre who made money out of the destruction of the environment, careless of the effect his operations had upon the local communities. This is a false picture, though the destruction was real enough. If a by-product could possibly be sold or recycled, then that course was taken. The sight of the white cloud emerging from the chimney, or the barge carrying the alkali waste out to sea, reminded the manufacturer of a dire waste of money. In fact, to satisfy the urgent demand for soda, much more hydrochloric acid was produced than could be sold; and in chemistry, product and by-product are linked by inexorable laws of nature. The most ingenious chemist cannot change the ratio of salt to hydrochloric acid or tank waste; all he can do is to devise more responsible ways of utilising the by-products. The search for such methods occupied much time and effort. Among uses for alkali waste which were explored and largely rejected were remedies for dry-rot and for potato and vine blight, in-fill for road making, cement and mortar, and glass-making.

## (a)   Remedial measures

Sometimes the relief of one kind of nuisance simply gave rise to another. In 1836 William Gossage, a salt and alkali manufacturer at Stoke Prior in Worcestershire, invented a tower which would trap the acid gas before it entered the chimney. Where the tower was used, it prevented the emission of the familiar white cloud but resulted instead in the production of unsaleable hydrochloric

*William Gossage (1799–1877) invented a tower for absorbing hydrochloric acid gas in water*

*Walter Weldon (1832–1885) invented a process for recovering manganese used in chlorine making*

acid which was discharged into the nearest river or canal. The tower reduced the draught through the saltcake furnace, so the potmen who found their output threatened often took steps to manipulate the dampers in order to by-pass the tower.[15]

At one time or another nearly every alkali manufacturer had to face action for damage to vegetation and livestock. Most pleaded that they were doing all that they could to find solutions to the problem. A solution did exist but it was not at first economically feasible. In 1770 the Swedish chemist Scheele had made chlorine from hydrochloric acid by boiling it with manganese dioxide, and in 1784 Berthollet in France had demonstrated the bleaching power of chlorine. The gas was difficult and dangerous to handle, so it was dissolved in water to which soda or potash had been added. In 1799 Tennant of Glasgow used milk of lime to dissolve chlorine, and in the same year his partner Charles Macintosh absorbed chlorine in dry slaked lime, so producing a bleaching powder or 'chorine of lime'.

## (b)  Utilisation of waste products

Though this seems like a solution to the problems of both alkali manufacturers

---

[15] J.F. Allen, 'Industrial celebrities: William Gossage', *Chem. Trade J.*, 1889, **2**, 111; D.J. Adam, 'William Gossage', *Educ. Chem.*, 1977, **44**, 46–47; D.W.F. Hardie, 'William Gossage', *Chem. Age*, 1958, **79**, 363.

and their hard-pressed neighbours, there was an obstacle to the general adoption of the scheme. All the manganese and half the chlorine was lost, and the high cost of wasted manganese made the bleaching powder too expensive.

$$MnO_2 + 4HCl = MnCl_2 + Cl_2 + 2H_2O;$$
$$Cl_2 + CaO = CaOCl_2 \text{ (bleaching powder)}$$

The answer to this difficulty was supplied not by an industrial chemist but by a journalist. Walter Weldon, founder and editor of *Weldon's Fashion Journal*, *Weldon's Patterns*, and *Weldon's Household Encyclopaedia*, began his experiments at St. Helens and completed them at Walker Chemical Works in 1869. He mixed a slurry of lime with the waste manganese liquors and blew a current of air through the mixture. The manganese was precipitated as 'Weldon mud' which was used to generate more chlorine. The 'Weldon blowers' soon became salient features, especially of Tyneside works.[16]

$$MnCl_2 + Ca(OH)_2 = Mn(OH)_2 + CaCl_2;$$
$$Mn(OH)_2 + \tfrac{1}{2}O_2 = MnO_2 + H_2O$$

In Widnes, St. Helens and Runcorn a different mode of operation prevailed. Henry Deacon, a Sandemanian protégé of Michael Faraday with experience in the iron and glass trades, devised a catalytic method of converting hydrochloric acid gas into chlorine by mixing it with air and passing the mixture over lumps of clay impregnated with copper salts ('Deacon marbles'). The resulting chlorine was diluted with nitrogen from the air. This led to regional differences in the construction and working of bleaching powder chambers, those for Weldon chlorine having thicker layers of lime and shorter periods of exposure to chlorine than Deacon chambers. Some unchanged hydrochloric acid was of course still required for metal finishing, copper extraction, and for generating hydrogen sulphide from alkali waste for separating copper from silver in burnt pyrites.

Where bleaching powder was made, there were still occasional escapes of acid gas due to leaks, poor working practices, and the general difficulty of managing a gas which quickly revealed its presence. In any case, the Leblanc factories had already earned their evil reputation. Muspratt had been forced to leave Liverpool for Widnes, and Clapham had been driven out of Newcastle to establish new works on the South side of the Tyne. Losh at Walker was accused of blighting everything in the vicinity of his works. 'The gardens yield neither fruit nor vegetables; many flourishing trees have become rotten naked sticks. Cattle and poultry droop and pine away.'

Of Widnes the *Victoria County History* commented:

> A district more lacking in attractive natural features it would be difficult to conceive.
> A great cloud of smoke hangs continually over the town, and choking fumes assail

[16] W. Weldon, 'On some recent improvements in industrial chemical processes', *J. Soc. Chem. Ind.*, 1882, **2**, 39.

*River Tyne at St. Anthony's: atmosphere over the chemical area of the River Tyne in the 1870s*

the nose from various works. In the face of such an atmosphere it is not to be wondered at that trees and other green things refuse to grow.

Morrison warned those of a gloomy temperament not to enter the place lest they be tempted to do away with themselves. Even the moderate prose of the Alkali Inspector conveys the same message. 'It is true that those coming to Widnes even from very dark and gloomy skies enter that town with a certain awe and horror ... and wonder if life can be sustained there'.[17] On Tyneside, once a place of sylvan beauty (as can be seen from the water-colours of J.W. Carmichael or T.M. Richardson), only one tree survived on the banks of the river. A local dialect song, suitably rendered into English, complained:

> The banks of the Tyne, I remember,
> Were covered with bonny green fields;
> But now there is nought but big furnaces
> Down from Newcastle to Shields.
> And what with their sulphur and brimstone,
> Their vapour, their smoke and their steam,
> The grass is all gone, and the farmers
> Can neither get butter nor cream.[18]

[17] *Northern Year Book*, Newcastle, 1829, p. 135; Alkali Inspector, *13th Report*, (c.2199), P.P. 1879, XVI, p. 11.
[18] W.A. Campbell, *A century of chemistry on Tyneside*, Society of Chemical Industry, Newcastle, 1968, frontispiece.

**Table 4** *Works on Merseyside and Tyneside registered under the Alkali Act of 1863*

| Merseyside | Tyneside |
|---|---|
| Borax & Alkali Works, Widnes | J. & W. Allen, Heworth |
| Crosfield Brothers, St. Helens | C. Allhusen, South Shore, Gateshead |
| Evans, McBride, St. Helens | T. Burnett, Bill Quay |
| Thomas Fleetwood, Widnes | T. Burnett, Dunston, Gateshead |
| Gaskell & Deacon, Widnes | T. Bramwell, Heworth |
| James Gamble, St. Helens | Blaydon Alkali & Manure Works, Blaydon |
| Greenbank Alkali Works, St. Helens | Carville Chemical Co., Wallsend |
| John Hutchinson No. 1, Widnes | Cook Brothers, Walker |
| John Hutchinson No. 2, Widnes | John Cook, Newcastle |
| Hazlehurst & Sons, Runcorn | Friars Goose Chemical Works, Gateshead |
| Andrew George Kurtz, St. Helens | Heworth Chemical Works, Heworth |
| Mersey Chemical Works, Widnes | Jarrow Chemical Works, Tyne Dock |
| James Muspratt, Liverpool | Jarrow Hill Chemical Co., Jarrow |
| J. Marsh & Co., St. Helens | Thomas Lomax, Jarrow |
| Frederick Muspratt, Widnes | Low Walker Chemical Co., Walker |
| New Road Chemical Works, St. Helens | H. L. Pattinson, Felling, Gateshead |
| Runcorn Soap & Alkali Works | Ebenezer Ridshaw Ridley, Newcastle |
| St. Helens Chemical Co. | Solomon Mease, Jarrow |
| | Tyne Chemical Co., South Shields |
| | Walker Alkali Works, Walker |
| | Washington Chemical Works, Washington, Co. Durham |

In the Alkali Inspectorate lists, both titles and addresses often differ from usual forms.

The local landowners on the South side of the Tyne formed themselves into an Association for the Prosecution of Alkali Manufacturers; meeting in the Blue Bell Inn in Felling, they kept daily records of the descent of white fumes on to the land. Witnesses to actual damage were not always easy to find, for the chemical workers purchased their butter, milk and eggs from the neighbouring producers, and were formidable defenders of their work places. It must not be forgotten, too, that mothers sometimes brought children afflicted with whooping cough into the alkali works so that they could breathe the tainted, but supposedly healing, atmosphere.

In 1863 private litigation by individual landowners was replaced by parliamentary action. Lord Derby, possibly discharging a political debt, had pleaded before the House of Lords that the courts were no longer able to restrain the alkali makers and legislation was necessary. The Alkali Act of that year compelled alkali makers to absorb 95% of the hydrochloric acid gas from the decomposition of salt, and set up an organisation of Chief Alkali Inspector and regional inspectors with rights to enter works and test flue gases. The Act did not restrain the manufacturers from pouring their compulsorily absorbed hydrochloric acid into the rivers and canals, nor did it regulate the emission of other gases. It did however generate new analytical methods for monitoring flue gasses, and it brought competent observers into the works.

At about the same time, a technical change in the paper industry possibly did more to induce alkali makers to convert their acid into chlorine for bleaching powder than did the Act itself. The introduction of Esparto grass into paper making set up a new demand for bleaching powder, and within a few years this by-product of the soda industry had become more valuable than soda; indeed, soda was often made at a loss for the sake of the more profitable bleaching powder.

Less intrusive, but just as economically damaging, was the loss of sulphur in the waste heaps. Gossage had tackled the issue as early as 1837, and several firms on Tyneside had recovered some sulphur by partial oxidation to thiosulphates which could be sold to paper makers for removing the last traces of chlorine from their bleached rags. Ludwig Mond brought his undoubted ingenuity to bear on the matter. Addressing fellow alkali makers he observed:

> I see here several gentlemen who have spent thousands of pounds in trying to turn this obstinate substance to advantage. I see many more who still spend many hundreds every year to have it carried away to the dumb fishes of the sea ... and probably there is nobody in this room who has not more than once had his nose offended and his appetite spoiled by coming into too close contact with the waste heap.[19]

Mond's suggestion was to follow the partial oxidation to thiosulphate with the precipitation of sulphur using waste hydrochloric acid. By this means he obtained a 50% recovery of sulphur, though others were not so successful.

Meanwhile the local populations had to continue to endure the annoyance, and worse, of the heaps. The condition of the River Don, a minor tributary of the Tyne at Jarrow Slake near South Shields, gave rise to mounting complaint. A cluster of small alkali and other chemical works had turned the little stream into a stinking ditch. The situation was made worse by a factory making chlorides of barium and strontium from heavy spar (barium sulphate) and celestine (strontium sulphate) by reduction to sulphides and treatment with hydrochloric acid. The liberated hydrogen sulphide was allowed to escape into the atmosphere without even the benefit of a tall chimney to dissipate the nauseating gas over a wider area.

$$SrSO_4 + 4C = SrS + 4CO;$$
$$SrS + 2HCl = SrCl_2 + H_2S$$

In 1887 the attitude of South Shields town council over this matter was described by the Alkali Inspector as one of 'masterly inactivity' though in the following year they did proceed against the firm.

Also in 1887 a fatal accident occurred which generated less outcry than might have been expected. Waste heaps were frequently drained into the nearest waterway, causing more than one kind of nuisance. At Friars Goose near Gateshead a group of waste heaps drained into the Tyne, and to prevent the

---

[19] L. Mond, 'On the recovery of sulphur from alkali waste', *Trans. Newcastle Chem. Soc.*, 1868, **1**, 75.

*Ludwig Mond (1839–1900) introduced the ammonia–soda process to England at Winnington in Cheshire*

*Alexander M. Chance (1844–1917) successfully recovered sulphur from alkali waste*

smell of hydrogen sulphide being carried down the river a number of ventilation shafts had been sunk into the alkali waste; the sulphide gases of course vented into the surrounding air. The hope that ventilation might assist in the oxidation of sulphides to less noisome sulphites was not realised:

> Although these holes had been properly fenced, some stupid boys could find no better amusement than clambering up and down one of them. The consequence was that they so stirred up the stagnant drainage water at the bottom of the shaft as to cause sufficient $H_2S$ to be evolved as to fill the shaft. One of them, overcome by the gas, fell to the bottom. Two men who gallantly went to his rescue were also overcome. The outcome was fatal in all three cases. During the investigation I was much struck by the want of knowledge among the men continually exposed to danger from this source, how to protect themselves.[20]

At St. Helens it was claimed that the whole neighbourhood was filled with the foul smell of rotten eggs.

> Nor was this horrible stench confined to St. Helens alone. The offensive thick, yellow drainings from the waste heaps ... flowed slowly down the Sankey valley. The

---

[20] Alkali Inspector, *24th Report*, 1887, (*c*.5417), XXVI, p. 35.

Rector of Winwick described the smell as ... like Harrogate water, doubly distilled.[21]

The process which finally displaced all others was that of A.M. Chance (1882) combined with that of C.F. Claus (1888). Carbon dioxide from the lime kilns was blown through a suspension of calcium sulphide waste to liberate hydrogen sulphide gas, leaving chalk as a by-product. The hydrogen sulphide was then partly oxidised to sulphur in a Claus kiln. Care was needed to prevent total oxidation to sulphur dioxide, but the use of a catalyst permitted oxidation to take place at a fairly low temperature.

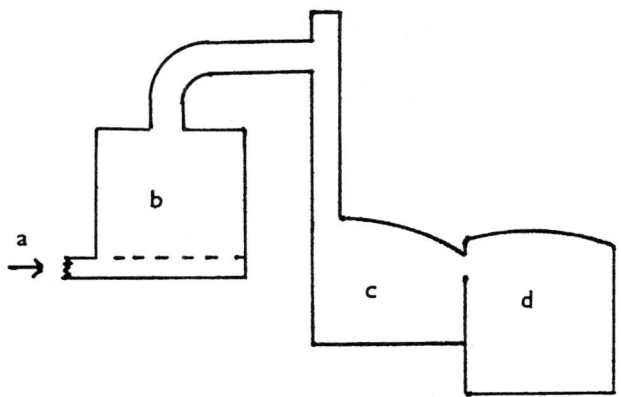

**Claus Kiln for Sulphur Recovery as used at Chance's Oldbury Works in 1887**

a. Inlet for hydrogen sulphide and air
b. Iron oxide chamber
c. Brimstone chamber
d. Flowers of sulphur

*Claus Kiln for sulphur recovery as used at Chance's Oldbury works in 1887*

Sulphur made in this way was extremely pure.[22]

$$CaS + CO_2 + H_2O = CaCO_3 + H_2S;$$
$$H_2S + \tfrac{1}{2}O_2 = H_2O + S$$

Such new additions to the Leblanc process called for new skills on the part of the workers, even if the picture painted by the Alkali Inspector was a little too rosy:

In one works there are four sets of seven carbonators, twenty-eight in all, connected by a series of pipes provided with cocks for directing the passage of the gases in and out of several vessels. Arranged in a single building there may be counted no less

[21] Barker and Harris (ref. 7), p. 353.
[22] A.M. Chance, 'The recovery of sulphur from alkali waste by means of lime kiln gas', *J. Soc. Chem. Ind.*, 1882, **2**, 162.

than a hundred cocks. The correct position of each of these is essential to the continuance of the operation; any misplacement would cause confusion. Yet ... it has been found possible to select from the ranks of the ordinary workmen those who can be trusted to control this complicated apparatus.

Duties were not limited to turning the right cocks at the right time, but the workman was also required to take gas samples and carry out analyses.

Thus the man is raised from the mere toiler valuable in proportion to his strength of muscle, to a skilled artisan, where his power of thought and judgement are brought into play as well as those of hand and arm.

This seems a far cry from the drunken potman!

## 5   Sulphuric Acid Manufacture

The preparation of sulphuric acid was not strictly part of the Leblanc process, but as that process relied so heavily on sulphuric acid all the larger manufacturers made their own acid. Liebig's memorable saying, that the economic health of a country might be measured by its consumption of sulphuric acid, is more than a clever quip. Sulphuric acid for superphosphate and for ammonium sulphate is the basis of the fertiliser trade; for making alum and Epsom salts it lies at the heart of the dyers' art; japanners and tinplate workers have used it since the 18th century; synthetic dyes called for the finest grades of the acid; and today much is used in making detergents, wrappings, pigments and dyes, and in water treatment. Thus Liebig's dictum is still true.

In 1736 a quack doctor named Joshua Ward made 'spirit of vitriol' at Twickenham (see also Chapter 3). His method was to burn a mixture of sulphur and saltpetre in closed flasks containing a little water; several charges were usually needed to produce an acid of useful concentration.[23] It was not at first realised that spirit of vitriol made from sulphur was identical in all but concentration with 'oil of vitriol' made by calcining iron sulphate or green vitriol. This latter method continued to furnish fuming (*i.e.* super-concentrated) acid until late in the 19th century; the principal seat was Nordhausen in Saxony, hence the common name of 'Nordhausen sulphuric acid', but it was also made in the coal-mining areas of Britain. Ten years after Ward's enterprise began, the process was made safer and its scale enlarged by John Roebuck who replaced the glass flasks by lead 'houses', later called lead chambers.[24]

### (a)   Lead chamber process

The lead chamber process in its earliest form mimicked closely Ward's method of working. Sulphur and saltpetre or nitre were ignited in the chamber which contained a few inches of water at the bottom. The nitre furnished oxides of

---

[23] W.A. Campbell, 'Portrait of a quack; Joshua Ward 1685–1761', *Univ. Newcastle Med. Gazette*, 1964, **58**, 118.
[24] Lead was one of the few constructional materials not strongly attacked by sulphuric acid.

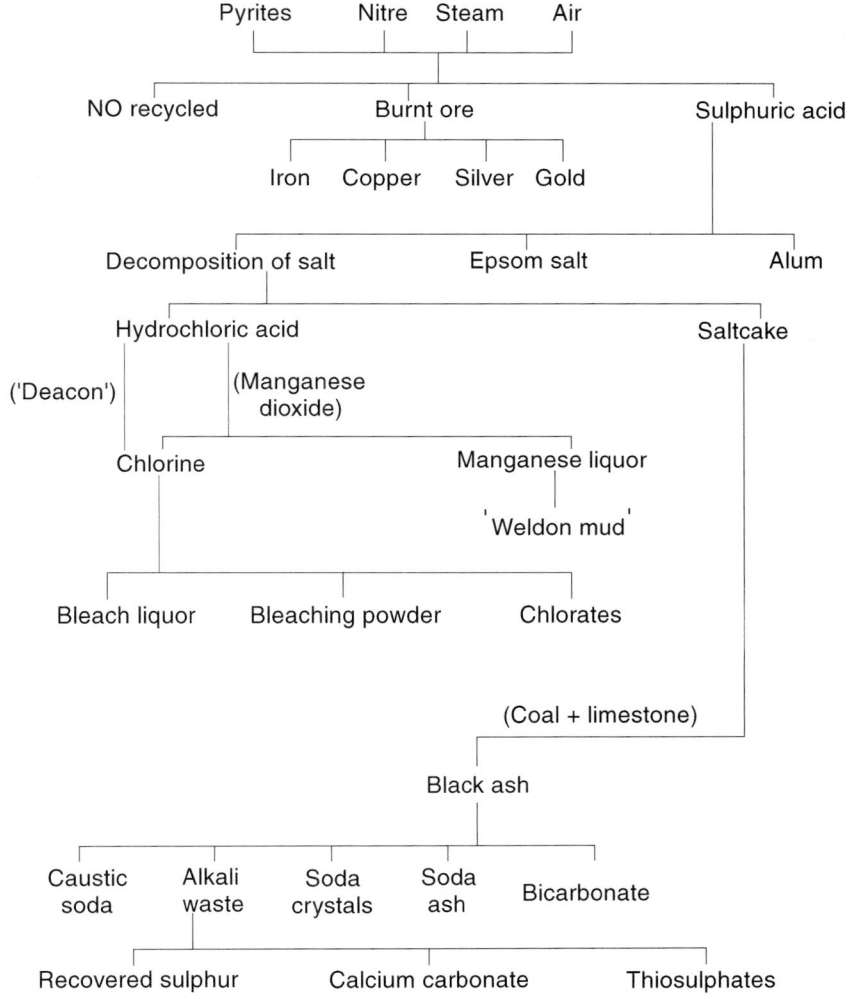

*Later flowsheet of the Leblanc process*

nitrogen which conveyed oxygen to the sulphur dioxide from the burning sulphur. The resulting formation of sulphuric acid does not occur easily with cold water, so time was taken in achieving solution; the product was a dilute acid contaminated with nitric acid. When the acid was concentrated by evaporating off most of the water in open vats, the nitrogen oxides were lost. In times of war, saltpetre was required to make gunpowder and its effective loss through concentrating sulphuric acid was serious. Moreover, the fumes of mixed sulphuric and nitric acids were as evil a pollutant as the hydrochloric acid from the Leblanc process. Isaac Cookson set up an acid works at South Shields on a slight eminence known locally as Paradise; after

*John Glover (1817–1902) invented a tower for denitrifying and concentrating sulphuric acid*

a few months the locals gave the place the more ominous name of Vitriol Hill.[25]

In 1827 the French chemist Gay-Lussac found that oxides of nitrogen would dissolve in moderately strong sulphuric acid and invented a tower in which this could be achieved. Now the question was how to get the nitrogen oxides (chiefly nitric oxide NO) out of the acid for re-use in the chambers. A solution was found in 1858 by John Glover. He had entered the industry as a plumber constructing lead chambers and had risen to become manager of Hugh Lee Pattinson's works at Washington, County Durham; thus he knew the industry from the bottom upwards. Glover's tower allowed hot gases from the sulphur burners, which now were outside the chamber, to sweep the nitric oxide out of Gay-Lussac's nitrified acid and carry it to the chamber, together with sulphur dioxide and hot air. Apart from purifying the Gay-Lussac acid, this operation concentrated the acid and so dispensed with the open vats and the pollution which they brought.[26]

The construction of a lead chamber was a difficult operation owing to the weight of lead involved and the floppy nature of the lead sheets. Sheets of lead were fixed to a horizontal frame of stout wood, using clout nails dipped in molten lead to protect them from the acid. The lead sheets were then joined by a skilled plumber called a lead burner, often using a hydrogen blow-torch. When all the frames had been thus prepared they were carefully hoisted into a vertical position and secured together. The lead burner then joined the corner sheets. In use the sheets became thin, and the chamber was stripped down and the lead melted for re-use, an operation which often led to complaints of nuisance.

[25] Campbell (ref. 14), p. 31.
[26] E.E. Aynsley and W.A. Campbell, 'John Glover and the Clean Air Acts', *Chem. Ind.*, 1959, 1540.

Sulphur for the chamber process came from Sicily and cost £7 per ton in this country. A clumsy attempt by Tennant and Muspratt to buy the sulphur mine led to defensive action and doubled the price of Sicilian sulphur to Britain. Matters were soon adjusted by the arrival in Sicilian waters of a gunboat, but confidence in foreign sulphur was shaken, and a new raw material was sought. As early as 1818 a patent for making sulphuric acid from iron pyrites had been granted to Thomas Hills and Uriah Haddock in London.[27] The first to work the process successfully was Thomas Farmer, owner of an old-established vitriol works in Kennington, in 1838 after the Sicilian *débâcle*.

New furnaces for burning pyrites had to be designed, and new ways found of usefully disposing of the residue of mixed metal oxides. In 1841 Longmaid's process for making saltcake directly from salt and pyrites seemed to offer a way of by-passing the sulphuric acid stage, but the process presented great technical difficulties on a commercial scale, and few who tried it continued with it. One advantage that the process offered was the easy extraction of copper from the burnt pyrites. Ten years later Henderson's wet process for copper extraction made pyrites an altogether more attractive raw material for the sulphuric acid manufacturer, though acid made in this way often contained arsenic. In 1900 beer drinkers in Lancashire began to fall ill in their hundreds, the outbreak being traced to arsenic in the beer arising from the use of arsenical sugar which in turn had been contaminated by arsenical sulphuric acid used in the sugar refining process.[28]

The lead chamber process was essentially:

$$S + O_2 = SO_2;$$
$$NO + \tfrac{1}{2}O_2 = NO_2;$$
$$SO_2 + NO_2 = SO_3 + NO;$$
$$SO_3 + H_2O = H_2SO_4;$$
$$SO_3 + H_2SO_4 = H_2S_2O_7;$$
$$H_2S_2O_7 + H_2O = 2H_2SO_4$$

## (b)    Contact process

The chamber process remained awkward and capricious. While chemists argued about the mechanisms of the chamber reactions, chambers would go out of service only to be coaxed back into operation by means bordering on the magical. Yet as long ago as 1831 Peregrine Phillips, a Bristol vinegar maker, had received a patent for a process which was easier to interpret. He proposed passing air and sulphur dioxide through a tube packed with platinum wire, the resulting sulphur trioxide being absorbed in dilute acid in a stone jar:

$$2SO_2 + O_2 = 2SO_3$$

[27] C.T. Kingzett, *History, products and processes of the alkali trade*, Longman Green, London, 1877, p. 24.
[28] *Report of Royal Commission on arsenical poisoning*, 1902–1903, 94. P.P. 1904, IX, p. 94.

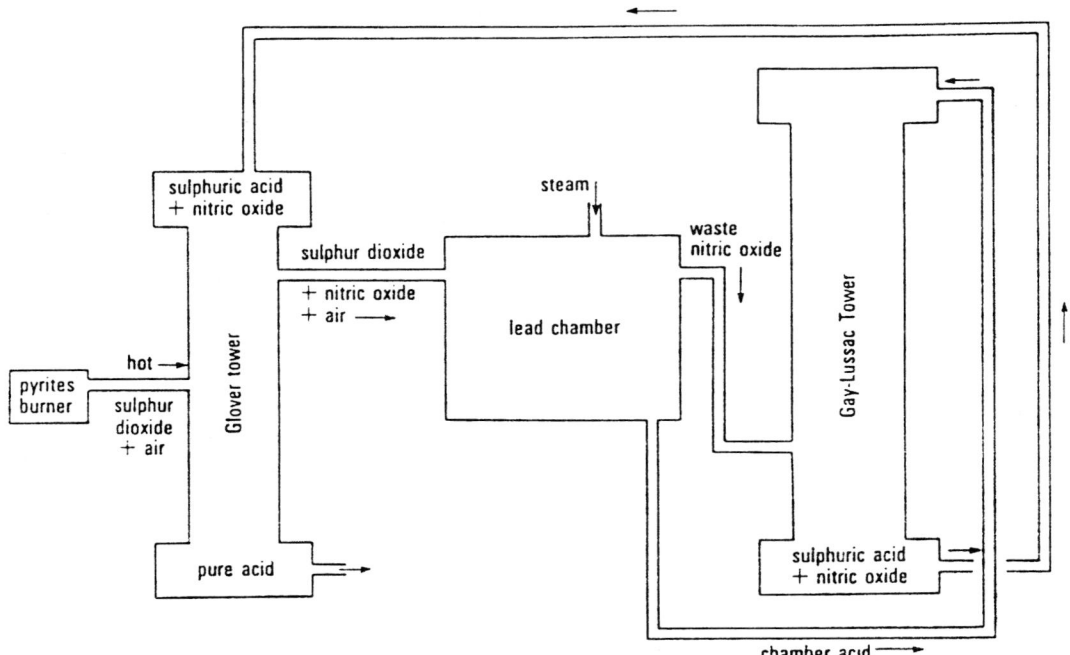

*Circulation of gases and acid in the lead chamber process*

The process did not at first succeed because the concept of gas kinetics lay in the future, as did knowledge of the factors which might inhibit catalytic reactions.[29] Apart from gaps in the appropriate chemical knowledge, there was then little chemical engineering know-how about handling gases.

In 1876 Rudolph Messel and W.S. Squire had some success at their Silvertown chemical works, largely because Messel had researched the poisoning of catalysts. Twenty years later, a team at BASF on the Rhine discovered that the reaction was assisted by an excess of oxygen, and for the first time the contact process became a serious rival to the chamber process. One inherent weakness was the need for a very pure sulphur to avoid catalyst poisoning, whereas crude sulphur ores would suffice for the lead chambers.

In broad terms, Leblanc alkali makers continued with chamber acid, while the purer contact acid was used in making dyestuffs and drugs. Even the fertiliser industry remained loyal to the chamber process, to the extent of introducing new designs of chambers.

[29] E. Cook, 'Peregrine Phillips, inventor of the contact process for sulphuric acid', *Nature*, 1926, **117**, 419.

# 6 Demise of the Leblanc Process

As the sources of waste in the Leblanc process were tackled one by one, so the process became too complicated for easy management. An early Leblanc works would have taken in sulphur, salt, limestone, nitre and coal, turning them into a single saleable produce, soda. By 1890 the same works would take in pyrites, salt, nitre, limestone, coal and manganese ore, producing soda, caustic soda, bicarbonates, thiosulphates, bleaching powder, hypochlorites, chlorates, sulphuric, nitric and hydrochloric acids, pure sulphur, sodium sulphide, calcium sulphate and Epsom salts. Such a trade was economically unstable, for the demand for one product might go up at the time when demand for a related product went down.[30]

Faced by this daunting situation, the Leblanc manufacturers sought safety in solidarity; in 1890 the United Alkali Company was formed by amalgamation of the largest and most powerful Leblanc firms. The following year saw the new company buying up and closing down the smaller and less efficient plants, in fact rationalising the industry.

Although the Company's brave prospectus brought almost the whole of industrial chemistry, as well as mining and electricity supply, within its remit, the UAC was primarily committed to the Leblanc process. Its chief chemist, Ferdinand Hurter,[31] was emotionally as well as scientifically involved with the process to the extent that he was blind to the merits of newer ways of making alkali.

Hurter was born in Switzerland and studied under Bunsen at Heidelberg. He became chief chemist to Gaskell and Deacon in Widnes where he worked out the practical details of Deacon's chlorine process (p. 92). He acquired a reputation as one of the two most learned alkali chemists in Europe, the other being Georg Lunge (manager of the Tyne Alkali Company at South Shields and eventually professor of technical chemistry at Zürich Polytechnic).[32] Hurter's major contribution to the UAC was the founding in 1891 of the central research laboratory in Widnes; here every aspect of the Leblanc process could be examined. It was a bold gesture but it came too late to save the Leblanc process from its rivals.

## (a) Ammonia–soda or Solvay process

The first rival had been known since 1838 when John Hemming and his partner Harrison Dyar were granted a patent for making soda from ammonium carbonate and brine. The by-product was ammonium chloride which was heated with calcium carbonate, chalk or limestone, to form ammonium carbonate again. This recycling of the ammonium salt was only partly effective, and the partners lost a great deal of money. The underlying idea was never-

[30] J.W. Mellor, *Comprehensive treatise on inorganic and theoretical chemistry*, Longman Green, London, 1926, p. 728.
[31] D. Adams, 'A chief among chemists', *Chem. Brit.*, 1998, **34** (6), 43–44.
[32] P. Bedson, *J. Chem. Soc.*, 1923, **123**, 948.

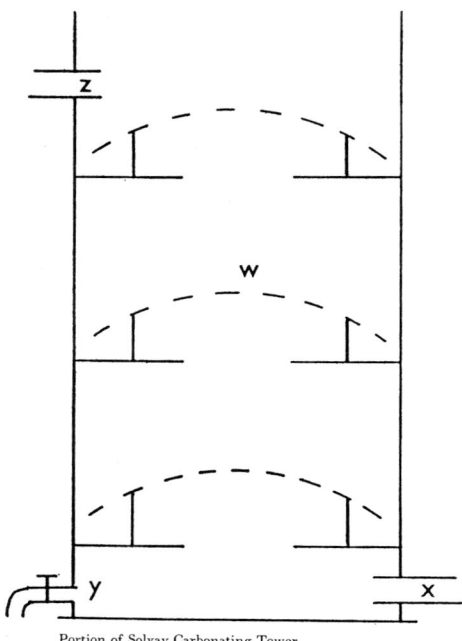

Portion of Solvay Carbonating Tower

w. Perforated dome to disperse carbon dioxide through liquid
x. Inlet for carbon dioxide
y. Stopcock for running off carbonated liquor
z. Inlet for ammoniacal brine

*Portion of Solvay carbonating tower*

theless tempting, and several others – including Muspratt, Gossage, and Gaskell & Deacon – also lost heavily while trying to make it work on an industrial scale. During the 1860s, backed by experience in the salt and coal gas industries, Ernest Solvay[33] patiently experimented with a modified process using the same chemical reactions near Charleroi in Belgium. His method was to saturate the brine with ammonia and to pass carbon dioxide through the ammoniacal solution when sodium bicarbonate separated. Finally the bicarbonate was heated to yield soda and carbon dioxide. The most difficult part of the process was the recovery of ammonia from the by-product ammonium chloride:

$$NaCl + NH_3 + CO_2 + H_2O = NaHCO_3 + NH_4Cl;$$
$$2NaHCO_3 = Na_2CO_3 + H_2O + CO_2;$$
$$2NH_4Cl + CaO = 2NH_3 + CaCl_2 + H_2O$$

Success came to Solvay at his works opened near Charleroi in 1873, and in that year he granted a licence to Ludwig Mond to work the process at Winnington in Cheshire. Mond had learned his chemistry from Bunsen at Heidelberg, and had

[33] D.W.F. Hardie, 'Ernest Solvay', *Chem. Age*, 1958, **80**, 234.

already experienced the alkali trade at Hutchinson's works in Widnes. There were still many technical problems to be solved and Mond brought all his energy and enthusiasm to these, even sleeping in a camp-bed on a catwalk over the blowing engine in order to be on hand when trouble arose.[34]

Joining with John Brunner, he created the firm of Brunner Mond, the first real threat to the Leblanc trade. That threat was limited by the fact that the new ammonia–soda plant could not make bleaching powder. The commercial stability of the new soda industry was strengthened by their ability to buy cheap ammonia liquors from near-by coal gas works; later, as explained in Chapter 5, they were further helped by Fritz Haber's discovery of a way to make ammonia from nitrogen in the atmosphere.

Brine pumping led to the collapse of houses and larger buildings through subsidence. Among the buildings which stood, many were askew or had metal plates and tie-bars to hold them together. The firm acknowledged its share in causing damage to property, and when John Brunner became a Liberal member of parliament he introduced a bill to compensate those affected by the extraction of brine.[35]

The social consequences of the rise of the ammonia–soda trade were considerable. The new factories did not emit acid fumes from their chimneys, nor did they leave mountains of smelly waste. They also called for a different type of workmen; in place of the burly but skilful men with rakes and slices, men were needed who could read dials and take appropriate action by manipulating taps and valves. Mond however remained puzzled by the attitudes of some of his men whom he described as foolish, obstinate and lazy. Nevertheless, the firm built workmen's houses, opened a co-operative store and provided changing rooms with showers at the works. A free school followed in 1886 and a recreational club in 1890.

### (b)  Electrolysis of brine

Meanwhile the secure confidence of the Leblanc manufacturers in their unique ability to make bleaching powder was to be shattered. In 1895 the Castner–Kellner Alkali Company was formed, commencing production of caustic soda and chlorine by the electrolysis of brine at Runcorn in 1897. These two products easily react together so the major technical difficulty was to keep them apart.

A cell with a porous diaphragm had been patented in 1888, but Hamilton Young Castner,[36] a skilled chemical engineer, who had managed the Aluminium Company's works at Oldbury near Birmingham, was working on entirely different lines. He designed a cell in which the sodium formed at the cathode was dissolved in mercury and the resultant sodium amalgam allowed to react with water to yield caustic soda. The chlorine formed at the anode was available for

[34] L. Mond, 'The origin of the ammonia soda process', *J. Soc. Chem. Ind.*, 1885, **5**, 257.

[35] W.F.L. Dick, *A hundred years of alkali in Cheshire*, ICI, London, 1973, p. 96.

[36] A. Kidner, in D. J. Jeremy (ed.), *Dictionary of Business Biography*, vol. i, Butterworth, London, 1984, pp. 614–616.

making bleaching powder, formerly the trump card of the Leblanc process. Environmentally the new electrolytic process was a great improvement as there were no unsaleable by-products.

The UAC Board was inclined to try out the new process, but Hurter advised against it on the ground that electric power would prove too expensive. This advice effectively took the UAC out of the mainstream of industrial development and consigned it to the backwater of declining Leblanc prospects.[37] In fact, the invention of the steam turbine generator in 1884 by Charles Parsons, and the concept of 'electricity in bulk' due to John Theodore Merz in 1900, brought the cost of electricity within the reach of alkali manufacturers, making possible the development of what is now known as the chlor-alkali trade.

## 7   Coda

Why does alkali merit a whole chapter in a book on the social and environmental history of the chemical industry? The answer lies in the nature of the chemical input into the Industrial Revolution. The mechanisation of the textile trade and the gathering together of the spinning and weaving machines into mills powered first by water wheels and then by steam engines permitted a vastly increased output of cloth. The cloth, however, could not be finished without soap, dyestuffs, mordants and bleach; the mills had large windows, thus setting up an increased demand for glass; and the movement of people from small rural communities engaged in agriculture to large urban communities engaged in industry meant that a small proportion of rural workers had to feed a larger proportion of town dwellers, thus calling for more fertilisers. In all this activity, three chemicals occupied key positions, namely alkali, sulphuric acid, and chlorine bleach. The effects on family and community life of the arrival of plentiful soap, inexpensive glassware, household bleaches, and cheap paper for printing are obvious. These social improvements were a gift from the alkali industry.

---

[37] W.J. Reader, *Imperial Chemical Industries: a history*, : vol. i, *The forerunners, 1870–1926*, Oxford University Press, London, 1970, pp. 116–119.

Chapter 5 **_The Nitrogen Industry_**

W.A. CAMPBELL

## 1  Chemical Manure

The middle of the 18th century marks a watershed in British agriculture. Earlier, Britain had been a grain-exporting country on a large scale but later found some difficulty in feeding its increasingly urbanised population. Around this time, leading figures in science, especially in Scotland, saw the need for a closer collaboration between chemistry and agriculture.

In 1757 Francis Home, Professor of Materia Medica in Edinburgh, and a pioneer of the use of sulphuric acid in bleaching, had published his *Principles of agriculture and vegetation*. In this work he advocated the use of soot and nitrates, both nitrogenous fertilisers though chosen on empirical grounds.

Archibald Cochrane, ninth Earl of Dundonald (p. 77), brought out his *Treatise on the connection of agriculture with chemistry*, in 1795, suggesting that mineral fertilisers should be based on the composition of plant ashes, putting back what the plant had taken out from the soil. This book was commended by Humphry Davy in his *Elements of agricultural chemistry* of 1813.

The great thrust to the fertiliser industry however came from Justus von Liebig, Professor of Chemistry at Giessen, and described in his own day as 'the Pope of chemistry'. In 1837 Liebig was invited to address the Liverpool meeting of the British Association for the Advancement of Science, giving him the opportunity to meet a number of prominent chemists. Supposedly in response to a request from the BAAS, he published his *Chemistry in its application to agriculture and physiology* in 1840. Two years later, Lyon Playfair took Liebig on a kind of 'royal progress', introducing him to titled landowners and leading industrialists. Liebig was entertained by the Manchester chemical entrepreneur William Henry; here he first encountered an English butler, and was over-awed by the splendour of the house and the standard of living of a British chemical manufacturer.

### (a)  Liebig and superphosphate

It had been known for a long time that bones were an effective fertiliser on some soils but not on others. James Murray, Inspector for Ireland under the Anatomy Act, had employed 'vitriolised bones' near Belfast in 1808, but it was Liebig who explained the action of sulphuric acid in liberating the phosphate in bones to produce a faster acting fertiliser named 'superphosphate'.

*Justus von Liebig (1803–1873), founder of the research school at Giessen and pioneer in both organic analysis and agricultural chemistry*

*Sir James Murray (1788–1871), pioneer in superphosphate manufacture*

*Sir John Bennet Lawes (1814–1899), founder of Rothamsted*

*Sir Joseph Henry Gilbert (1817–1901), Lawes's collaborator at Rothamsted for 57 years*

John Bennet Lawes, an Oxford graduate with some chemical knowledge who was heir to the Rothamsted estate, had begun to experiment with crop rotation and liming in 1832 and was well placed to test Liebig's conclusions. In 1841 he built the first superphosphate factory in this country at Deptford Creek on the Thames. Here he used as raw materials bones, bone ash from the potteries, spent animal charcoal from sugar refineries, and Cambridge coprolites (fossilised dung from prehistoric animals, the significance of which had been explained to Liebig by the geologist William Buckland). Many buyers stipulated that their superphosphate must be made only from coprolites, presumably on the ground that they were dug from the soil to which they were to be returned. Lawes's list of raw materials indicates that the chemical industry was a user as well as a creator of waste products.[1]

Where he got his bones from was the manufacturer's own problem, though Muspratt speaks darkly of 'bones from the butcher and bones from the battlefield'. Sampson Langdale, one of the largest manure manufacturers in the country, entered into an advantageous agreement with a meat-canning establishment on the Danube to ship their waste bones back to England. In warm weather, seamen refused to sail in the bone ships and Langdale was compelled to build a calcining factory at the canning site, converting the noisome cargo into bone ash. The supply of bones was not sufficient to satisfy a growing trade, and Lawes soon began to employ such mineral phosphates as

[1] Sir J. Russell, *Chem. Ind.*, 1931, 823; Sir W. Ogg, *Chemistry and crop nutrition*, RIC Lecture Reprints, 1955, No. 5; J. Fenwick Allen, *Some founders of the chemical industry*, Sherratt & Hughes, Manchester, 1907, p. 94.

*Rothamsted Manor, home of J.B. Lawes and agricultural research centre*

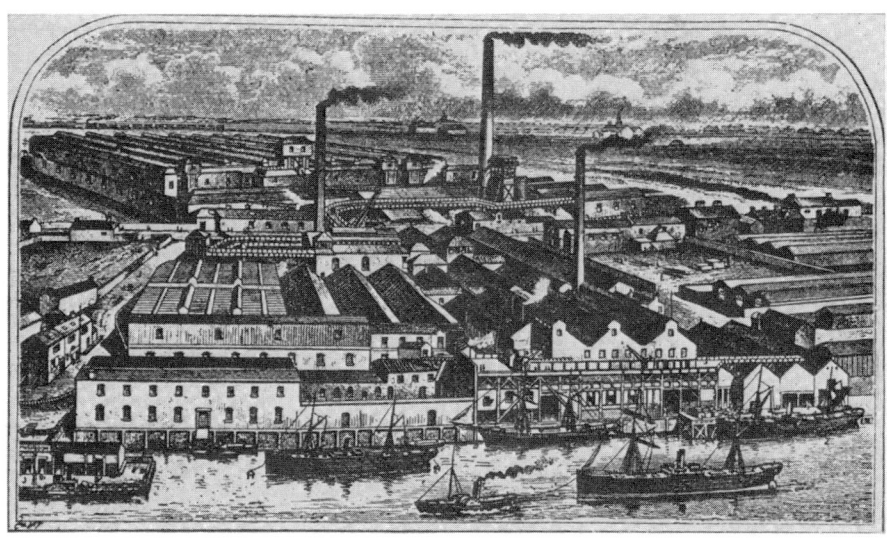

*Langdale's Manure Works, Newcastle, 1875*

*Blaydon's Alkali and Manure Works: an advertisement from 1950*

apatite, a fluorine-bearing calcium phosphate. Continuing the Liebig connection, Lawes engaged Joseph Henry Gilbert, one of Liebig's PhD graduates, to manage the research side of the works. Lawes left the business early to continue his scientific farming, and in 1872 his factory was acquired by Sampson Langdale; the works closed in 1969.

Though Lawes and Gilbert owed much to Liebig in the matter of soluble phosphates, this did not deter them from successfully disputing Liebig's claim that plants got all their nitrogen requirements from the air and not from the soil. They showed at Rothamsted that crops of wheat rose from twenty to forty bushels per acre when the land was treated with ammonium salts.

In 1844 the second chemical manure factory in England was opened at Blaydon on the banks of the Tyne to the west of Newcastle. The founder was Thomas Richardson who had gained his PhD at Giessen under Liebig. Richardson went on to become Reader in chemistry at the Newcastle College of Medicine, and to run an analytical and consulting service for local industry as well as conducting private classes in chemistry.[2]

There was little sophistication in these pioneering manure factories. A 'den'

---

[2] W.A. Campbell, *A century of chemistry on Tyneside*, SCI, Newcastle, 1968, p. 29.

was constructed by building a circular wall of ashes about two feet in height. Crushed bones were thrown into the enclosed space, and sulphuric acid poured on to them. Twenty-four hours were allowed for the reaction to occur, during which time the heap was turned repeatedly by a man with a spade. The wall of ashes was then raked into the heap and thoroughly mixed with the treated bones.

Superphosphate as produced in these factories was a pasty mass which had to be mixed with absorbent material to render it more manageable by the farmer. Murray, who had transferred his operation from Belfast to Dublin, specified in a patent 'bran, sawdust, dust of malt, husks of seeds, brewer's and distiller's grain, ground rags, dry tan, siliceous sand, peat or other sandy mould, dry dust, earthy clay, fine sifted cinders and the like'. All this suggests a certain desperation in the face of an intractable material.[3]

A better way of dealing with superphosphate was to mix it with other fertilising substances yielding nitrogen and potassium to make a balanced fertiliser. This brought the manure manufacturer into direct conflict with the surrounding communities. Even the simple treatment of mineral phosphates with sulphuric acid was unpleasant enough, for among the gases evolved were the highly corrosive hydrogen fluoride and silicon tetrafluoride. But the desire to mix superphosphate with nitrogenous manures brought in the horse slaughterer, bone boiler, and glue maker with predictable consequences. At Blackburn the rakings from the town middens were carried to the chemical manure works, while in Newcastle Sampson Langdale, a larger-than-life character, offered to build a public convenience at his own expense if the effluent could be piped into his factory; a suspicious Town Council rejected the offer.

## (b)    Manure and the environment

As late as 1882, the environmental impact of a chemical manure factory was graphically set out by the Alkali Inspector:

> The manure is made from carcasses, shoddy, leather, slaughterhouse refuse, and some mineral phosphate. The method is as follows: a heap of twenty to thirty tons of shoddy is made in the shed, and onto this is poured blood and refuse from the slaughter-house. Any carcasses that the owner may buy are, after being skinned, buried in this heap, the heap being allowed to stand and rot for five or six months. This is then shovelled into the mixer with some leather, crushed bones and acid. After mixing, this is let into an open den and a man shovels onto it a certain quantity of mineral phosphate.

The inspector described the stench during mixing as 'simply intolerable'.[4] Some firms even compounded the offence by including fish offal and the sweepings from the fish market.

In spite of the comments above, this kind of nuisance was not subject to

[3] W.A.L. Alford and J.W. Parker, *Chem. Ind.*, 1953, 852.
[4] Alkali Inspector, *19th Report*, (c.3715), P.P. 1882, XVIII, p. 74.

control by the Alkali Inspector whose responsibility was limited to that portion of the works where sulphuric acid was added to mineral phosphate. Fortunately, the first Chief Alkali Inspector, Robert Angus Smith, had taken a broad view of his duty; where he could not interfere under the Act, he was ready to offer unofficial advice as suggested by his motto '*Nihil chymicum ad meum alienum puto*' (*I consider nothing chemical as alien to myself*).

Local authorities made by-laws to regulate the trade relying on their Nuisance Inspectors to enforce them. Particularly they placed limits on the proximity of manure works to dwelling houses. The lax manner in which such by-laws were interpreted is revealed by the-following exchange at a nuisance enquiry:

> *Mayor*: At what distance are the works from inhabited houses?
> *Witness*: There are inhabited houses all round, but they are occupied almost entirely by their own workmen.[5]

Fertiliser was for many decades the mainstay of the lead chamber process for sulphuric acid; indeed, the last lead chambers to continue in operation were those in the manure trade where the stringent purity criteria of, say, the dyestuff industry did not apply. The smaller operators usually bought in their acid from the Leblanc alkali factories. This, with the dependence on slaughterhouses, tanyards, sugar refineries *etc.*, illustrates again the network principle which governed so much of 19th century industry (Table 1). The manure works was usually situated near to a tanyard and a slaughter-house at least, and often some of the others would be present as well. This reduced transport costs, but also minimised the potential for nuisance through carrying smelly materials in open carts. Having regard to the very large production of superphosphate (Lawes alone was making 40,000 tons per year in the 1860s) this was not a trivial consideration.

John Morrison, a Tyneside manure manufacturer, whose writings reveal a deep distrust of workmen, introduced mechanisation into his factory. The mixing with sulphuric acid was carried out in an iron cylinder resembling the 'revolver' of the Leblanc alkali works. He also designed a tower for absorbing those acid fumes which caused so much complaint and litigation, and Morrison towers soon became common features in manure works. Being dependent on machinery, Morrison extended the network principle by siting his factory between those of the railway pioneer Robert Stephenson and Hawthorns the marine engineers.

## (c)   The analysis scandal

The chemical manure trade was widely distrusted by farmers. Much of the manure which arrived at the farm was of a low standard, though it seems that the fraud was perpetrated more by the wholesaler than by the manufacturer. Readers of *Barchester Towers* will remember that an archdeacon who had

[5] Campbell (ref. 2), p. 30.

**Table 1** *Industrial network based on livestock*

| Source | Primary outlets | Secondary outlets |
| --- | --- | --- |
| Hides | Tanning | White lead |
|  | Fertiliser |  |
|  | Glue and gelatine |  |
| Blood | Prussian blue | Paint |
|  | Fertiliser |  |
| Bones | Cutlery |  |
|  | Fertiliser |  |
| Bone ash | Pottery |  |
|  | Phosphorus | Matches |
|  | Superphosphate |  |
| Bone black | Sugar refining |  |
|  | Printing ink |  |
| Grease | Lubricant |  |
|  | 'Butterine' | Butter substitute |
| Ammonia | Dyeing |  |
|  | Bleaching |  |
|  | Fertiliser |  |
| Sal ammoniac | Batteries |  |
|  | Metal finishing |  |
| Dippel's oil | Heterocyclic bases | Dyestuffs, medicinals |
| Hair and wool | Textiles |  |
| Wool-fat (lanolin) | Pharmacy |  |
| Tallow | Soap |  |
|  | Candles |  |

purchased a ton and a half of guano complained that there was not five hundredweight of genuine product in it. There were, however, deeper reasons for concern than simple adulteration, no matter how gross.

Manure was sold on the basis of chemical analysis for nitrogen, phosphorus, and potash content, or any one of these. The analysts often contradicted each other's results, and it was easy for both buyers and sellers to identify 'high' and 'low' analysts, and to make money by buying on a low analysis and selling on a high figure. A faulty analysis could easily lead to a manufacturer being branded as a swindler.

In order to bring some order into a chaotic situation, it was necessary to standardise the analytical methods in use, and this proved difficult. Commercial analysts were unwilling to disclose their analytical procedures. In 1874 the BAAS voted £10 to a committee led by A.H. Allen, Public Analyst for Sheffield, to investigate analytical methods for chemical manure. In answer to a thousand letters of enquiry, only thirty-seven replies were received. One prominent London firm of analysts answered: 'We do not think ourselves called upon to publish our methods of analysis which we have perfected after long and careful investigation for the sake of those who have not taken the trouble'. Another analyst stated that his analytical methods were his only capital.[6]

[6] *BAAS Reports*, 1875, 24; 1877, 9; W.A. Campbell, *Endeavour*, 1979, ser. 2, **3**, 83.

The confusion was further compounded by the habit of attaching money values to analytical results. Two analysts, examining the same manure, placed values of £5 and £8/10 (£8.50) per ton upon it. Such evaluation was of course quite outside the competence of the chemist, especially as the putative price was based only on the raw materials, taking no account of the value to the farmer. These scandals led to the passing of the Fertilisers and Feeding Stuffs Act of 1893, permitting County Councils to appoint Official Agricultural Analysts whose function in the agricultural field was similar to that exercised in the foods and drugs field by the Public Analysts. The 1906 amendment to the Act led to the publication of official methods for the analysis of agricultural products. Knowledge that their results might be scrutinised by the Official Agricultural Analyst impelled the commercial analysts to adopt the official analytical methods.

Concern about the poor quality of agricultural analysis also played a part in the organisation of chemists in general, especially through the founding of the (Royal) Institute of Chemistry in 1877.

Lest this paints too gloomy a picture of the relation between the chemical manure trade and its customers, it should be stated that there were excellent analysts who devoted their careers to improving the scientific knowledge available to landowners and farmers. Two examples will suffice.

Farmers were at the mercy of manure manufacturers when it came to selecting a fertiliser, for they had no knowledge, other than crop behaviour, about the nature or extent of mineral deficiency in their soils. For this reason, a group of Midlothian farmers in 1843 formed an Agricultural Chemistry Association. Out of ten applicants, James Finlay Weir Johnston was chosen to be the Association's chemist. He was already Reader in Chemistry in the University of Durham where he was neither overworked nor overpaid; the Association added £400 a year to his income. Johnston had studied chemistry with Thomas Thomson in Glasgow and with Berzelius in Sweden, and his best-selling *Chemistry of common life* showed his ability to communicate chemical principles to a wide audience.

In 1848 the Association was taken over by the Highlands and Agricultural Society who provided Johnston with a laboratory in Edinburgh and three assistant chemists earning £300 a year. Through public lectures he popularised the ideas of Liebig, and provided soil analyses for the members. Johnston was active in the foundation of the BAAS in 1831 and of the Chemical Society ten years later. Thus agricultural chemistry had a significant presence in these national bodies, and Scottish farmers had access to scientific information about their needs.[7]

South of the border, one of Johnston's students was active in a similar way. Augustus Voelcker became Professor of Chemistry in 1849 at the Royal Agricultural College, Cirencester, a bold venture aimed at combining academic and commercial interests. He also became chemical adviser to the Royal

[7] A. and N.L. Clow, *The chemical revolution*, Batchworth, London, 1952 [reprinted Gordon & Breach, Philadelphia, 1992], pp. 505, 507, 510; G.R. Clemo and N.S. Brown, *J. Roy. Inst. Chem.*, 1956, **80**, 14.

Agricultural Society. Voelcker introduced British farmers to the use of potassium salts from the Stassfurt deposits in the 1860s. He was a founder member of the (Royal) Institute of Chemistry, and a formidable critic of the state of commercial analysis in Britain. He sat on the Parliamentary Commission on Food Adulteration in 1872. Like Johnston, Voelcker was elected FRS. Thus two well-trained and highly articulate chemists raised their voices to protect the farmer from the incompetent analyst and the unscrupulous supplier.[8]

Agricultural chemistry was to make a further, though unexpected, contribution to national well-being. Liebig had hoped that farmers would not merely make use of chemistry but would study the subject, and the notion gained ground that a knowledge of chemistry would inevitably lead to an increased yield from the land. Samuel Parkes, author and sulphuric acid manufacturer, wrote: 'Is your son heir to a Large estate? Make him an analytical chemist'.

As they had done with the Royal Institution at its founding in 1799, influential landowners eagerly espoused the idea of a college where agricultural chemistry would be taught and from which new knowledge of use to the farmer would emanate. As plans took shape, the post of director was offered to Carl Remigius Fresenius, arguably the most learned analyst in Europe and a powerful contributer to the debates on fertiliser analysis, but he declined it. Next, Heinrich Will, inventor of a method for determining nitrogen in manures, was approached, with the same result. In the end, August Wilhelm Hofmann, a brilliant organic chemist with absolutely no interest in agriculture, became Professor at the Royal College of Chemistry. From that appointment sprang the whole field of coal-tar chemistry and the synthetic dyestuffs industry, but only cold comfort for the landowners.

It is clear that the manufacture of fertilisers gave rise to pollution and nuisance, but we must now consider what effects fertilisers had upon British farming and the rural landscape. Fertilisers allowed the cultivation of scrubland and the upgrading of poor grazing land, and the increased area devoted to wheat brought the country nearer to self-sufficiency than it had been for a century. Chemical manure allowed a field to bear the same kind of crop year after year, removing the visual variety provided by crop rotation and the fallow periods.

The situation changed in the early years of the 20th century when the import of cheap grain from Canada and America came in. Mechanisation and cultivation on a huge scale in those countries brought advantages in which the smaller British farms could not share. This competition led many farmers, tempted into wheat growing by cheap fertilisers, to turn from arable to dairy farming, while others simply left the land.

The important fertiliser ammonium sulphate is dealt with under ammonia below.

---

[8] G. Taylor, *The fertilizer and feeding stuffs Acts and their analytical applications*, RIC Lecture Reprints, 1948 (30th Streatfield Mem. Lect.).

## 2 Explosives

It is often said that the three greatest technological advances of the Renaissance period were moveable type, the mariner's compass, and gunpowder. The manufacture of black powder (p. 62) from sulphur, charcoal and saltpetre is not a chemical process, though the social and environmental impact of the product on warfare, civil engineering, and country pursuits has been far-reaching in the extreme. Here we are concerned only with those explosives which contain nitrogen and are made by a process of chemical change. The nitration products of cellulose and glycerol are dealt with in Chapter 8, but two important nitrated explosives remain.

### (a) Picric acid

Picric acid was discovered by the eccentric recluse Peter Woulfe in 1771 when he observed the action of nitric acid on indigo. The fact was reported in the French chemical literature, and eight years later Jean Joseph Welter was able to use picric acid to dye wool and silk a brilliant yellow shade. Picric acid was so named on account of its intensely bitter taste; its chemical name is 2,4,6-trinitrophenol. For about a century it was used in 1% aqueous solution as a treatment for burns; bright yellow burn dressings were a feature of first-aid kits.

In the 1840s picric acid began to be made as a dyestuff by nitrating phenol, a coal tar product. The process was improved by first sulphonating the phenol in an iron pot heated by steam. Exactly a hundred years after Woulfe's discovery, Hermann Sprengel demonstrated the explosive nature of picric acid, and it was quickly taken up by the military, especially in France. Picric acid is a strong acid which reacts easily with metals and their oxides, the resulting metal picrates being extremely sensitive explosives. For this reason the nitration stage was carried out in small earthenware vessels, and the insides of shells to be filled with picric acid were varnished.[9]

Lead picrate or a mixture of picric acid and lead oxide is a particularly violent explosive. In 1887 an appalling explosion occurred at the dyestuff works of Roberts Dale near Manchester. An accidental fire had melted a heap of picric acid which flowed on to some lead, creating a detonator for the entire stock. A similar explosion happened in 1900 at Read Holliday's dye works at Huddersfield. So great was the fear of lead picrate that all traces of lead paint were removed from the vehicles in which picric acid was transported.

During the First World War, picric acid was made in very large quantities under the name of 'Lyddite' (after Lydd in Kent where early tests had been done). The women who worked on production and those who filled the shells elsewhere found their hair and skin stained yellow. Perhaps fortunately, they tended to live in closed communities of 'munition cottages', often little bigger than garden sheds, hastily put up in the neighbourhoods of the new factories. Some of these emergency housing estates survived until the 1950s. The nitration

---

[9] W. Taylor, *Modern explosives*, RIC Lecture Reprints, 1959, No. 5, p. 22.

*Nitration plant for picric acid, at Brookes Chemicals, Halifax, 1915*

stage of the synthesis produced copious brown fumes of poisonous nitrogen dioxide, so it was usually arranged that nitration should take place only under cover of darkness. An explosives factory was an unwelcome neighbour, quite apart from the danger of explosion. Siegfried Sassoon, recovering from a war wound, spent some time at a training camp in Cheshire, close to Brotherton's explosives factory 'which flared and seethed and reeked with poisonous vapours a few hundred yards away'.

Picric acid was usually fired by a detonator of mercury fulminate, but for a time a French priming charge of potassium chlorate mixed with coal-tar was employed. A terrifyingly sensitive composition of picric acid mixed with potassium chlorate was tried as a shell filling, but proved too dangerous to handle.

## (b)  TNT

As the war progressed, picric acid was largely replaced by the less sensitive TNT, trinitrotoluene. The raw materials were toluene (also a coal-tar product), nitric acid, and oleum (fuming sulphuric acid). Before the war, Britain had been almost totally dependent on Germany for supplies of oleum. The lead chamber plants as used in the fertiliser and alkali trades could not make oleum directly, so contact process plants were built at the TNT works. When nitric acid was made in the old way by acting on sodium nitrate from Chile with sulphuric acid, large amounts of 'nitre cake' (sodium hydrogen sulphate) were generated. As this salt possesses strongly acid properties, much of it was used in place of

*TNT washing plant, at the Government factory at Queensferry, 1917*

sulphuric acid in such industries as metal finishing. Otherwise it was dissolved in water and run into the sea; much nitre cake from the Merseyside factories was dumped off the Welsh coast.

TNT was used alone as a bursting charge in shells into which it was poured in the molten state. It was also used in such compound explosives as Amatol (20% TNT and 80% ammonium nitrate). The mining explosive Ammonal contained 30% TNT, 47% ammonium nitrate, 22% aluminium powder and 1% charcoal. The manufacture and quality control of such compound explosives was a highly sophisticated matter, for which the expertise emerged only slowly. Much of the research was entrusted to the long-established government factories such as Waltham Abbey with its centuries old experience of gunpowder manufacture, and to the ordnance factories like that of Armstrong at Elswick, Newcastle. Lord Armstrong had established a testing ground deep in the Northumberland moors at Ridsdale; after Armstrong's death, his partner Sir Andrew Noble FRS, a former captain in the Royal Artillery, conducted research into the whole field of ballistics.[10]

Unless the strictest hygiene could be enforced in the TNT factories, the health of the workers suffered from both dust and fumes. Personal susceptibility seemed to play a large part, and workers who were prone to chestiness or dermatitis could soon be identified and transferred to other jobs.

The largest TNT plant was established at Queensferry near Chester, with a capacity of 100 tons per day. Other important plants were built at Gretna, Avonmouth and Greenwich, each with its own sulphuric acid facility.

[10] Sir A. Noble, *Artillery and explosives*, John Murray, London, 1906.

Though the First World War gave a powerful thrust to the chemistry and engineering involved in TNT manufacture, production did not cease entirely when hostilities ended. Mining, quarrying, tunnelling, road and harbour construction all demanded high explosives, often with highly specific characteristics, and this kept some of the Royal Ordnance factories in work. Quantities of underwater explosive were used in blowing up submerged wrecks, which were a continuing hazard left over from the war. Totally new uses have been found, ranging from exploration for oil to separating the stages of rockets as they journey into space.

Safety in the manufacture of explosives was first regulated by the Explosives Act of 1875 which classified explosives according to the ease with which they could be set off. Amendments to the Act, together with Home Office rules, regulate the lay-out of factories, requiring statutory distances between buildings, and the provision of protective mounds of earth so that a minor incident does not become a major disaster. Routes have been devised for the movement of chemicals through the factory, so that, for example, chlorates do not come into accidental contact with either sulphur or charcoal.

## (c)   **Detonators**

Small copper tubes containing cap-composition are known as detonators. A detonator can be fired by percussion, ignition (including electrical), or more commonly by a fuse. Bickford's safety fuse (1831) consisted of a flexible tube containing gunpowder; the time of burning was proportional to the length of fuse, and specially slow-burning fuses were available; the rate of combustion of the gunpowder was reduced by mixing in up to 10% of rosin. Bickford and Smith, the Cornish developers and manufacturers of the fuses, were absorbed into Nobel's industrial complex in 1918.

For detonating explosives, fulminates and azides are chiefly used. Fulminating gold and silver have been known for centuries. Samuel Pepys witnessed an experiment by a Dr Allen involving 'something made of gold, which they call in chemistry aurum fulminans, a grain of which put into a spoon and fired will give a blow like a musket, and strike a hole through the spoon downwards'. Pomet's *History of drugs* (1712) recommended mercury fulminate as a 'sudorifick very proper in the small pox, being given two grains to six. It is likewise good to stop vomiting'. Silver fulminate was extensively studied by Liebig in 1826 in Gay-Lussac's Paris laboratory, a courageous piece of work which did much to gain him his chair at Giessen.

Mercury fulminate was discovered in 1800 by Edward Charles Howard. In spite of serious injury sustained in the course of his research, Howard continued to experiment with this dangerous material. As a result of Howard's persistence, musketry underwent a change around 1824, the flint lock being replaced by the percussion cap. A typical cap composition contained mercury fulminate, antimony sulphide, and potassium chlorate. Later, a little gunpowder was added to give a more durable flash to set off the main charge. The small paper percussion caps for toy guns usually contained potassium chlorate and red phosphorus. For Christmas crackers, silver fulminate was preferred.[11]

The manufacture of mercury fulminate is simply a scaling up of the laboratory preparation. One part by weight of mercury is dissolved in ten parts of nitric acid, and 8.3 parts of ethanol are added to the solution. A vigorous reaction occurs (for which reason a large vessel is used) and grey-white crystals of fulminate separate out. Care must be exercised through control of temperature to avoid the formation of large crystals which are especially sensitive to shock or friction.

Lead azide is a salt of the phenomenally sensitive and rarely encountered hydrazoic acid $HN_3$; even a sample-containing 30% water will detonate. Instead of the free acid, sodium azide is used as the starting material for the preparation of lead azide. The sodium salt is made on a large scale by passing nitrous oxide $N_2O$ over sodamide at 190 °C. An alternative method is to treat sodium nitrite with hydrazine.

$$NaNH_2 + N_2O = NaN_3 + H_2O$$
$$H_2NNH_2 + NaNO_2 = NaN_3 + 2 H_2O$$

Sodium azide is easily soluble in water, and thus the lead salt can be prepared by precipitating with lead acetate. Although lead azide has certain advantages over mercury fulminate, these have to be weighed against its tendency to react with other metals, especially the copper tubing in the detonator, forming the much more sensitive copper azide.

Since the quantities of explosive required for detonators have always been far smaller than those for shell-filling, the impact on the neighbouring community has been correspondingly reduced. For the workers, there has been the hazard of lead or mercury poisoning, or the inhalation of nitrous fumes, but these have largely been controlled by the conventional methods of industrial hygiene.

## 3   Cyanides and Their Derivatives

### (a)   Prussiates

Although nitrogen was recognised only in 1766 by the Scottish chemist Daniel Rutherford, its chemistry is much older. In industrial terms, it may be dated from the beginning of the 18th century when compounds containing the CN group began to be made. The source of nitrogen for these purposes was always some sort of animal matter. The manufacture of Prussian Blue and the prussiates (hexacyanoferrates) is explained later (see p. 163). Calico printers made early use of the yellow prussiate of potash (potassium ferrocyanide or hexacyanoferrate(II)); cloth was soaked in a solution of the prussiate and then treated with a ferric salt when Prussian Blue was re-formed within the fibres.

The earliest process for prussiates worked in this country was cumbersome

---

[11] J. Read, *Explosives*, Penguin, Harmondsworth, 1942, p. 1312.

but ultimately effective. A mixture of saltpetre and tartar from the wine vats was ignited to yield potassium carbonate which was calcined with dried blood. The product was extracted with water and boiled with alum and iron(II) sulphate, the solution being concentrated and left to crystallise. It was soon recognised that flesh, wool, or horns could be substituted for blood, and Edward Turner, first Professor of Chemistry at London University, added guano to the list of nitrogen providers.

The largest manufacturer of prussiates in Britain was Thomas Bramwell of Heworth on the south bank of the River Tyne. His business can be traced back to the Germany of Dippel and Diesbach. Diesbach's process for Prussian Blue was kept secret until 1711 when it was described in the Archives of the Berlin Academy, recently established in 1700. No account in English appeared until 1724 when the Royal Society published a paper by Woodward. Yet the manufacture of Prussian Blue had commenced on Tyneside in the very early years of the century when a German Jew had set up a plant in Gateshead, eventually settling at Corbridge in Northumberland where there was a community of dyers. There he was assisted by a Mr Atkinson.[12]

When the Corbridge business failed, it was bought by Thomas Simpson 'a gentleman of extensive knowledge in Chemistry, and of unwearied and persevering industry' who transferred the manufacture to Elswick, Newcastle. Again, Mr Atkinson was in charge of the practical running of the works. On the death of Simpson in 1758, Thomas Bramwell bought the business and moved it to Heworth under the style of Bramwell and Atkinson; the managing partner was the grandson of the original Atkinson who had learned his trade from the German who first introduced it to this country.

Early in the 19th century the Gas Light and Coke Company made Prussian Blue from cyanides obtained from their ammoniacal liquors; in this they were followed by several smaller gas companies. Indeed, the Prussian Blue often appeared spontaneously under gas-works conditions, as industrial archaeologists who have excavated old gas-works will know.

As other uses for animal products developed, manufacturers of prussiates found themselves starved of nitrogen. It was natural therefore that experiments should be instituted on the use of atmospheric nitrogen. In 1839 Lewis Thompson of Byker, Newcastle, drew attention to the fact that cyanogen, $(CN)_2$, was produced when coke, potash and iron were heated in a stream of air. The mechanism probably involved the formation and decomposition of hexacyanoferrates. The Society of Arts was sufficiently impressed to award a medal to Thompson for his observation. Meanwhile Lyon Playfair, acting on advice from Bunsen, made cyanides by passing nitrogen over a heated mixture of potash and charred sugar. Bramwell's firm took up the quest, but at the high temperatures needed for success the reactants attacked the materials of the plant.[13] Only in 1899 was Scottish Cyanides able to operate a similar process

[12] R. Surtees, *History and antiquities of the County Palatine of Durham*, Gateshead section, Andrews, Durham, 1830, p. 40.
[13] G. Martin, *Industrial and manufacturing chemistry*, Part II, *Inorganic*, Crosby Lockwood & Son, London, 1925, pp. 483–495.

successfully, using the electric furnace invented by Héroult in 1887. The importance of cyanides to the gold mining trade is explained on p. 312.

### (b) Cyanamide

While nitrogen does not combine easily with carbon, it reacts much more readily with calcium carbide to form calcium cyanamide $CaCN_2$. To make calcium carbide very high temperatures are needed, such as can only be attained in electric furnaces, so the manufacture of carbide is confined to countries where cheap electric power, usually from hydroelectric installations, is available. Although carbide was made by the Acetylene Illumination Company in 1895, the coal-fired power stations of the time could not compete on price with hydroelectric stations, and carbide production in this country ceased in 1914. Thereafter the principal sources of carbide and cyanamide for use as a nitrogen-rich fertiliser were Norway, Canada and America. The value of cyanamide to the chemical industry was enhanced by the discovery that it can be hydrolysed by superheated steam to yield ammonia.

## 4 Leather and Glue

Since collagen, the chief constituent of hides, contains some 18% of nitrogen, the conversion of hides to leather, and the associated manufacture of glue and gelatin, must be considered part of the chemical industry of nitrogen.

The tanning trade is one of the most efficient in the natural products sector of the chemical industry, with its careful attention to the utilisation of waste products. In 1889 a tanner told the annual meeting of the BAAS:

> All the waste products of a tannery are capable of utilisation. The horns go to the comb maker, the hair and wool to the carpet thread spinner, the waste parts of skin and feet are made into glue and gelatin. Even the spent bark and sewage of the factory, dried and ground together with some charcoal from the same source, form a manure with a sensible quantity of ammonia and phosphate.[14]

In fact, the leather trade stood at the centre of an efficient industrial network. The tannery would often be situated next to a slaughterhouse, a bone-processing plant, a pottery which would take the bone ash, a chemical manure factory, and a white lead works which could use the spent bark from the lanyard. Such clusters were to be found, among other places, on the east, west, and north sides of Newcastle-upon-Tyne.

The oldest method of making leather was practised in Britain by the Romans, much of whose leather has been recovered from excavations at civilian sites such as Vindolanda in Northumberland. The hides were stripped of hair and simply steeped in pits filled with a 'soup' of bark, leaves or fruits of suitable plants. The tannic acid in the plant material combined with gelatin in the hide to form a durable leather. The tanning process was slow, and as weeks passed the pits

[14] J.W. Richardson (ed.), *Handbook to local industries*, BAAS, Newcastle, 1889, p. 177.

developed a smell which was a serious nuisance. Another ancient method involved treating the hides with alum and salt, producing a leather of different texture.[15] In 1858 the use of chromium salts was introduced, making a tough leather for the soles of boots and shoes. In chrome tanning the hide was steeped in an acid solution of a dichromate and then transferred to a bath of hypo (sodium thiosulphate) which reduced the absorbed chromate to a green chromium salt.

This further stimulated the trade in chromates, originally made for the pigment industry. Since 1830 the firm of J. & J. White had made chromates at Rutherglen near Glasgow, and as chrome leather gained in popularity they opened a subsidiary, the Eglinton Chrome Tannery in Glasgow. After a time, the tannery began to make its own chromium salts. In 1927 an old-established fertiliser and sulphuric acid firm at Eaglescliffe, County Durham, turned their attention to the manufacture of chromates on a large scale. After amalgamation with J. & J. White the firm became British Chrome and Chemicals, closing down in the 1960s. Working with chromates was hazardous, the workers being prone to 'chrome ulcers' on the hands.

Apart from this, work in a tanyard was not particularly dangerous although the cutting and trimming of hides did lead to accidents. As was the case with lead (p. 303), users were exposed to more danger than producers. Thus in boot and shoe factories where soles and heels were stamped out from whole hides of leather on a piece-work basis, many cutters lost fingers in an attempt to squeeze one more sole from a hide; indeed, losing a finger-end was regarded as some kind of baptism into the trade.

The word 'tanning' was used when no tannins were involved, as in the production of soft absorbent wash leathers by treating the hide with some easily oxidisable oil, and allowing a pile of hides to heat up through the heat of aerial oxidation. Since fish oils were frequently employed, these establishments were often judged a nuisance.[16]

Tanneries were formerly sited near to the forests from which the bark of various sorts was obtained. From the 1860s, tannins began to be extracted at source, and concentrated tanning juices began to be imported into this country from America, Canada, and India. By this time the major centre of the tanning trade had been established in the neighbourhood of Leeds, and the old geographical connection between spent bark and white lead works (p. 159) was severed. The diminishing leather trade on Tyneside continued to use bark, but also made much use of the ground leaves of the shumac tree imported from Palermo. Leather tanned with shumac had a pleasant spicy smell.

The trimmings from hides before tanning were converted into glue or gelatine, either in the tanning establishment or in a separate glue factory. The process entailed the partial degradation of collagen by simmering the pieces in

---

[15] B. Blount and A.G. Bloxam, *Chemistry for engineers and manufacturers*, C. Griffin & Co., London, vol. 2, 1896, pp. 367ff; C. Singer, *The earliest chemical industry*, Folio Society, London, 1948, p. 74.

[16] G. Dodd, *Days at the factories*, Charles Knight & Co., London, 1843 [reprinted A.M. Kelley, New York, 1967], p. 161.

water at 60 °C for six to eight hours. As soon as a test portion would set, the liquor was run off for concentration and drying. There was a close connection with this trade and the manufacture of paper, for large quantities of gelatine were required for sizing writing paper to prevent the ink running.

Crushed and degreased bones were also made into glue in a similar manner to hide offcuts. This branch of the trade was a cognate of the bone meal and fertiliser industry; it therefore shared the opprobrium which the manure trade invited. Williamson and Corder at Walker-on-Tyne (later British Glues and Chemicals) earned the pejorative title of the 'Bone Yard' on account of the persistent smell which hung in the fog on the river.

Grease extracted from bones by steam treatment, mixed with milk and chalk, was sold as a cheap butter substitute under the name of 'butterine'. Public Analysts' reports indicate that purchasers were often led to believe that butterine was a cheap grade of butter, and that the courts regarded this deception lightly.

Even less popular as a neighbour was the manufacturer of fish glue. Offal of cod, haddock and hake was treated in the same way as skins and bones. Fish glue was not taken to dryness, being sold as a viscous liquid. The popularity of such hobbies as fretwork and model making, and the launch of such periodicals as *The Woodworker*, gave a fillip to all sections of the glue trade. Whether from hides, bones or fish, the trade provides another example of the determination to find useful outlets for even the least promising kinds of waste.[17]

## 5   Ammonia

Until the early years of the 20th century there were three sources of volatile alkali or ammonia. The first was the deposits of sal ammoniac (ammonium chloride) found in desert regions of Egypt. This salt became an important reagent for the preparation of metallic chlorides from the days of Paracelsus to the 1650s when Glauber pointed out a more direct route. We shall consider the manufacture of this material below.

### (a)   The trade in urine

The second source was stale or 'fermented' urine in which urea had degraded into ammonium carbonate. Charles Macintosh (Sir Walter Scott's 'Webster Charlie') required ammonia for bleaching cloth, and for the manufacture of a lichen dye named cudbear after Cuthbert, brother of its inventor George Gordon. Macintosh acquired a fleet of horse-drawn carts equipped with barrels and sent them out to purchase the contents of chamber pots from Glasgow and Edinburgh. This in itself had a social consequence, for the urine was no longer poured out from tenement windows on to the heads of passers-by, with or without the traditional warning cry of 'Gardy loo' (*Gardez l'eau*).[18]

[17] G. Martin, *Industrial and manufacturing chemistry*, Part I, *Organic*, Crosby Lockwood & Son, London, 1913, p. 593.
[18] A. and N.L. Clow (ref. 7), p. 211; F.M.L Thompson, *Econ. Hist. Rev.*, 1968, 2nd ser., **21**, 62.

The urine was at first purchased solely by volume, leading to a mean, but widespread, adulteration with water from the nearest standpipe. To counteract this deception, Macintosh commissioned a Glasgow instrument maker called Twaddell to make robust hydrometers with which the carters could test the gravity of the urine. Until very recently, densities of such liquids as sulphuric acid and caustic soda were expressed in degrees Twaddell. Macintosh's operations called for some two thousand gallons of urine per day, so to supplement the supply from the carts he encouraged local factory owners to install casks in their works. After he had distilled off the ammonia he was left with a residue containing microcosmic salt ($NH_4KHPO_4$) which he sold to metal finishers as a cheap substitute for borax as a flux. The cudbear trade became an important part of Scottish economy, soon leading to an eight-fold increase in the price of the lichen.

### (b)   Gas-works ammonia

With the spread of gas-lighting in the first two decades of the 19th century there arrived a new source of ammonia. The crude gas was washed by passing it through water; the resulting ammoniacal liquor was usually converted to ammonium sulphate for the fertiliser trade. These liquors also contained significant amounts of hydrogen sulphide which led to complaints of nuisance from nearby residents. If the sulphide was absorbed on lime ('gas-lime') or moist iron oxide then the transport of that material through the streets was another cause of complaint.[19] Nevertheless the ammonia section was an important item in the economy of a town's gas-works. In 1810 Macintosh undertook to buy all the tar and ammonia from the Glasgow gas-works, a move which led him to the eponymous waterproof cloth.[20]

Even with ammonia from every town's gas-works, the demands of the fertiliser, dyestuffs, and explosives industries were barely met. The prospect of making ammonia from the nitrogen of the air by direct combination with hydrogen had attracted chemists since 1784 when Berthollet had shown ammonia to be a compound of these two gases. Insufficient knowledge of the theory of gas reactions and lack of experience in the industrial manipulation of gases combined to frustrate efforts at ammonia synthesis throughout the 19th century.

### (c)   The Haber process

In 1904 Fritz Haber at Karlsruhe began to study the equilibrium between nitrogen, hydrogen and ammonia at various temperatures and pressures, and in the presence of various catalysts. Success on a laboratory scale came with the use of the rare metal osmium. As a result the German firm of Badische Anilin

[19] W.A. Campbell, *Archaeologia Aeliana*, 1985, 5th ser., 13, 200; *Newcastle Council Reports*, Jan. 7, 1870, p. 53; Jan. 11, 1871, p. 123.
[20] A. and N.L. Clow (ref. 7), p. 251.

*Fritz Haber (1868–1934)*

und Soda Fabrik (BASF) bought up the entire world supply of osmium, only to be informed by one of Haber's collaborators that a far cheaper iron catalyst worked much better.[21]

In 1913 Carl Bosch brought Haber's laboratory process into full industrial production, and the fixation of nitrogen as ammonia was firmly established. During the First World War it became clear to the British authorities that the Haber process was enabling Germany to continue the war, and in 1917 the Government decided to set up a synthetic ammonia plant. A 100 acre site at Billingham on Tees was selected; hydrogen was to be made from water-gas and also by Lavoisier's process of passing steam over red-hot iron. In conjunction with the Billingham enterprise, the Castner–Kellner plant at Runcorn was selected for ammonia synthesis since they already made hydrogen as a by-product of their electrolytic chlor-alkali process.

The underlying reason for the Government to interest itself in ammonia production was the demand for ammonium nitrate explosive and ammonium sulphate fertiliser. The latter was to be made at Billingham, not from sulphuric acid but from the large deposits of calcium sulphate (anhydrite) which lay beneath the site. The reaction takes place in two stages; saturated ammonia solution is treated with carbon dioxide to form ammonium carbonate which then reacts with solid anhydrite to yield ammonium sulphate.

$$2NH_3 + H_2O + CO_2 = (NH_4)_2CO_3$$
$$(NH_4)_2CO_3 + CaSO_4 = (NH_4)_2SO_4 + CaCO_3$$

[21] D. Stoltzenberg, *Fritz Haber. Chemiker, Nobelpreisträger, Deutscher, Jude. Eine Biografie*, VCH, Weinheim and New York, 1994; M.F. Perutz, *New York Review*, June 20, 1996.

*Hydrogen plant at Billingham*

Once more this seemingly simple transformation ran into enormous engineering difficulties. The Billingham affair brought the British chemical industry into the political arena, and into the popular press. In April 1920 the Government offered its Billingham factory for sale, when the only firm to show any interest was Brunner Mond. The package was to include all relevant knowledge in Government hands, including that acquired by the Allied Commissioners who had investigated the Haber–Bosch plant at the BASF factory at Oppau on the Rhine. A newspaper reported that 'all enemy patents have been placed by the Government at the company's disposal', and much popular feeling was aroused against Brunner Mond who were thought to be making private profit out of secret information.[22]

In fact, at the time of sale the Billingham 'factory' was a collection of derelict buildings on an undrained site. The development into production in 1924 was entirely due to Brunner Mond's subsidiary Synthetic Ammonia and Nitrates. Long after this enterprise had been absorbed into ICI, older inhabitants of Billingham continued to refer to the whole establishment as 'The Synthetic' or 'The Ammonia'.

During the Second World War, concern about the vulnerability of the well-known Billingham plant to air attack led to the establishment of an ammonium

[22] D.W.F. Hardie and J.D. Pratt, *A history of the modern British chemical industry*, Pergamon, Oxford, 1966, p. 106; G. P. Pollitt, *J. Soc. Chem. Ind*, 1927, 291.

*Synthetic ammonia plant at Billingham*

sulphate plant at Prudhoe, fourteen miles west of Newcastle-upon-Tyne. The factory was owned by the Ministry of Supply with ICI in a managerial role.[23] The by-product of calcium carbonate was piled into elongated mounds to shield the factory from the nearby railway line to Cumbria and South-West Scotland (and *vice versa*). The mounds were planted with turf in which the employees carved a large white elephant.

### (d)  Ammonium chloride or sal ammoniac

The manufacture of ammonium chloride in Britain was originally a Scottish enterprise. The first successful operation was that of James Hutton at Edinburgh in 1756. Round-bottomed flasks of green glass were charged with soot and arranged in oblong furnaces with their necks protruding. The arrangement was so similar to that in early sulphuric acid plants, such as that of Joshua Ward at Twickenham in 1736, that some transfer of technique seems likely. The sal ammoniac sublimed on the cool necks of the flasks which were sometimes broken off to retrieve the salt. Hutton and his partner Davie took all the soot that the Edinburgh sweeps could furnish. The profits from the manufacture of sal ammoniac enabled Hutton to leave the trade and to devote the rest of his life to those pioneering geological studies for which he is famous.

In the 1780s Charles Macintosh began to make sal ammoniac from a mixture of soot and urine, following a method devised by Baumé in Paris. Like Hutton he employed green glass bottles, possibly obtained from Roebuck's oil of vitriol

[23] G.I. Higson, *Chem. Ind.*, 1951, 750.

works at Prestonpans. The sal ammoniac was sold to dyers, braziers and tinplate workers.[24]

Lord Dundonald engaged in ammonium chloride manufacture, drawing his ammonia from the ammoniacal liquors at his coal tar works. He bought the necessary hydrochloric acid from Roebuck's successor at Prestonpans. He also continued to use the older method based on soot and urine.

By 1814 there was only one sizeable factory in Scotland still engaged in the sal ammoniac trade. This was Joseph Astley's establishment at Bo'ness where woollen rags were heated with the bittern left in the salt pans after common salt had crystallised out and consisting mainly of magnesium chloride solution. Thus two more waste products were being transformed into a saleable material. Astley, like the earlier manufacturers, continued to use the soot process, in pursuit of which he attempted to bribe the Edinburgh sweeps who were then under contract to supply soot to a manufacturer in Beverley named Walker.

As gas lighting became more widespread, the sal ammoniac trade moved from dedicated factories to the gas works with their ready supply of ammonia.[25] The invention in 1867 of the Leclanché cell using moist ammonium chloride as electrolyte provided a new outlet to the trade, and instituted a number of social changes based on the portable dry battery and the flashlamp.

## 6   Nitric Acid

That *aqua fortis* or nitric acid was important in very early commerce can be gleaned from the *Probierbüchlein* (little testing book) of 1556, where the use of nitric acid for separating gold from silver is described. At that time it would be made by the dry distillation of a mixture of saltpetre and copperas [iron(II) sulphate crystals]. A century later nitric acid was regularly made by the action of sulphuric acid on saltpetre; the residue, nitre cake or potassium sulphate, was sold to alum makers or to the glassworks.[26]

In spite of competition for saltpetre from the gunpowder mills, this method sufficed until well into the 19th-century. The discovery of vast deposits of sodium nitrate on the eastern slopes of coastal hills in Chile and the import into this country of this Chile nitre in 1830 opened a new route to the manufacture of nitric acid. Thus when the demand for nitric acid suddenly escalated owing to the establishment of the synthetic dyestuffs industry the new raw material was securely in place. The larger Leblanc alkali factories, where there was always spare sulphuric acid, became the chief suppliers.

It is clear that the transport of Chile nitre would be slow and expensive, and so distant a source would be open to political or military disruption. Attempts to make nitric acid from atmospheric nitrogen had therefore been undertaken over many years, without prospect of commercial success. Lord Rayleigh demonstrated that nitrogen and oxygen would combine to form nitric oxide

[24] A. and N.L. Clow (ref. 7), p. 419.
[25] Dodd (ref. 16), p. 411.
[26] E. Ronalds and T. Richardson (eds. and tr.), *Knapp's chemical technology or chemistry applied to the arts and manufacturers*, vol. 1, London, 1848, 423.

NO from which nitric acid could be made, but only at the temperature of the flame from a powerful electric arc. Efforts to scale up the reaction failed because the surface area of the arc flame was small, restricting the quantity of gas that could react.

In 1903 Kristian Birkeland and Samuel Eyde in Norway invented a disc-shaped arc by spreading out a conventional arc under the influence of an oscillating magnetic field. The large area of flame permitted a rapid oxidation of nitrogen, but the cost in terms of electrical energy was high so that the Birkeland–Eyde process was commercially possible only in countries with cheap hydroelectric power, especially Norway, Canada and America.[27]

Isaac Milner in Cambridge had demonstrated in 1788 that ammonia could be oxidised to oxides of nitrogen by passing it, mixed with air, over heated manganese dioxide. Many workers studied the oxidation of ammonia, employing a range of catalysts, but it was Ostwald who in 1908 established the correct procedure. Using a platinum catalyst in a nickel tube he was able to achieve a 90% conversion. By 1917 the Gas Light and Coke Company at their Beckton works in London were making nitric acid from their own ammonia using a catalyst of their own design. A similar plant was operated by Brunner Mond in 1917, and by the Widnes works of the United Alkali Company in 1922.

Haber and Bosch devoted half of the ammonia from their Oppau factory to the production of nitric acid and ultimately of ammonium nitrate, an important ingredient of explosives. In 1919 when peace was declared, Allied investigators visited the Oppau plant in connection with both reparations and war guilt; it was easy for the directors to claim that nitrates were made for use in fertilisers.

The huge stockpile of ammonium nitrate at Oppau furnished a terrifying example of the destructive potential of the salt. On September 21, 1921, at 7.30 a.m. there occurred an explosion which destroyed the ammonium nitrate silos, leaving a deep crater; 531 lives were lost and 7000 persons made homeless. The cause of the reaction is still argued, but it seems likely that the rock-hard mass was being broken down by blasting, experiments on a small scale having suggested that the procedure was safe.[28]

This disaster alerted the British authorities to the possibility of caking when ammonium nitrate is stored in bulk, with consequent danger when the hard mass is broken up. Two methods were adopted to minimise the hazard; the crystals were waxed, or the crystal habit was modified by changing the conditions of crystallisation. More recently, the practice of producing agricultural materials in small, hard, spherical granules known as prills has made caking almost impossible.[29]

[27] J.R. Partington and L.H. Parker, *The nitrogen industry*, Constable, London, 1922, p. 175.
[28] BASF, *In the realm of chemistry*, Econ-Verlag, Düsseldorf, 1965, p. 94.
[29] L.M. Dewhurst, in P.J.T. Morris, W.A. Campbell and H.L. Roberts (eds.), *Milestones in 150 years of the chemical industry*, Royal Society of Chemistry, Cambridge, 1991, p. 160.

*Crater after the Oppau explosion of 1921*

# Chapter 6 *The British Pharmaceutical Industry*

N.G. COLEY

## 1 Introduction

It is a popular perception that 'the chemist' dispenses medical prescriptions and sells proprietary medicines, cosmetics and related items from a shop on the high street. This view of the chemist as a pharmacist and retailer allows people to recognise a direct input of chemistry into their lives at a personal level. Thus the retail pharmaceutical industry is in a position to influence popular notions of the role of chemistry in everyday life in a different way from almost any other branch of the chemical industry. Pharmaceutical preparations are usually accompanied by seductive claims for their therapeutic properties, but the range of chemicals and drugs used is limited and perceived differences between rival preparations depend largely on the formulation of the mixture and the persuasive skills of the advertiser. Nevertheless, modern legal demands ensure that the chemical names of all the ingredients and their proportions in the mixture are printed on the package. Retail chemists who supply such products direct to the public are well-known through their advertising and distinctive presence in the high street and shopping precinct, but the great pharmaceutical corporations like SmithKline Beecham, Rhône-Poulenc, Glaxo-Wellcome, or Merck, which form the foundation of the modern international pharmaceutical industry, are less widely known.

The pharmaceutical industry works closely with the medical profession which makes specific demands, exerts constraints and promotes what it deems the most successful products. Medical control of the use and dosage of drugs is necessary to limit, or prevent undesirable side-effects. In these and other ways, most notably the immense costs of research and development which are higher than in any other branch of chemical industry, the pharmaceutical industry is unique, although it has much in common with the fine chemicals industry.

## 2 The British Pharmaceutical Industry in the 18th and Early 19th Centuries

In Britain the *manufacture* of pharmaceutical products became established during the 18th century at a time when the preparation and sale of drugs and medicines was still largely in the hands of apothecaries licensed by the Royal

*High street chemist's shop, 1810*

College of Physicians, or by experienced but unqualified tradesmen who called themselves 'chemists and druggists'. Medicines sold by both groups included many vegetable extracts such as lemon juice as an anti-scorbutant, extract of willow bark to relieve inflammation, digitalis as a heart stimulant and opium to relieve pain and induce sleep. An extract from the bark of the cinchona tree known as Peruvian or Jesuit's bark was used in treating fevers, even in the 16th century.[1] Indeed, medicinal vegetable extracts, or 'galenicals', the history of which goes back to ancient times, were used throughout the Middle Ages and remained the main therapeutic agents in the apothecary's armoury until well into the 19th century. Some were powerful and have become the focus of research in recent times, for example the ancient analgesic, myrrh.[2] From the 16th century there have been in addition 'chemical medicines' containing compounds of metals like mercury, arsenic or antimony which, although highly poisonous in anything more than minute amounts, were used in well-known mixtures, tinctures and potions. Thus, in Joshua Ward's 'White Drops' and Dr James's Fever Powders, the main therapeutic constituent was an antimony compound. Exaggerated claims were made for such remedies, but many of them were of doubtful value and in extreme cases may even have caused death – Dr James's powder is said to have killed Oliver Goldsmith.

In the 18th century the growing popularity of mineral waters for bathing and drinking gave rise to the revival of the spas as health resorts. The demand for mineral waters grew, but they were difficult to transport in wooden casks over long distances so as to arrive in a drinkable state. Chemical analysis revealed the salts present in common mineral waters and it seemed that they could be copied

[1] J.K. Borchardt, 'A short history of quinine', *Drug News and Perspectives,* 1996, **9**, 116–118.
[2] *Idem,* 'Myrrh: an analgesic with a 4000-year history', *ibid.,* 1996, **9**, 554–557.

by dissolving the dry salts in the correct proportions in ordinary water. This practice began at the end of the 17th century when Nehemiah Grew exhibited crystals of Epsom Salts at the Royal Society, but it spread rapidly when it was found that the sale of these salts provided a lucrative trade. When the demand outstripped the supply from natural waters, the salts were prepared from other sources. Epsom Salts (magnesium sulphate) were obtained from magnesian limestone, or from 'bittern', the liquor remaining in salt pans after the extraction of common salt. Such sources were virtually inexhaustible.[3] In the 18th century, when aerated mineral waters were in vogue, some pharmaceutical chemists prepared powdered mixtures consisting mainly of sodium bicarbonate and tartaric acid which effervesced when dissolved in water. Mixtures such as Eno's Fruit Salts and Andrews Liver Salts, antacid laxatives, were sold as 'health salts'.[4] Still available today, they are a reminder of the 18th-century vogue for drinking mineral waters.

By the 19th century there were some 700 apothecary and chemist shops in London alone; most sold unique mixtures for which they were locally known, in addition to common drugs and proprietary mixtures supplied by others. The Society of Apothecaries, founded in 1617, campaigned for the right to regulate the activities of unqualified competitors and the Apothecaries Act (1815) made some progress towards professionalisation. It specified that the qualified apothecary must possess the Licence of the Society, gained after a prescribed course of study, experience and examination. Some apothecaries also became members of the Royal College of Surgeons of England and these two qualifications became the normal requirement for the surgeon-apothecary or 'general practitioner', a title increasingly used after about 1830. Not all apothecaries took this route, however, and the profession began to divide between those who leaned towards the medical profession and the others who, along with some of the more reputable chemists and druggists, concentrated on dispensing medicines or manufacturing drugs. Leading members of the trade including Jacob Bell, Joseph Bevan, Luke Howard and others, many of them Quakers whose firms had prospered as they established a name for reliability, pressed for a new profession of pharmacy. In a trade abounding with charlatans and quacks, they recognised that it was essential to avoid becoming involved in the tempting business of making quick profits by trading on false claims.

After Bevan retired, his business at Plough Court, off Lombard Street, was taken over in 1797 by William Allen in partnership with Luke Howard. Allen took a firm line with his customers, expecting payment by stipulated times, but in return he undertook to supply only the purest and best-quality products. He applied careful quality control to materials bought in and to his products, a process which has remained a fundamental concern of pharmaceutical manufacturers ever since. Allen was himself a competent chemist and, through his

---

[3] Industrial ventures in this field are discussed by W.V. Farrar, K.R. Farrar and E.L. Scott, in a series of articles on 'The Henry's of Manchester' published in *Ambix* between 1970 and 1977. See especially *Ambix*, 1977, **24**, 10–14, 16–19.

[4] W.A. Campbell, 'James Crossley Eno and the rise of the health salt trade', *University of Newcastle upon Tyne Medical Gazette*, 1966, **60**, no. 3.

contacts with famous men like Humphry Davy, he became widely known in scientific circles. In 1807 he was involved with Davy and others in establishing the Geological Society of London and in the same year he was elected FRS after his experiments with William Hasledine Pepys had shown that diamond is pure carbon. In 1808–1809 Allen and Pepys worked on respiration,[5] and the laboratory at Plough Court where many of these experiments were carried out became one of the centres of scientific research in London. It was recognised far beyond the circle of London chemists; a collection of chemical reagents from Plough Court was exhibited to the National Institute of Paris in 1801 and Allen was a friend of the Swedish chemist Jöns Jacob Berzelius, who in 1820 sent to Plough Court a quantity of the new metallic element selenium which he had isolated in 1817. Allen's reputation carried weight among his contemporaries and in 1841, when the Pharmaceutical Society of Great Britain was established, he was elected its first President. This gave professional status to those pharmacists who were members of the Society and set up standards of achievement by examinations and practical experience. It also helped to set the pharmacist apart from those who taught chemistry in the universities, or who practised the science in industry.

Allen married his second wife Charlotte Hanbury in 1806 and it was through this marriage that the two families of Allen and Hanbury were linked and the famous pharmaceutical company was founded. The firm remained at Plough Court which underwent changes and extensions as the business flourished until it was completely rebuilt in 1872. The building had always been cramped and it was now becoming altogether too small. The demand for new products coupled with technological advances in drug manufacture meant that larger premises were essential for survival and in 1874 they leased part of an old match factory at Bethnal Green.[6]

The new factory provided the opportunity to develop special lines. One of the first of these was refined cod-liver oil, which by the 1850s had become an important product and remained one of their main lines for the next century despite competition from rivals including Duncan, Flockhart & Co. in Edinburgh,[7] and T.J. Smith (later Smith & Nephew) in Hull. Malt extract was another important early product which, like cod-liver oil, needed specialised treatment. In 1879 Allen and Hanbury's began to experiment with vacuum distillation at low temperatures to produce an uncaramelised malt extract rich in diastase. Used first in farinaceous foods this extract was soon found to mix well with cod-liver oil and in liquid form with a whole range of other medicaments. It was from this beginning that Allen and Hanburys entered the field of infant and invalid foods for which they were to become famous. The vacuum distillation process also proved invaluable for preparing dried milk, another baby-food product.

---

[5] W. Allen and W.H. Pepys, *Phil. Trans.*, 1808, **98**, 249–281; *ibid.*, 1809, **99**, 409–429.
[6] G. Tweedale, *At the sign of the Plough; Allen & Hanburys and the British pharmaceutical industry 1715–1990*, John Murray, London, 1990, p. 74.
[7] P.J.T. Morris and C.A. Russell, *Archives of the British chemical industry 1750–1914*, British Society for the History of Science, Monograph no. 6, Faringdon, 1988, pp. 84–85.

# 3 The British Pharmaceutical Industry from the Later 19th Century

During the 19th century, as the numbers of town-dwellers steadily increased, the problems and diseases of urban life gave rise to new markets for medicines. The public began to indulge in the delights of an emerging consumer society. New products with an air of luxury began to appear; sales of proprietary medicines more than doubled between 1852 and 1870 and trebled by 1890. Physicians' fees were far beyond the means of most ordinary people and it was usual to treat minor illnesses at home using preparations obtained from the local pharmacist. Many proprietary medicines which later became famous appeared on the household scene. Thomas Beecham in St. Helens advertised his 'Beecham's Pills – worth a guinea a box', although they cost only a few pence; James Crossley Eno, a Tyneside druggist, launched his famous 'Fruit Salt' and Thomas Holloway in London began to sell 'Holloway's Pills'. Later, Jesse Boot in Nottingham retailed a whole range of proprietary medicines and began to build up his pharmaceutical empire. Most proprietary medicines were compounded in the tradition of the apothecary and most were heavily advertised with extravagant claims. In the 1880s Beecham was spending £22,000 a year on advertising, while Thomas Holloway is said to have spent as much as £50,000 a year. The profits, which grew in direct proportion to the degree of advertising, were public benefactions. In 1876 Holloway bought a large estate at Egham in Surrey, 'to provide the best education suitable for women of middle and upper middle classes'. Holloway also donated a valuable collection of paintings to his new College, which received its 'Royal' accolade from Queen Victoria, ten years later. Jesse Boot too, donated land and a generous endowment for the establishment of University College, Nottingham.

From the mid-19th century a thriving pharmaceutical industry began to take shape, with numerous individual firms established in London, Manchester, Edinburgh and other important centres of population. Some of these went on to become famous. For example, Burroughs, Wellcome & Co.[8] had been formed in London in the 1880s by a partnership between Silas M. Burroughs and Henry Solomon Wellcome. Initially agents for several American firms specialising in tablets, they began manufacturing in 1893 at Dartford, Kent. When Burroughs died in 1895 Henry Wellcome became the sole proprietor. He opened a branch in Melbourne (Australia) in 1896 and by 1914 had associated houses in various overseas locations, including South Africa, Italy, Canada, USA, Shanghai, Argentina and India. Interested in research as well as production he set up the Wellcome Chemical Research Laboratories in 1896 and followed this with the Wellcome Bureau of Scientific Research and the Wellcome Historical Medical Museum in 1913. In 1942 he consolidated these enterprises in the Wellcome Foundation Ltd. which achieved world-wide renown for its benefactions to scholarship.[9]

[8] *Ibid.*, p. 49.
[9] In the early 1990s Wellcome research and production was taken over by rivals Glaxo to become a new pharmaceuticals giant, Glaxo-Wellcome.

*Henry Wellcome (1853–1936)*

*Burroughs Wellcome pharmaceutical products*

Like Allen and Hanbury's, Howard & Sons[10] also had Quaker origins. Luke Howard had taken charge of Allen's new laboratories in the East End, first at Plaistow and later at Stratford, where they made heavy chemicals like nitric and sulphuric acids, salts of ammonium and mercury. In 1807 the partnership was dissolved and Howard took over the Stratford factory as Luke Howard & Co. Here he began making fine chemicals for the pharmaceutical industry and was

[10] Morris and Russell (ref. 7), p. 103.

*Interior of typical late-Victorian chemist's shop: a reconstruction at Wilberforce House Museum, Hull*

particularly known for cinchona alkaloids. In 1888 Howards bought up Hopkin & Williams, who made laboratory and photographic chemicals at Wandsworth; they also had interests in camphor manufacture and thorium nitrate for gas mantles. Between 1898 and 1923 both gradually transferred their activities to a new site at Ilford. After the war the company took another direction and began to manufacture solvents such as isopropyl alcohol, cyclohexanol and lauryl alcohol. The business was acquired by Laporte Industries Ltd. in 1961.

Another well-known British pharmaceutical manufacturer, Evans Medical Ltd., was established in 1809 by John Evans, a local druggist in Worcester.[11] Evans then went to London and formed several consecutive partnerships, culminating in Evans, Lascher & Webb in 1879. He had three sons who established a side company in Liverpool which, in 1902, was amalgamated with the main firm to form Evans, Sons, Lascher and Webb Ltd. They prepared biological products (sera and vaccines) and with Liverpool University established the Incorporated Institute of Comparative Pathology (1902), which became the Evans Biological Institute in 1929. After the war the Liverpool and London Branches joined to become Evans Medical Supplies Ltd. and in 1961 the firm became part of the Glaxo Group.

May & Baker Ltd., based at Battersea, London,[12] were well-known for the manufacture of fine chemicals, but became more involved in pharmaceuticals

[11] *Ibid.*, p. 86.
[12] *Ibid.*, p. 137.

*Clinical cleanliness in the barbiturates plant at May & Baker (1953)*

during the first world war through the production of salvarsan.[13] In 1927 May & Baker was partly acquired by Poulenc Frères which itself merged in 1928 with Usines de Rhône and completed the take-over of M&B by 1934. Through this French connection May & Baker became involved with the sulphonamides.[14] Whiffen & Sons Ltd. was another London chemical manufacturer, based in Battersea, Southall and Fulham.[15] This firm was started in 1854 when a London pharmacist, Thomas Whiffen, joined a small firm manufacturing fine chemicals in Battersea. The chief products were medical chemicals and poisons including strychnine. Whiffen acquired full control after 1868 and by the 1890s Whiffen & Sons had become one of the leading fine-chemical manufacturers in Britain. The firm was acquired by Fisons in 1947.

In Edinburgh there were several famous firms connected with the pharmaceutical industry. Duncan, Flockhart & Co., already mentioned, opened in Edinburgh in 1820 as a retailer of drugs with a small laboratory for testing manufacturing processes. They began to manufacture chloroform and in 1847 supplied J.Y. Simpson, whose pioneering use of this anaesthetic was widely known. Chloroform became one of the firm's chief products and large amounts

[13] Judy Slinn, *A history of May & Baker, 1834–1984*, Hobsons, Cambridge, 1984, pp. 91–94.
[14] *Ibid.*, pp. 122–125 and *passim*.
[15] Morris and Russell (ref. 7), p. 203.

were supplied to Florence Nightingale. In 1876 they began to supply vaccines as well as anaesthetics. T. & H. Smith,[16] another Edinburgh pharmaceutical manufacturer, began in 1827 when Thomas Smith, an Edinburgh medical graduate, took over his brother William's pharmacy. With his brother Henry he manufactured galenicals (medicines prepared from plant sources) and fine chemicals. They later turned to opium alkaloids, especially morphine which they began extracting from opium in 1837. John Fletcher Macfarlane, an Edinburgh surgeon, took over an apothecary's shop in 1815 and established a large trade in laudanum. Later he made morphine which with the development of the hypodermic needle increased the effectiveness of the drug by direct injection into the blood stream. Demand increased and his trade flourished. He also made anaesthetics (ether and chloroform) and surgical dressings used by Lister. In 1840 he opened a factory and by the 1900s J.F. Macfarlane & Co.[17] had become one of the largest suppliers of alkaloids in the country. Developing by take-overs and extensions T. & H. Smith united with Macfarlane in 1960 to form Macfarlane Smith Ltd., part of Edinburgh Pharmaceutical Industries which had been established in 1956 and was taken over by the Glaxo Group in 1963.

In Manchester James Woolley, a druggist, began a small pharmaceutical firm in 1831.[18] After his death he was succeeded by his son George Stephen, also a pharmacist. The firm developed in manufacturing and wholesaling until In 1872 it was joined by another son, Harold, with responsibility for scientific apparatus and surgical instruments. The firm became famed for equipment and medicines, notably galenicals; in 1936 they acquired J.C. Arnfield & Sons Ltd., a Stockport pharmaceuticals company, and continued to thrive until after the Second World War when it went public in 1950 and was acquired by British Drug Houses Ltd. (BDH) in 1962. It subsequently became part of Vestric, a joint company formed between BDH and Glaxo for their wholesale pharmaceutical business.

The story of the British pharmaceutical industry is one of 'rationalisation' and consolidation as the smaller family firms found it impossible to compete with the increasingly complex scientific, legal and economic demands placed upon them. Through mergers and take-overs they gradually amalgamated to form larger chemical and pharmaceutical manufacturers and then, especially after 1945, they were taken over by much larger international corporations until there were only one or two pharmaceutical giants operating in Britain.[19] The main reason for these moves has been the ever-increasing need for research and development with mounting costs unsustainable except by the largest multi-national corporations.

[16] *Ibid.*, p. 174.
[17] *Ibid.*, p. 134.
[18] *Ibid.*, p. 213.
[19] By 1990 Glaxo had become by far the largest of these and after taking over the Wellcome business in the early 1990s is now a world-class corporation.

## 4 Organic Chemistry and Fine Chemicals

Nineteenth-century developments in organic chemistry resulted in the discovery, and synthesis, of many new substances with medicinal properties or potential. In 1812 the French chemist Bernard Courtois isolated iodine, later shown to be an element by Davy and Gay-Lussac. Bromine was discovered in 1826 by Antoine Jerome Balard, benzene by Michael Faraday in 1825. Important alkaloids were discovered, including morphine in 1804 and quinic acid in 1806. Prior to this the existence of organic *bases* was unknown, but following these discoveries the search for other organic bases began. Pierre Joseph Pelletier and Jean Bienaimé Caventou isolated strychnine and brucine from *nux vomica* in 1818–1819, quinine and cinchonine from Peruvian bark and veratrine from *hellebore* rhizomes in 1820. Pierre Jean Robiquet isolated narcotine in 1817 and caffeine in 1821; atropine was obtained in 1831. Discovery of the chemical and physiological properties of these compounds extended the range of substances that could be used for medicinal purposes.

Anaesthesia using nitrous oxide, ether and chloroform was introduced into surgery in the mid-19th century, beginning in America. Chloroform, described in 1831 by Samuel Guthrie, an American physician, as 'chloric ether' or 'sweet whisky', was recognised almost simultaneously by Liebig in Germany and the apothecary Eugène Soubeiran, professor at the École de Pharmacie in Paris. Phenol (carbolic acid) was identified by the German pharmacist Friedlieb Ferdinand Runge in 1834 and in the hands of Joseph Lister in the 1860s it vastly increased the safety of surgical operations. Lister used a solution of carbolic acid as a fine spray in the atmosphere of the operating room and antiseptic dressings also using carbolic acid, applied to surgical wounds. Together these discoveries reduced the risks and extended the possibilities of surgery from the middle of the 19th century.[20] However, the greatest care had to taken to ensure that the chemicals used for these purposes were absolutely pure. It was this search for purification techniques, by which even minute traces of foreign compounds could be eliminated, that improved the quality of pharmaceuticals while at the same time increasing production problems and costs.

The discovery in coal-tar distillates of aromatic compounds such as benzene, phenol, toluene, naphthalene and many others stimulated the search for new methods of synthesis of therapeutic compounds. In 1838 salicylic acid (2-hydroxybenzoic acid), a component of the glycoside salicin, extracted from willow bark, was shown by Rafaelle Piria, professor of chemistry at Turin, to be a derivative of phenol. This compound was later found to provide a number of useful derivatives, several of which have become important medicinal substances. These, and other organic compounds, were produced in quantity by the fine chemicals industry from the 1880s and in 1899 the analgesic and anti-inflammatory properties of acetylsalicylic acid (aspirin) were recognised. Aspirin was found to be less toxic than salicylic acid itself and, without the

[20] Frederick Crace Calvert established a large chemical works making coal-tar products, especially phenol (carbolic acid) in Manchester in 1865.

depressant effects of sodium salicylate, it rapidly became a major product of the pharmaceutical industry in the early years of the 20th century.[21] The drug was manufactured by Burroughs Wellcome Ltd., and Boots Pure Drug Co., pharmaceutical manufacturers in Nottingham, also became a major supplier of aspirin. Shortly afterwards the barbiturates, derivatives of phenobarbitone were introduced for the treatment of epilepsy.

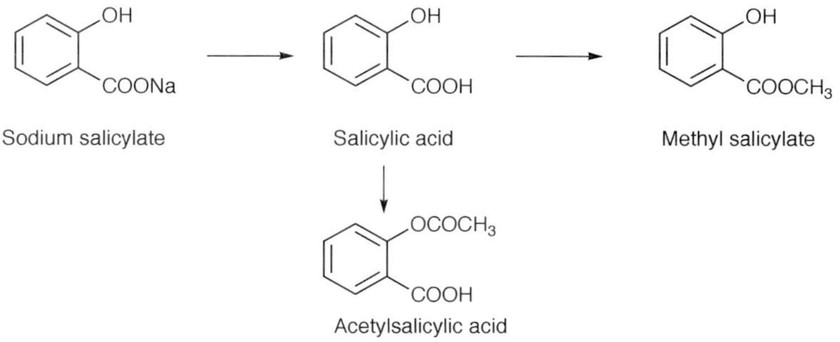

Sodium salicylate          Salicylic acid          Methyl salicylate

Acetylsalicylic acid

*Medicinal derivatives of salicylic acid*

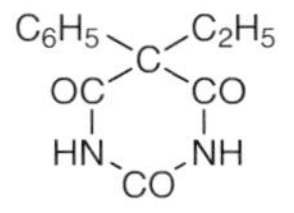

Phenobarbitone

These synthetic organic compounds quickly found their way into the pharmacopaeias and there were one or two other chemical additions, substances of known therapeutic value such as ethyl nitrite or 'sweet spirit of nitre', a soporific, and nitroglycerine, an anti-coagulant. The latter was introduced by Alfred Bernhard Nobel who accidentally discovered that it gave him relief from angina pectoris and began to dose himself with this product of his explosives factory at Ardeer in Scotland, about 1895. It is now used as a vasodilator in medicine and is perhaps the drug with the strangest origins of all.

In general, however, there was very little change in ideas about the nature of medicinal compounds from the late 18th century until in 1881 Paul Ehrlich, a medical student at Leipzig, used the organic dye Methylene Blue to stain bacteria for microscopic examination. Observing that certain dyes had a specific affinity for particular micro-organisms Ehrlich was led to the idea that it might be possible to devise synthetic drugs with specific affinities which could be used to destroy particular micro-organisms and parasites without affecting the host tissues. The most important result of Ehrlich's work was the synthesis of salvarsan (arsphenamine), first used to treat syphilis in 1911. This compound, popularly known as Ehrlich's 'magic bullet', was first manufactured by Hoechst

---

[21] For a history of this important drug see John R. Vane, *Aspirin and other salicylates*, Chapman & Hall, London, 1993.

in Germany and rapidly gained ground. Supplies were cut off during the war; the Board of Trade suspended the patent and granted a licence to Burroughs Wellcome & Co. to make the product under a different name. At the same time another licence was given to Poulenc Frères to sell Ehrlich's compounds through their British agents, May & Baker, who were also required to begin manufacture as soon as possible. Production began in 1916. Arsphenamine is a strong acid and must be carefully neutralised immediately before injection, a defect which was later overcome in an improved form of the drug (neosalvarsan) which was still in use in the 1940s.

Arsphenamine, 'Salvarsan'                Neoarsphenamine, 'Neosalvarsan

The successes achieved by these new drugs mark the beginning of the era of chemotherapy, a new approach to pharmaceuticals, based on organic chemical synthesis. Britain also witnessed the beginnings of drug testing procedures, supervised by the Medical Research Committee established in 1913 (later named the Medical Research Council). The financial resources of this Committee were initially small and the testing procedures were of necessity paid for by the manufacturers.[22]

## 5   19th Century Drug Synthesis

The basic principle on which the chemists' search for new drugs operated from the late 19th century onwards was the notion that chemical compounds with similar molecular structures are likely to show similar physiological effects on living organisms. Based on this assumption one approach to the problem of discovering a new drug has been to prepare large numbers of compounds related to a compound of known therapeutic properties and to test these on animals in the expectation that they may have similar effects and some may be even more powerful. The method, which was impracticable until there was a well established system for organic synthesis, is often very lengthy and expensive as a large proportion of compounds prepared in the initial stages of the search may well have to be discarded. It is also somewhat haphazard, though it can lead to unexpected results. For example W.H. Perkin, working at the Royal College of Chemistry in London, was searching for a method of synthesising quinine by oxidation of allyl toluidine when he discovered the aniline dye mauveine in 1856. The problem was more difficult to solve than Perkin had imagined. It took more than 50 years before the molecular structure of quinine could be fully worked out and a successful chemical synthesis devised.[23] From the late 19th century to the First World War

---

[22] Slinn (ref. 13), p. 92.
[23] P. Rabe, *Berichte*, 1923, **64**, 2487; R.B. Woodward and W.E. Doering, *J. Am. Chem. Soc.*, 1944, **66**, 849.

Germany held the lead in this field, owing in large part to the large number of organic chemists who had been trained there by Liebig and his successors. There was a new attitude to organic synthesis supported by the development of acylation techniques and the availability of synthetic agents such as acetoacetic ester. But, if Germany led the way, British chemical firms rapidly sought to supply the urgent need for drugs created by the loss of German supplies after 1914. The manufacture of anti-syphilitic drugs by May & Baker, already mentioned, is a good example.

Among organic compounds which showed anti-pyretic properties the first to be recognised about 1886 was acetanilide ($C_6H_5NHCOCH_3$), a compound which dissolves readily in hot water or alcohol and was used in 'Daisy powders' as a febrifuge under the name 'antifebrine'. It is now a scheduled poison. Phenacetin (4-ethoxyacetanilide), a related compound with similar antipyretic properties, was also synthesised in the 1880s. Like acetanilide it was used to reduce body temperatures in fevers, but it was also found to act as a remedy for neuralgia.

A still more important febrifuge called 'antipyrine' is a derivative of pyrazole, and its 4-dimethylamino derivative, known as pyramidone, is also a febrifuge and has been used as a substitute for antipyrine. These examples bear out the notion that compounds with similar molecular structures may produce similar physiological effects.

Acetanilide          Phenacetin          Antipyrine          Pyramidone

There is no certain way of knowing in advance all the likely physiological effects of a new type of compound on animals, but once a substance of known molecular structure has been found to have certain properties it may be promising to synthesise and test related analogues. The sulphonic hypnotics researched by Eugen Baumann between 1886 and 1889 illustrate this principle. The soporific effects of sulphonal were first observed in a dog fed with this drug in an attempt to trace the metabolic pathway of sulphur. Baumann synthesised a series of compounds with the sulphonal structure, but with different groups at $R^1$, $R^2$, $R^3$ and $R^4$. All were found to be soporifics, but they showed different degrees of effectiveness and toxicity. This research helped to set the pattern for the modern pharmaceutical industry.

Otto Philipp Fischer prepared tetrahydroquinoline compounds in the hope of discovering a substitute for quinine. One of these had a similar antipyretic effect, though not on malaria. Called 1-ethyl-5-hydroxytetrahydroquinoline, or 'kairine', it is a powerful antipyretic, but is useless as a drug because, like its analogues, it is is toxic to the red blood cells.

Kairine

Nevertheless, Fischer's work is important as one of the first *deliberate attempts* to synthesise a drug. He set out with a plan to produce a synthetic antipyretic, but although he succeeded in producing a compound with the required properties, the result was not favourable. This early example highlights one of the major difficulties which chemists still face in the search for new drugs, namely unacceptable toxicity. The need for thorough physiological and clinical testing becomes apparent at once.

# 6   The Early 20th Century Drugs Industry

When German supplies of drugs and fine chemicals were cut off during the First World War, England, France and America each began to expand their own pharmaceutical and fine chemicals industries out of necessity and they continued in competition with the German industries after the war, though the latter remained dominant in size and research input up to the 1960s.

Allen and Hanbury was the largest British pharmaceutical manufacturer before 1914, but Glaxo had already begun to compete in the manufacture of milk products. Glaxo owed its origins to a Londoner, Joseph Nathan, who had emigrated to New Zealand where he dealt in groceries and ironmongery imported from Britain. In the 1890s, with the development of refrigerated ships, he switched to the export of frozen meat and butter and by 1900 his company was registered in London. It was at this point that Nathan began to produce dried milk and by 1908, after a shaky start, he had launched a dried milk business with the brand name 'Glaxo' backed by an extensive advertising campaign. Although the new company lacked Allen and Hanbury's standing in the pharmaceuticals field, its advertisements were addressed directly to the public as well as to doctors, nurses and retail pharmacists and by 1921 Glaxo's sales had reached £1.5 million per annum, far outstripping Allen and Hanbury's total business.

In 1923 Glaxo's aggressive marketing policy began to change under a newly appointed chief chemist (Sir) Harry Jephcott.[24] While at the International Dairy Congress in Washington in 1923, Jephcott heard about American research on 'accessory food factors' (vitamins). At that time no vitamins had been isolated, although it had long been known that traces of unidentified substances present in natural foods were essential for a healthy diet. At Wisconsin University Elmer McCollum discovered substances in milk which he called 'fat-soluble A' and 'water-soluble B' (vitamins A and B) and later he identified the anti-rachitic vitamin D in cod-liver oil. Shortly afterwards Theodore Zucker at Columbia University discovered a process by which the proportion of vitamin D in treated cod-liver oil could be increased a thousand-fold. At Jephcott's instigation Glaxo secured a licence for Zucker's process and began to manufacture baby food reinforced with the 'sunshine vitamin D'. In 1924 they began to manufacture 'Ostelin' the first vitamin-concentrate to be made on a commercial scale in Britain. These developments preceded others which, by giving Glaxo the marketing edge, led to increased competition based on research. Jephcott began the process of rationalisation of the British pharmaceutical industry as Glaxo progressively acquired other companies in the 1960s and extended the search for new drugs. A new stage in the development of the British pharmaceutical industry had begun.[25]

## (a)   The sulphonamides

One of the most important developments in the German pharmaceutical industry was the sulphonamide compound Prontosil Red, the most active of a large number of sulphonamide dyes which were tested for therapeutic properties following Ehrlich's use of Methylene Blue for malaria in 1911. Discovered in 1932, prontosil was found to be effective against streptococcal infections such as blood poisoning and wound infections as well as a large group of major diseases like scarlet fever, pneumonia, cerebrospinal meningitis and gonorrhoea. This discovery was an important advance in chemotherapy and led to further research in other European countries. Research at the French pharmaceutical company Poulenc Frères showed that prontosil breaks down in the human body to form sulphanilamide (4-aminobenzenesulphonamide) which was found to have similar powerful therapeutic properties. The difference was that sulphanilamide was neither a new substance nor a patented product. Moreover, it was relatively cheap and could be used in place of the highly-priced prontosil. When it was found that removal of the 4-amino group

---

[24] Tweedale (ref. 6), pp. 145–148.
[25] Allen and Hanbury's merged with Glaxo in 1961. *Ibid.*, pp. 193–221.

destroyed the therapeutic properties, research was directed towards derivatives of sulphanilamide.

Prontosil

Sulphanilamide

Sulphapyridine

Sulphathiazole

New compounds with greatly improved properties were developed by researchers at May and Baker who had entered the sulphonamide field in 1936 following their take-over by Poulenc Frères two years earlier.[26] May & Baker researched several new sulpha-drugs of which sulphapyridine (M&B 693) and sulphathiazole (M&B 760) were the most important. These new drugs were very successful in treating wounds and they undoubtedly saved the lives of many servicemen and women during the war. They could be administered in comparatively large doses both orally and as a powder to be dusted on wounds. In addition they were used in the treatment of a wide variety of other diseases.

Sulphapyridine (M&B 693) was described as a 'wonder drug' and successes were hailed in the popular press as miracle cures. It gained world-wide fame during 1938 and 1939 as a cure for previously fatal diseases, but the outbreak of war brought even greater need for such a drug and in 1941 it was used on a large-scale to prevent gangrene amongst the wounded troops of the British Expeditionary Force brought back from the Dunkirk beaches. It was also used with signal success when Winston Churchill suffered two attacks of pneumonia. There was such a call for it that May & Baker faced problems in manufacturing it in sufficient quantities. A plant producing twelve tons of the drug a month was installed, but its capacity was soon overtaken by demand. M&B 693 also reduced the cost of treating pneumonia quite dramatically from £5 to a few shillings a time. The discovery of the sulphonamide drugs heralded the beginning of the chemotherapeutic era and the disappearance of the old manufacturing chemist. After the Second World War the pharmaceutical industry was dominated by large international pharmaceutical companies for whom research, innovation and patent activity was of paramount importance.

But, successful as they were, the sulphonamides had their drawbacks. Their therapeutic properties, though powerful, were limited in range and there were many bacterial diseases which still could not be treated successfully. They also showed a distinct tendency towards unacceptable toxicity, a fact which limited

[26] Slinn (ref. 13), p. 124.

*Upper: making sulphonamides at May & Baker, Dagenham; lower: research laboratory where M&B 693 was discovered*

their usefulness in medicine, though many became invaluable in veterinary practice.

## (b) Penicillin and other antibiotics

The term 'antibiotic' is used to describe metabolic products produced by the activity of microbes which inhibit or destroy other micro-organisms. The first antibiotic to be prepared on a large scale was *pyocyanase* isolated in Munich in 1901 from the microbe *Pseudomonas aeruginosa* and successfully used in treating several hundred patients. However, poor control of the production process led to variable results and its use was discontinued. Penicillin was discovered about 1927 by (Sir) Alexander Fleming at St. Mary's Hospital in London, but it was a decade later before further work was done by Ernest Boris

*Commercial production of penicillin: the fermentation stage*

Chain and Howard Walter Florey at Oxford leading to clinical trials in 1941 which showed great promise. Penicillin is the generic name for a whole group of compounds. The commercial production of penicillin began in America in 1943, the earliest methods requiring a large surface of nutrient media in shallow trays. Quantities were severely restricted and it was not until the method of submerged fermentation was developed that penicillin could be made in bulk. Penicillin, a strong acid, is usually supplied in the form of its sodium salt which is crystallised and dried out in sterile conditions prior to packing in polyethylene bags or in stainless steel or glass containers.

The formula of the penicillins was published early in 1946. Several penicillins were known in the late 1940s, their molecular structures differing only in the substituent R.

$$\text{RCONH}-\underset{\underset{H}{|}}{C}-\underset{|}{C}\begin{matrix} OC-N \\ \\ S \end{matrix}\begin{matrix} COOH \\ CH_3 \\ CH_3 \end{matrix}$$

| | |
|---|---|
| Penicillin I (or F) | R = $CH_2CH{=}CHCH_2CH_3$ |
| Dihydropenicillin I | R = $CH_2CH_2CH_2CH_2CH_3$ |
| Penicillin II (or G) | R = $CH_2C_6H_5$ |
| Penicillin III (or X) | R = $CH_2C_6H_4OH\text{-}4$ |
| Penicillin IV (or K) | R = $CH_2(CH_2)_5CH_3$ |
| Penicillin V | R = $CH_2OC_6H_5$ |

Although the sulpha-drugs had shown great promise in the treatment of septicaemia, penicillin was found to be even more effective and to succeed where sulphanilamide or the other sulpha-drugs had failed. It proved so effective that other antibiotics were sought and the modern pharmaceutical industry now supplies a whole range of these drugs. They include streptomycin and related compounds, chloramphenicol, the tetracycline group and many others.

Chloramphenicol

| | $R^1$ | $R^2$ |
|---|---|---|
| Tetracycline | $R^1$ = H, | $R^2$ = H |
| Oxytetracycline | $R^1$ = H, | $R^2$ = OH |
| Chlortetracycline | $R^1$ = Cl, | $R^2$ = OH |

## 7 The Modern British Pharmaceutical Industry

The discovery of new drugs by chemical research provided the most significant advances in the pharmaceutical industry between the wars, but economic factors at work in Britain and elsewhere caused important and far-reaching changes in the industry. Traditionally, therapeutic products had been sold only by chemists and pharmacists, but in the 1930s there were increasing numbers of multiple stores which sold pharmaceuticals and proprietary medicines over the counter. Boots employed qualified dispensing chemists to manage their shops and thus began to take trade away from the independent pharmacists. Taylors, Glaxo, Allen and Hanburys, Beechams and Boots were all pharmaceutical manufacturers, patenting their own lines and retailing them. The development of the British pharmaceutical industry was significantly affected by the two world wars, owing to the loss of supplies from the German-based industry and to the urgent need to find ways of manufacturing new and established drugs and medicines to treat the troops, whether wounded in action or suffering from tropical diseases in Africa or the Far East. Between the wars the successes achieved by the sulphonamides, penicillin and anti-malarials, encouraged fresh research to find new synthetic drugs.[27]

These changes, along with growing foreign competition and especially the introduction of the National Health Service in 1948, created demands which have revolutionised the British pharmaceutical industry. British firms which were large enough to take part in this activity gained greatly from the experience and a research-based pharmaceuticals industry was established. Then in the 1940s and 1950s, the industry changed quite dramatically with the rise of research laboratories such as those of the Beecham Company where the basic

[27] B.J. Price and M.G. Dodds, 'The quest for new medicines', in P.J.T. Morris, W.A. Campbell and H.L. Roberts (eds.), *Milestones in 150 years of the chemical industry*, Royal Society of Chemistry, Cambridge, 1991, Chapter 7, pp. 27–53; see also S.M. Roberts and B.J. Price (eds.), *Medicinal Chemistry: the role of organic chemistry in drug research*, Academic Press, London, 1985.

core of penicillin was isolated and new modifications of the substance were developed to broaden its therapeutic range. Patents ensured massive profits from the applications of these discoveries. Imperial Chemical Industries (ICI) which had not been much involved in the manufacture of pharmaceuticals before the war, began research on anti-malarials and, in 1946, an anti-malarial compound of novel type called 'paludrine'[28] was announced by ICI. It was found to be more effective than quinine and did not give rise to nausea, nor stain the skin yellow like mepacrine (atebrine). Moreover, it was more easily prepared than other anti-malarials. The success of 'paludrine' led to the founding of ICI's Pharmaceutical Division in 1957 and a range of new drugs was researched and put into production. The vast resources of ICI changed the face of the British pharmaceutical industry. By the 1970s when other parts of British industry were failing, the pharmaceutical industry was able to compete successfully with the best in the world. Drugs-related research by British companies has often led the field and it has a long and highly successful history.

Paludrine                                        Mepacrine

While ingenuity in research placed the British pharmaceutical industry at the forefront of the developments, demand for drugs created by the establishment of the National Health Service in 1948 was also highly significant. The cost of drugs used by the NHS rose from £6.8 million in 1947–1948 to £55.6 million in 1955–1956 and has continued to rise at much the same rate. This escalating demand further stimulated the pharmaceutical industry, heralding the demise of the old proprietary drug trade and encouraging the search for new synthetic drugs. In the 1960s there was a period of rapid growth in the number of products and new companies including foreign drug firms like Pfizer, Smith Kline & French, Merck, Parke Davis and Wyeth were attracted to this lucrative field. The nature of the industry as a high risk business increasingly dominated by large-scale multi-national firms was becoming apparent. Large companies formerly engaged in the manufacture of fine chemicals, oils, food, cosmetics and perfumery also entered the field and the industry grew rapidly. These firms, which were often able to supply new drugs where the British industry was wholly lacking, brushed aside the British suppliers who soon began to realise the scale of research and development which would be needed if they were to produce their own patented products and compete successfully. They were greatly helped by pricing agreements with the NHS which kept the price of drugs high, supported by Government subsidies. This provided British pharmaceutical companies with the profits necessary to engage in fresh research and so stimulate growth in the industry.

[28] F.H.S. Curd and F.L. Rose, *Chem. Ind.*, 1946, 75–77.

*Modern drug plant at Burroughs Wellcome*

# 8 The Search for New Drugs

There are several kinds of pharmaceuticals in modern use. First there are products used for minor ailments which are well established as *safe and simple remedies* without serious side effects and can be used with no more medical guidance than can be given in print on the side of the packet. They contain substances which are not the subject of current patents; they are usually mild-acting remedies for common complaints. These products are well known from long familiarity or as a result of advertising; they can be manufactured and marketed without enormous capital outlay and offer good prospects of a steady, though not spectacular return.

A second category of pharmaceuticals is the so-called *ethicals*. These are controlled drugs available only on medical prescription. They include drugs which are subject to current patents and can be manufactured only by the patent-holders or their licensees. In addition there are the *generics* – drugs which have passed out of patent control and can be manufactured by any firm provided it is licensed as a pharmaceutical manufacturer. To produce these drugs requires more sophisticated chemical plant and careful quality control, but minimal research effort because the high-risk research and development stages have already been carried out by others.

A further group of pharmaceuticals, the *new chemical entities* or NCEs, are the ones which require the expensive research and development effort. In recent years diversification has led to the search for remedies for arthritis, heart disease, cancer, gastric ulcers, mental disorders, fertility regulation and other conditions. It is now usual to study the molecular configuration of the disease agent and new work on the chemistry of the living cell has provided a deeper understanding of heredity, growth and ageing, while viruses entering the cell disrupt its chemistry. Derangements in cell chemistry are now thought to be the underlying causes of cancer, heart disease and mental illness and future drugs may be aimed at correcting faulty cell chemistry and the wasteful 'grape-shot approach' of the former method may thus be avoided. It may also be possible to improve lives as well as saving them. A drug that would stimulate RNA formation in the cells could improve memory, so increasing human intellect and preventing senility. Another might be able to affect the genes, suppressing undesirable traits and enhancing those that are favourable to health and long life.

The greater part of the cost of developing a new drug for general medical use comes from the necessary chemical, physiological, toxicological and clinical studies over a long period of time extending up to a decade or more. While there are differences in the legal requirements of various countries, the problems accompanying the development and marketing of a new drug are common to all pharmaceutical manufacturers world-wide. They have affected the rise of the British pharmaceutical industry no less than others and it is in order to sustain the vast capital outlay needed that so many industrial take-overs and amalgamations have been required since the 1950s.

The comprehensive coverage and stringency of tests has increased enormously over the past half-century. In the early 1960s a new drug could be marketed in about 3–4 years from initial discovery. By the late 1980s this had risen to 10 or 12 years and today the time-lapse from the initial identification of a likely compound to the marketing of a successful new drug is about 15 years. During all this time the risk of failure remains high since it is never certain that the substance will prove satisfactory at every stage of the process and it need only fail one crucial test to be considered unsuccessful. The new drug would then be discarded and the time and money spent on it would have been wasted. Clearly the financial risks involved in this sector of the industry can only be supported by the largest and wealthiest companies able to sustain such losses. British pharmaceutical firms spend about 15–16% of their annual turnover on research and development.[29] Most of this expenditure occurs before a new drug is patented, so that by the time it reaches the market the drug represents a massive capital investment by the company. It is not surprising that the number

---

[29] In 1968 the average research and development costs were around £15 million. By 1990 this had risen to £1082 million, an eightfold rise in real terms.

of NCEs marketed annually fell from 50–60 in the 1960s to less than half that number in 1990.

The pricing of drugs is an important factor which influences the degree of research undertaken by pharmaceuticals manufacturers. In Britain the NHS is the largest purchaser of medicinal drugs and prices are controlled by the Department of Health through the Pharmaceutical Price Regulation Scheme, first set up in 1957. The scheme aims to limit the rate of return on capital expenditure devoted to sales promotion and the figures show that this rate has fallen from 20% to about 5–6%. This may be compared with an increase from 15% to 20% over the same period for manufacturing industry generally. Other countries have their own regulations, based on different criteria. The same drug may be available in various countries at different prices even at launch and these prices diverge further with time. Thus the development of new drugs remains a high-risk business. Boots spent £100 million in developing and testing a heart-treatment compound, before running into problems which meant that the drug would never be safe and had to be scrapped. The project was abandoned and the costs written off. On the other hand the profits to be gained from just one really successful drug can be astronomical. Glaxo, known in the 1970s for the manufacture of antibiotics, turned to research on an anti-ulcer compound. 'Zantac' was launched on the American market in 1980 and is now the world's most successful drug with sales in 1994 of £2.5 billion and profits of about £1 billion. With profits of that order available, it is not surprising that competition between multinational pharmaceutical companies is fierce.[30]

## 9   Some Modern Problems

There is no branch of the chemical industry which is more prone to controversy than the pharmaceutical industry and its products, every one of which carries risks as well as benefits.[31] Despite the thoroughness of the testing procedures, mistakes and accidents seem inevitable. There are so many contributory factors in human physiology that it is impossible for any new drug to be tested in every conceivable situation in which it may be used whether singly or in conjunction with other drugs. If it is used in conditions for which no tests were made the results can be disastrous as in the case of thalidomide, the sedative prescribed for pregnant women in the 1950s. In 1960–1962 unusually large numbers of babies were born with severely deformed or merely vestigial limbs and extremities. The cause was traced to thalidomide and the tragedy, most marked in the UK and Germany, led to a reappraisal of testing procedures to include tests on pregnant animals. Even this is not foolproof, as there is no certainty that the effects of the drug in animals would be the same as that in human mothers.

The thalidomide tragedy gave rise to a negative view of the pharmaceutical industry, but one positive outcome was the development of much more

---

[30] Recently, Japanese companies which entered this market late have been expanding rapidly and are now becoming significant.

[31] L. Caglioti, *The two faces of chemistry*, MIT Press, Cambridge, Mass., 1985, pp. 44–66.

sophisticated prostheses which could be operated by a pneumatic system even by very young children. Another valuable outcome was that the ethics of experimenting on human subjects were reconsidered. During the 1960s some other drugs were found to have adverse effects in pregnancy leading to damage to offspring. These included the tranquilliser meprobromate, which was found to cause mental impairment in young rats. Salicylates were also found to cause malformations in rodents, and meclozine, a drug commonly prescribed for pregnancy sickness, the antibiotic tetracycline, phenmetrazine, a drug used for weight reduction, and the long-standing chemotherapeutic agent sulpha-methoxypyrimidine were all found to have adverse effects on pregnant animals and their offspring. The risks of using these drugs in human pregnancies remain uncertain and direct causal relationships are difficult to establish.

Public awareness of the possible long-term effects of other drugs increased and cases of toxicity in drugs which had long been considered safe were reported in the 1960s. Chloroquin, an anti-malarial drug which had also been found valuable in rheumatoid arthritis, was reported to damage the lens and retina of the eye, and cortico-steroids, introduced in 1945, were recognised as a cause of cataracts. The anaesthetic halothane, which had been found quite safe in laboratory tests and had been given to many millions of patients in Europe and America, was reported in 1963 to cause liver damage, although the picture was complicated by side issues including the fact that in many cases other drugs had been administered at the same time. Cases of thrombophlebitis in women taking oral contraceptives also began to be reported about this time. Thus after the great expansion of chemotherapy in the two preceding decades, the 1960s witnessed growing doubts.

Another worrying aspect of medication by drugs arose from the fact that many general practitioners in Britain, bearing an ever increasing workload, were prone to prescribe pills too readily for insistent patients. Stimulants like the amphetamines or tranquillisers such as the barbiturates began to find their way in large quantities into the nation's medicine cabinets and people began to take pills rather than seeking other means of relieving the stresses of modern urban life. This led to cases of drug dependency in people who were otherwise healthy. It also meant that many homes possessed supplies of drugs which could be abused for non-medical purposes by members of the household other than the patient. There was the strong possibility of drug addiction and entry into the downward spiral of drug abuse, especially among the young. While the pharmaceutical companies cannot be held responsible for the inappropriate use or abuse of their products, the world-wide problem of criminal drug trafficking inevitably rebounds to some extent on the drug companies.

Pharmaceutical companies are also prone to strident objections from animal rights campaigners demanding a ban on the use of laboratory animals for testing drugs. The protesters are rarely able to offer adequate alternative procedures to replace the use of animals, but, unrealistic as their demands may appear, they pose a real dilemma for the pharmaceutical manufacturer who must achieve the highest possible safety standards and reliability for new medicinal products whilst at the same time maintaining good public relations.

Chapter 7  # General and Fine Inorganic Chemicals

W.A. CAMPBELL

The roots of preparative inorganic chemistry on a commercial scale lie principally in the system of medical chemistry known as iatrochemistry, and in the production of artists' pigments. Techniques evolved in one field were transferred to the other, often by the same operators. For example, Sir Theodore Turquet de Mayerne was instrumental in introducing new metallic remedies into the first London *Pharmacopoeia* of 1618, but he also advised Rubens and van Dyke on the preparation of their pigments.

## 1  Iatrochemistry

Led by the wayward genius of Paracelsus, many alchemists in the 16th and 17th centuries turned away from gold-making, seeking instead new chemical remedies based largely on mercury, arsenic, antimony, iron, tin and lead. This potentially dangerous innovation was in part a reaction against two prevailing fashions in medicine, the polypharmacy of such Renaissance preparations as mithridate with its hundred ingredients, and the exaggerated reverence for the botanical remedies of Galen and Avicenna.[1]

Of major importance among the new medicines was corrosive sublimate or mercury(II) chloride with which the iatrochemists had achieved a good deal of success in treating syphilis. Robert Boyle recommended it wholeheartedly as 'a certain rough Emetick with which the French Pox has often been cured', and it soon made inroads against the alternative cure based on guaiacum wood. This use of mercurials may be the source of the expression 'quack doctor' from *Quecksilber*, the German name for mercury.

The process of sublimation (heating a solid and condensing the vapours on a cooled surface) was central to this period of practical chemistry, and suitable apparatus in glass, cast iron or earthenware was a salient feature of the iatrochemical laboratory. Corrosive sublimate was made by subliming a mixture of mercury, common salt and alum, though sometimes the salt was replaced by ammonium chloride and the alum by green vitriol (iron(II) sulphate).

Although mercury(II) chloride had been prepared by treating mercury with chlorine gas as early as 1841, the most widely used process in the 19th century

[1] W.A. Campbell, 'Marvellous Paracelsus', *Chem. Brit.*, 1994, **30**, 39.

again involved sublimation. Mercury was dissolved in sulphuric acid (with copious emission of sulphur dioxide gas) and the resulting mercury(II) sulphate mixed with salt and packed into round bottomed flasks. The flasks were heated in a sand-bath with their necks protruding, and corrosive sublimate collected in the cooler necks.

The iatrochemists made the popular purgative calomel (mercury(I) chloride) by subliming four parts of corrosive sublimate with three parts of mercury. It was later manufactured by dissolving mercury in nitric acid (when the dangerous brown fumes of nitrogen dioxide were given off) and precipitating calomel by adding common salt. Calomel kept its place in the materia medica long enough to be lampooned by W.S. Gilbert in *Patience* (1881). Corrosive sublimate on the other hand found new uses in preserving timber from dry-rot, and in taxidermy. Mixed with metallic mercury, corrosive sublimate was used in the process of felting hair for the hat trade. It thus contributed to the spread of that form of mercury poisoning called 'hatter's shakes' from which the Mad Hatter in *Alice in Wonderland* suffered.

Another significant remedy was butter of antimony or antimony trichloride. This was prepared from native antimony sulphide (stibnite) by sublimation with corrosive sublimate, until Glauber in 1668 showed how to make it by boiling stibnite with concentrated hydrochloric acid. The atmosphere in laboratories where such preparations were carried out can barely be imagined, particularly as these were often in the back premises of retail shops. On the continent, preparative pharmacies were frequently set up by cities, states or guilds; in this country the only comparable establishment was that at Apothecaries' Hall in London.

Thus the preparative techniques of commercial inorganic chemistry were being forged. The druggist with his knowledge of herbs was forced to become a 'chemist and druggist', a title which he was to retain for three centuries.[2]

The virtues of antimony were recorded by the pseudonymous 'Basil Valentine' in *The triumphal chariot of antimony* (1604). The most assiduous user in England was the quack doctor Joshua Ward, who is important in the history of the chemical industry because he made sulphuric acid on a factory scale, first at Richmond and then at Twickenham (p. 58). After an abortive political career which led to his exile in France, he began to make and sell antimonial and other chemical medicines in London in 1733.

In addition to antimony trichloride, Ward made extensive use of 'glass of antimony', a mixture of sulphide and oxide prepared by carefully controlled roasting of stibnite. This was the basis of Ward's Pills and Drops, of which it was said:

> Before you take his Drop or Pill,
> Take leave of friends and make your will.[3]

[2] R.P. Multhauf, *The origins of chemistry*, Oldbourne, London, 1966, p. 22.
[3] W.A. Campbell, 'Portrait of a quack; Joshua Ward (1685–1761)', *Univ. Newcastle Med. Gaz.*, 1964, **58**, 118.

Ward's grateful patients, apart from a stream of poor folk who obtained free treatment, included Horace Walpole the Whig politician and Henry Fielding the novelist and magistrate. The scale of the preparative operation can be judged by the fact that Ward needed three dispensaries, in Pimlico, Whitehall and Threadneedle Street, to satisfy the demand for his nostrums. During the period of iatrochemistry, the physicians of course continued to avail themselves of the whole range of 'galenical' or herbal remedies.

## 2   Pigments

The flowering of artistic enterprise that came with the Italian Renaissance depended on a ready supply of pigments. Although many artists prepared their own, there soon emerged a new trade of colourman. Some pigments were prepared simply by grinding some mineral, for example ultramarine from lapis lazuli, or the ochres, umbers and siennas from ferruginous earths. But chemical preparations such as white lead and verdigris were established in Roman times. By the 17th century there was considerable scientific interest in colour making. The Royal Society, as part of a general survey of industries in the second half of the 17th century, intended to publish an account of pigment manufacture. The project came to nothing however, perhaps because the philosophers and the artisans had little common ground for discussion. A century later the Society for the Encouragement of Arts, Manufactures and Commerce (now the Royal Society of Arts) offered prizes for improvements to the manufacture of specific pigments.

### (a)   Vermilion

Natural vermilion, a highly prized red, was expensive since it was made by grinding the somewhat scarce mineral cinnabar (native mercury sulphide). A cheaper form was synthesised by a particularly hazardous process. Sulphur was melted in a shallow vessel, and mercury was run in, the mix being stirred continuously. The black mass so formed was broken up and sublimed to yield the red sulphide. An average batch took 500 kg of mercury, so the amount of mercury vapour escaping at temperatures above 100 °C must have been considerable. In 1828 a wet method of manufacture was introduced. Sulphur and mercury were fully mixed and then digested under caustic potash solution; when the desired shade of red was reached, the liquid, which consisted largely of polysulphides, was decanted and the pigment washed and dried. Artists, however, preferred the vermilion made in the dry way.[4]

### (b)   White lead

White lead had been made in England by the stack process since Elizabethan times, usually under royal monopoly. A thin strip of lead was rolled into a loose

---

[4] W.A. Campbell, 'Vermilion and verdigris', *Chem. Brit.*, 1990, **26**, 558.

spiral and suspended over vinegar in a pot which was loosely closed with a lead plate. Four hundred such pots were packed in a square and covered with animal dung. Boards were placed over the layer of pots and a fresh square started. The stack consisted of four such layers accommodating 1600 pots. After some weeks the stack was dismantled and the white encrustation shaken and scraped from the lead strips and cover plates. The product was a basic lead carbonate.

In 1787 Richard Fishwick of Newcastle-upon-Tyne substituted spent bark from the tanyards for dung in the stack process, and from that time it was common to find white lead works situated near to tanneries, often with a manure works and a paint works to complete the network.

Late in the 19th century a quicker chamber process was introduced, but many believed that the pigment produced in the older way possessed better covering power when made up into paint. The chamber process for white lead differed from the stack process in that carbon dioxide was generated outside the chamber either from a lime kiln or a coke fire. Sometimes the vinegar, instead of being exposed in pots, was also boiled outside the chamber and the acetous vapours piped in. At about the same time, a precipitation process originated in France and was used for a time in this country. Lead acetate solution was boiled with litharge (the yellow monoxide of lead) to produce a basic lead acetate which was treated with carbon dioxide. Although a poorer pigment resulted, this had the advantage of avoiding the stripping of white lead from unchanged metal during which the workers were exposed to dust which inevitably was ingested. Women were often employed in this part of the works.[5]

## (c)   Red lead

Vitruvius has described how red lead or minium (lead oxide $Pb_3O_4$) was discovered when some Roman cosmetic jars containing white lead were damaged in a fire; since Tudor times the pigment has been made by roasting white lead. A more direct process consisted in roasting metallic lead in a reverberatory furnace which Richard Watson described as being 'like a baker's oven'. The lead became coated with the yellow oxide litharge, which was scraped off as it formed; about 24 hours sufficed to convert all the lead. The litharge was taken to the colouring furnace where it was exposed to a temperature of between 500 and 600 °C. Constantly raked to present a new surface, the litharge was converted into red lead in about 48 hours. Here again, women were often employed in the raking process. It later became common to engage a doctor to inspect the workers, but many evaded the examination, fearing dismissal if they were found to be ill. The women, particularly, would boast of having 'dodged the doctor'.

[5] T. Oliver (ed.), *Dangerous trades*, London, 1902; Anon., 'A day at the lead works', *Penny Magazine*, 1844, **13**, 357.

*Stirring red lead: Locke, Blackett & Co., Newcastle, 1844*

### (d)   Verdigris

Verdigris, a basic copper acetate, was made by a process dating from antiquity and resembling that for white lead. The English mode of manufacture involved placing alternate layers of woollen cloth and copper strips in a wooden box, the cloth being moistened with vinegar every day. After some weeks the copper was coated with the green pigment which was flaked off by bending the strips. During the 18th century this manufacture became a by-trade of the destructive distillation of wood, the pyroligneous acid (impure acetic or ethanoic acid) being used in place of vinegar. A century later, verdigris was made by precipitating from copper sulphate solution with calcium acetate, or by dissolving copper oxide in pyroligneous acid. Neither product was identical with that from the older process. Basic copper carbonate, sold as green verditer, was made in a similar manner by mixing solutions of copper sulphate and sodium carbonate. Apart from these ancient processes, many pigments were made in ways that seemed to owe more to serendipity than to science. Verditer for example was made by calcining boric acid with potassium dichromate to produce a crude chromium borate which was hydrolysed with boiling water. Schweinfurt Green was made by roasting together arsenic(III) oxide with verdigris.

### (e)   Scheele's Green

In 1778 the Swedish chemist C.W. Scheele prepared a new and vivid pigment by dissolving white arsenic oxide in hot aqueous potassium carbonate, and running

*C.W. Scheele (1742–1786), Swedish chemist, co-discoverer of oxygen and pioneer in gas chemistry*

in copper sulphate solution until precipitation was complete. The product, copper arsenite, became known as Scheele's Green. Scheele was well aware of its poisonous nature, but many decades elapsed before that knowledge reached colourmen and confectioners. In 1848 a number of persons became seriously ill and one died after eating a spectacular green blancmange at a public banquet in Northampton.[6]

The employment of Scheele's Green as a wallpaper pigment led to much illness and a number of deaths; it was once believed that Napoleon had died from this cause on St. Helena, but this is now doubted. Queen Victoria selected green arsenical wallpaper for the bedrooms at Osborne House on the Isle of Wight, a choice which resulted in sickness among her guests and, apparently, the death of a parrot. Poisoning only occurred if the bedroom was damp; moulds growing on the starch paste behind the paper could exude volatile arsenides to be inhaled by the sleeper. The fact that many of the most popular shades of wallpaper were achieved through poisonous pigments created a hazard to the health of the workers. The Commission on the Employment of Children in 1843 discovered that the boys who handled the pigments were known, not without reason, as 'blue boys', 'yellow boys' and 'green boys'. Since it was held to be uneconomical to stop the machinery for meal breaks, the boys

[6] A. Normandy, *Commercial handbook of chemical analysis*, Crosby Lockwood & Co., London, 1850, p. 78.

ate their food whilst serving the machines with compounds of mercury, arsenic, copper, lead and chromium.

Scheele's Green was so popular that everyone wanted to use it. In 1854 the *Medical Times and Gazette* carried an article by the toxicologist A.S. Taylor, entitled 'How families are poisoned'.[7] He described how, while cutting the bread for breakfast, he had observed some green streaks on the crust. Investigation revealed that the baker had brightened up his shop by painting the shelves with Scheele's Green. He of course was quite unaware that the paint was poisonous, but the colourman who supplied it knew very well but pleaded that without arsenic it was impossible to get a good green.

### (f)   Prussian Blue

Prussian Blue was discovered by accident in 1704. A Berlin colourman named Diesbach was in the habit of preparing a crimson lake from a mixture which contained iron(II) sulphate, alum and potash. Finding himself out of potash, Diesbach bought some from his neighbour Conrad Dippel, a slippery character who made and sold a universal medicine known as Dippel's Animal Oil. Dippel's potash was old stock which had been used in the destructive distillation of animal refuse in the preparation of the animal oil. It was therefore contaminated with cyanides, and when Diesbach boiled it with his iron sulphate he obtained the colour which was at first called Berlin Blue.[8]

Although no official publication of the method occurred in England until 1724, the process had been brought to this country nearly twenty years earlier by a Jewish immigrant who established the manufacture on Tyneside. His business was bought up by Thomas Bramwell of Heworth on the Tyne, who became one of the largest manufacturers of the pigment in Britain. Bramwell also did pioneering work on the fixation of atmospheric nitrogen, and prepared prussiates of potash. The older method for making potassium ferrocyanide was to fuse potash in an iron pot, stirring in iron turnings, blood, hide trimmings, and other slaughterhouse waste. Extraction with water followed by evaporation yielded yellow crystals of the salt. If the solution of ferrocyanide was made acid and treated with iron(III) sulphate, a precipitate of Prussian Blue was obtained. Blue factories caused much distress to nearby residents on account of the smell of the animal refuse. Later the ferrocyanide was made by the less objectionable process of boiling iron salts with potassium cyanide. In places where coal gas was purified by passing it over moist iron oxide, prussiates and even Prussian Blue itself were formed in the gas-works. This was never a source of the best quality pigment, for sulphur in the gas caused contamination with thiocyanates which form a red complex with iron.

Apart from its use in the paint trade, Prussian Blue was required in large quantities for calico printing. The cloth was soaked in iron sulphate solution followed by alkali, thus forming iron hydroxide in the fibre. The parts to be

---

[7] A.S. Taylor, 'How families are poisoned', *Med. Times & Gaz.*, 1854, new ser. **8**, 320.
[8] E.E. Aynsley and W.A. Campbell, 'Johann Konrad Dippel (1673–1734)', *Med. Hist.*, 1962, **6**, 281.

dyed blue were then treated with an acid solution of prussiate. The American Civil War of 1861–1865 dealt a fatal blow to the Prussian Blue trade by cutting off supplies of cotton. By the time that the supply was restored, coal tar colours had begun to replace the older pigment.

A significant quantity of Prussian Blue was exported to China for the purpose of 'facing' green tea. Much of the China tea brought into this country contained up to ten percent of Prussian Blue. When questioned, the Chinese exporters admitted that they did not drink the 'faced' tea, but the foreigners seemed to like it and were prepared to pay a good price for it.[9]

## (g)    Chromium pigments

Chromium was isolated by Vauquelin in 1797, and over the following decade he worked out the chemistry of chromates and dichromates. Two of these salts were useful pigments, barium chromate or lemon yellow and lead chromate or chrome yellow. The first step in manufacturing these was to make potassium chromate by roasting chrome iron ore, $FeO.Cr_2O_3$, with a mixture of potassium carbonate and potassium nitrate; formerly potassium nitrate was used alone, but the product was too expensive. The sintered mass was ground to a fine powder and extracted with boiling water; wood vinegar was added to precipitate alumina and silica, and the clear liquid concentrated until crystals of potassium chromate separated. The salt was widely used as an oxidant in dye-making, as well as a starting material for the chrome colours. In the 1880s most of the country's potassium chromate was made by John & James White of Rutherglen, but the later use of chrome chemicals in tanning brought more firms into the trade. Chrome leather was a very durable, though more expensive, product.

Barium chromate was prepared by precipitation from barium chloride solution and potassium chromate. Its use as a pigment dated from 1835 when George Field included it in his treatise on colours. Lead chromate occurs in nature as the mineral chrocoite, but Vauquelin prepared it by adding a soluble lead salt to potassium chromate. The pigment was manufactured in London around 1816 by J.E. Bollmann, but the larger and more famous operation was that conducted by Andreas Kurtz in Manchester in 1831. Princess Charlotte, daughter of George IV, had chosen chrome yellow for her carriage, and Kurtz used this fact to popularise the shade. The pigment was widely used as a food and confectionery colour, often with disastrous results. In the examination of coloured confectionery carried out by *The Lancet* in 1851, lead chromate was found to be the most popular colouring matter for sweets and cake decorations.[10] On a happier note, the shade was popular with some railway companies including the Lancaster and Preston Junction Railway for coach panels.

A later, and slightly strange, method of preparation involved digesting sparingly soluble lead sulphate with potassium dichromate solution, the resulting lead chromate being even less soluble than lead sulphate. The latter was a

[9] Normandy (ref. 6), p. 550.
[10] Report of the analytical sanitary commission, *The Lancet*, 1854 (**i**), 316, 425, 524, 581.

by-product of the manufacture of aluminium acetate mordant (p. 46) by the reaction between lead acetate and aluminium sulphate.

The chrome colours provided a useful alternative to arsenic trisulphide, orpiment or King's Yellow. Although this occurs in mineral form, it was manufactured by sublimation of a mixture of white arsenic, $As_2O_3$, and sulphur, or by bubbling hydrogen sulphide gas into a solution of white arsenic in hydrochloric acid. In 1860 a pastry-cook in Clifton, Bristol, used arsenic sulphide to colour some Bath buns in order to give the illusion of a cake-mix rich in eggs: several schoolboys became ill and one barely escaped with his life. The colourman who had supplied the pigment said he was very sorry; he had really intended to supply lead chromate![11]

Another yellow pigment which was popular for several decades was also made from a by-product. Turner's Yellow, a nonstoichiometric lead chloride, arose from Scheele's process for making caustic soda from lead monoxide and salt solution (p. 77). The pigment was patented in 1781 by James Turner, a colour manufacturer, and was made by the Losh brothers at Walker near Newcastle-upon-Tyne, and on a larger scale by Hugh Lee Pattinson at Washington in County Durham.[12]

Naples Yellow or lead antimoniate, $Pb_3(SbO_4)_2$, only became popular in Britain in the second half of the 18th century. It was made in a manner reminiscent of the early techniques of medical chemistry by heating together glass of antimony with lead monoxide.

### (i) Ultramarine

In Tudor times the costliest pigment was ultramarine, made by grinding the semiprecious stone lapis lazuli. An ounce of the pigment cost £7.50 in London in 1600. For this reason it was reserved for special subjects such as the robes of the Madonna in nativity paintings; miniaturists who required only tiny quantities found it valuable. Similarly it had been used five hundred years earlier by manuscript illuminators, chiefly for elaborate Gospel books. For a long time analysis of natural ultramarine was far too difficult for the growing science of chemical analysis, but by the end of the 18th century it was clear that it contained soda, alumina, silica and sulphur.

Prizes were offered by the Society of Arts in 1817 and by the French Societé d'Encouragement pour l'Industrie in 1824 for a method of synthesis which would produce a cheaper product which was comparable with the natural material. Meanwhile a blue encrustation had been noted in a black ash furnace at the Government glass and alkali works at St. Gobain near Paris. Goethe had seen a similar phenomenon in lime kilns in Italy and had observed its use there in place of lapis lazuli. Vauquelin soon showed the St. Gobain pigment to have the same composition as lapis lazuli. The British offer of a prize brought no result, but the French offer led to simultaneous solutions to the problem by J.-B.

[11] W. Crookes, *Chem. News.*, 1860 (**i**), 48.
[12] R. Harley, *Artists' pigments, 1600–1835*, Butterworth, London, 1979, pp. 91–92.

*Advertisement for Reckitt's Blue*

Guimet and Christian Gmelin. Guimet began to manufacture ultramarine in Lyons in 1828. A year later the Royal Porcelain factory at Meissen also began to make the pigment.[13]

Ultramarine was brought to this country by Isaac Reckitt, a starch manufacturer in Hull. Made up into little packages called 'blue bags' it was used to restore whiteness to yellowing fabrics.

In 1863 the firm began to manufacture its own ultramarine, previously bought from Guimet. The success of the blue bag led the firm into other household chemicals such as Brasso and Silvo metal polishes. In 1938 Reckitts joined the mustard manufacturer J. & J. Colman to form Reckitt–Colman who, in addition to such well-known products as Windolene, Harpic and Dettol, make chemicals for the pharmaceutical and veterinary markets.[14]

Recent developments in health and safety regulations have resulted in a diminished use, if not total abandonment, of mercury, lead and cadmium colours. These have largely been replaced by 'lakes' made by precipitating aluminium hydroxide from solutions of aluminium sulphate containing soluble organic colours. As the precipitate falls it brings down with it the colour firmly fixed in the aluminous base. White lead and the barium whites (pp. 159 and 183)

[13] S. Muspratt, *Chemistry applied to arts and manufactures*, London, 1862, vol. ii, p. 1072.
[14] P.J.T. Morris and C.A. Russell, *Archives of the British chemical industry, 1750–1914*, British Society for the History of Science, Monograph no. 6, Faringdon, 1987, p. 165.

**Table 1** *Common Inorganic Pigments*

| Colour | Pigment | Chemical constituents |
|---|---|---|
| White | Bismuth White | Bismuth carbonate |
| | Blanc fixe (permanent white) | Barium sulphate |
| | Chinese White (zinc white) | Zinc oxide |
| | Lithopone | Zinc sulphide + barium sulphate |
| | Paris White (whiting) | Calcium carbonate |
| | Titanium White | Titanium dioxide |
| | White lead | Lead carbonate |
| Red | Antivermilion | Antimony sulphide |
| | Burnt Sienna | Calcined ferruginous clay |
| | Cadmium Red | Cadmium sulphide |
| | Red lead | Lead oxide $Pb_3O_4$ |
| | Venetian Red | Iron oxide $Fe_2O_3$ |
| | Vermilion | Mercury sulphide |
| Yellow | Cadmium Yellow | Cadmium sulphide |
| | Cassel Yellow | Basic lead chloride |
| | Chrome Yellow | Lead chromate |
| | King's Yellow (orpiment) | Arsenic sulphide |
| | Naples Yellow | Lead antimonate |
| | Turner's Yellow | Basic lead chloride |
| Green | Brunswick Green | Lead chromate + Prussian Blue |
| | Chrome Green | Chromium oxide $Cr_2O_3$ |
| | Cobalt Green (Rinman's Green) | Cobalt zincate |
| | Scheele's Green | Copper arsenite |
| | Schweinfurt Green (emerald green) | Copper arsenite + copper acetate |
| | Verditer | Hydrated chromium oxide |
| | Verdigris | Basic copper acetate |
| Blue | Cerulean Blue | Cobalt stannate |
| | Cobalt Blue (Thenard's Blue) | Cobalt aluminate |
| | Prussian Blue | Potassium iron(III) hexacyanoferrate(II) |
| | Smalt | Powdered cobalt glass |
| | Ultramarine | approx. $3NaAlSiO_4.Na_2S_2$ |

have largely been replaced by titanium dioxide. This is made from ilmenite, $FeTiO_3$, by extraction with sulphuric acid; the iron sulphate by-product is used in water treatment.

Common inorganic pigments are listed in Table 1.

# 3 Glass and Glazes

Speculation about the discovery or invention of glass is not relevant to this book. Neither is a consideration of Egyptian and Roman glass, nor the first essays in the use of window glass by the Venerable Bede and Benet Biscop in the 8th century. It is sufficient to state that glass was made by fusing together sand, limestone, and soda or potash over a wood fire.

*Non-chemical outlets for wood*
Building material for ships, Building material for houses, Roof supports for mining. Fuel for brewing, Fuel for glass-making, Fuel for pottery kilns, Fuel for smelting metals, Fuel for foundry and forge work, Fuel for brick and tile making, Fuel for soap-boiling, Fuel for baking, Fuel for domestic heating.

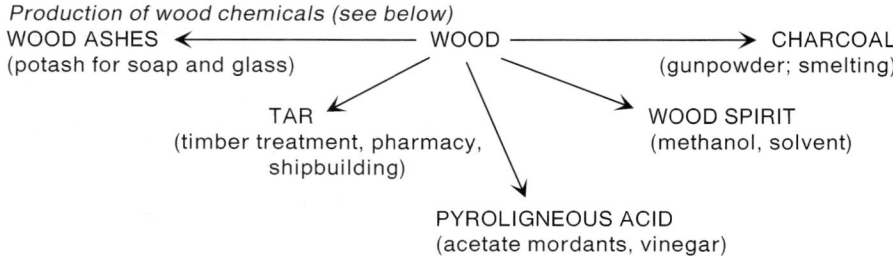

*Production of wood chemicals (see below)*

WOOD ASHES ⟵——————— WOOD ———————⟶ CHARCOAL
(potash for soap and glass)                                    (gunpowder; smelting)

TAR                                            WOOD SPIRIT
(timber treatment, pharmacy,              (methanol, solvent)
shipbuilding)

PYROLIGNEOUS ACID
(acetate mordants, vinegar)

**Figure 1**   *Early wood-based economy in Britain*

From the 13th century, glass-making in England was concentrated in the heavily wooded areas of the South East. It was a nomadic trade, glass furnaces being worked until the surrounding forest became denuded, and then moved to a new site. The glass-makers were often French, and were joined in the 16th and 17th centuries by Protestant refugees from Lorraine.

The modern glass industry in Britain dates from the edict of 1615 forbidding the use of wood as a fuel in glasshouses – an early example of legislation directed towards conservation. The necessity for such legislation is shown in Figure 1.

The restructured industry settled in Bristol, in Stourbridge and on Tyneside, but many technical changes were necessary before a satisfactory switch from wood to coal-fired furnaces could be achieved. The first significant operator after the edict was Admiral Sir Robert Mansell who set up glasshouses at Newcastle upon Tyne in 1619. There was on Tyneside a plentiful supply of cheap coal together with fireclay from the coal measures; sand and limestone were brought back as ballast by the wooden sailing collier ships which carried coal to London and the continental ports. Mansell was further protected by a monopoly from Charles I, a fact which led to Newcastle being the only port on the East side of the country to declare for the King during the Civil War.[15]

An influential change occurred in 1673 when George Ravenscroft took out his patent for lead glass, often made from ground flints and known as flint glass. Some or all of the limestone was replaced by lead oxides to provide a glass which was easier to work and more highly refractive. This gave a boost to the red lead trade, as well as allowing better lenses for telescopes and microscopes to be made. The influence of this development on 17th century astronomy and microscopy can hardly be over-stated.

By the 18th century London was the major centre for flint glass, followed by Stourbridge and Bristol for the glass bottles which had replaced earthernware and leather for containing wine, and Newcastle for window glass.

Coloured glass has a long history, and in the earliest cases it is not possible to

[15] E. Barrington Haynes, *Glass through the ages*, Penguin, Harmondsworth, 1959, p. 250.

*Glass-blowing furnace*

be certain if the colouring agent was added deliberately or if it was introduced as an impurity in one of the ingredients. Scientific control of coloured glass can be judged to begin in the 16th century when the production of the vitreous substance called zaffre was introduced. Both Biringuccio and Agricola refer to zaffre as a colourant for glass and pottery glazes. A similar compound called smalt was made by fusing cobalt ore with sand. Georg Brandt in 1742 showed that the blue colour of zaffre and smalt was due to cobalt. In Britain this fine sapphire blue glass became especially associated with Bristol, but it was made in all the glass-making centres.

Red glasses were first made by adding iron(III) oxide to the melt. This oxide was cheap, being obtained either by roasting iron sulphate or by collecting scale from the forge, but the colour was poor. Red copper oxide $Cu_2O$ had been used in the 16th century to impart a brilliant red so dense that thick glazes became almost opaque. The usual procedure was to coat a clear glass with a very thin layer of red copper glass. The best ruby glass was made by adding small quantities of gold. The sand was moistened with 'gold chloride' (chlorauric acid, $H_3AuCl_6$) before mixing with the other ingredients.

Addition of copper(II) oxide to the melt produced a fine emerald green glass. Care was always needed however to ensure that no reduction to copper(I) occurred, and for this reason a little potassium nitrate was added. Chromium(III) oxide also gave a good green, but on account of cost it was used only for special purposes. Sand containing a little iron would impart a dingy and less attractive green such as is found in wine bottles. If this colour was accidentally introduced into the melt, very small quantities of manganese dioxide would discharge it. For this reason manganese dioxide was sometimes known as 'glass-maker's soap'. It was made by grinding the mineral pyrolusite, or much later obtained through the Weldon manganese recovery process described on p. 92.

During the 18th century artificial gems became popular, and indeed the (Royal) Society of Arts offered premiums for the successful copying of precious stones. The base was a special kind of glass named 'strass' after its inventor. A typical strass formulation was:[16]

| | |
|---|---|
| Ground rock crystal | 100 parts |
| Red lead | 156 parts |
| Caustic potash | 54 parts |
| Boric acid | 7 parts |
| Arsenic(III) oxide | 0.3 parts |

The high proportion of lead conferred upon strass a refractive power close to that of diamond, and 'paste' diamonds were made from strass alone. Since traces of impurity might affect the colour and clarity, the rock crystal was washed with hydrochloric acid to extract iron, and the potash, boric acid and arsenic oxide were all recrystallised.

The musician Georg Berg, director of music at the Ranelagh pleasure gardens in London, experimented assiduously on the production of artificial gems from 1759 to 1774. Using several famous London glasshouses, and operating at weekends, he worked with a borosilicate glass suggested by Peter Shaw. Since borax cost £700 per ton in London in 1750, it is clear that products of the experiments would be expensive.[17]

For colourants Berg employed sulphates of copper and iron, verdigris, manganese ore, zaffre, smalt, iron oxide, silver and gold. Several of Berg's glasses were tested by members of the Society of Arts for use in enamelling precious metals. Saltpetre cost 6d a pound, sal ammoniac 3d and red lead 2d. This episode reveals that coloured glasses and enamels presented a free-for-all field in which the necessary expertise could easily be acquired. It also indicates the range of chemicals which could be purchased in mid-18th century London.

## 4   Matches and Pyrotechnics

At the State trial of Guy Fawkes for attempting to blow up the Houses of Parliament in 1605 it was said 'the fellow had about him a lantern and six matches'. These matches were merely spills of wood, or perhaps coarse strings of tow, dipped in melted sulphur; their function was to transport and not to initiate a flame. In Guy Fawkes's day, as for another two centuries, the common way to get a light was to generate a spark from a piece of steel by striking it a glancing blow with a flint, allowing the spark to fall on some charred rags or

[16] *Thorpe's Dictionary of pure and applied chemistry*, 4th ed., Longman, London, 1956, vol. v, pp. 513, 605.
[17] W.A. Campbell, 'Musical glasses', *Chem. Brit.*, 1989, **25**, 145.

dried leaves in a tinder-box, and gently blowing the smouldering tinder into a flame. Sometimes combustion was assisted by soaking the tinder in potassium nitrate solution before drying.

*The discovery of phosphorus, imaginatively reconstructed in a painting by Joseph Wright*

### (a) Phosphorus

Chemical means of initiating a flame date from 1669 when Hennig Brand isolated phosphorus by distilling urine with sand. Brand's process was operated on a commercial scale in Robert Boyle's laboratory by Ambrose Godfrey Hanckwitz who for several decades was the major European supplier of phosphorus from premises in Southampton Street, London. Boyle called phosphorus 'the icy noctiluca', an early suggestion that it might be used as a night-light. When Krafft visited Boyle in 1677 he demonstrated how to ignite gunpowder with phosphorus, scorching Boyle's carpet while doing so.

In 1775 the Swedish chemist Scheele prepared phosphorus from bone ash (calcium phosphate). His method was to treat the bone ash with concentrated sulphuric acid and to reduce the resulting phosphoric acids with charcoal. Phosphorus was manufactured by this process until the 1840s when mineral phosphates from South America were substituted for bone ash.

The Quaker pharmacist Arthur Albright began to make phosphorus in

*A group of chemical workers at Albright & Wilson's phosphorus works, Oldbury, West Midlands (1890)*

Birmingham in 1840. At first he used bones left over from the meat canning operations on the Danube. Complaints about the stench from the cargo by shippers who transported the bones led Albright to set up a calcining unit at source. The phosphorus factory moved to Oldbury in 1851.[18]

The effect of phosphorus vapour on the workers was horrific in the extreme. The vapour entered the cavities in carious teeth and caused necrosis of the lower jaw. Moreover, phosphorus burns were slow to heal, and there was the ever-present risk of the phosphorus spontaneously igniting. In 1845 Anton Schrötter of Vienna found a way to transform the dangerous yellow phosphorus into a less reactive red form. Being a powder, it was easier to handle than the waxy solid, and it transformed phosphorus-based industries.

A further advance was made in 1888 when continuous production in the electric furnace was introduced. The body of the furnace was a large iron tank lined with firebrick and having carbon electrodes at opposite ends. Ground phosphate, carbon and silica were fed in through a hopper and yellow phosphorus was collected under water. To convert it into the red allotrope, the phosphorus was maintained at 250 °C in a covered iron pot under pressure.

Meanwhile the search for a convenient method of initiating a flame continued. The bottle coated internally with phosphorus into which a sulphur-tipped splint could be dipped appeared in 1886. It was both uncertain and dangerous, but it continued in use for another thirty years.

[18] R.E. Threlfall, *A hundred years of phosphorus manufacture*, Albright & Wilson, Oldbury, 1951, p. 8.

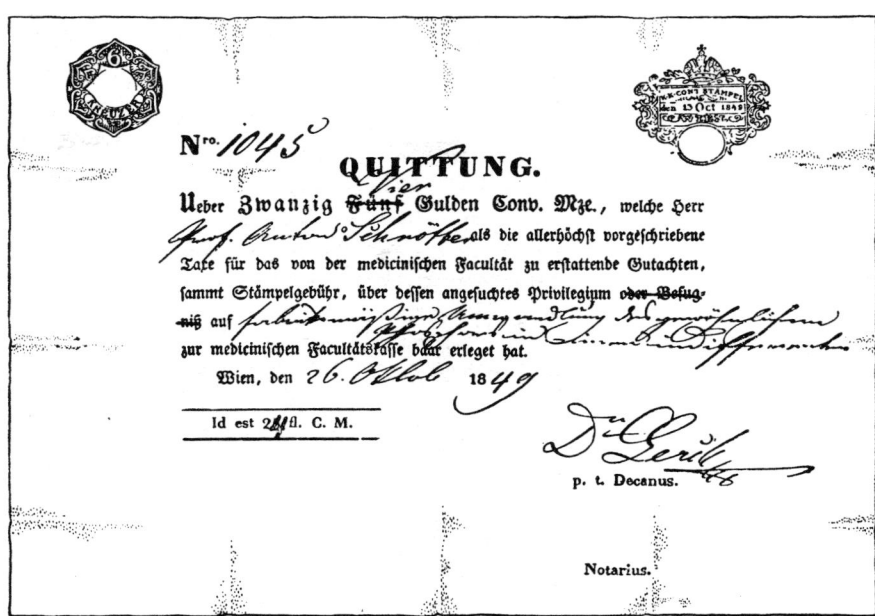

*Receipt for Anton Schrötter (1802–1875), discoverer of red phosphorus*

### (b) Incendiary mixtures

In 1784 the French chemist Berthollet had demonstrated that a mixture of potassium chlorate and sugar would ignite when touched with a glass rod dipped in sulphuric acid. From 1805 matches tipped with the chlorate and sugar mixture could be bought. These were dipped into a small bottle filled with asbestos soaked with concentrated sulphuric acid. A variant was a thin sealed glass tube filled with sulphuric acid and embedded in a roll of paper containing the incendiary mixture; breaking the tube led to ignition of the mixture. This device was known as the *Promethian* and was made and sold by Samuel Jones in the Strand, London; the firm supplied small pliers to break the glass tube, but bolder users broke it between the teeth.

Friction matches were invented by John Walker, a pharmacist in Stockton-on-Tees, in 1827. His match was tipped with a mixture of potassium chlorate and antimony trisulphide, and was ignited by drawing it between a folded sheet of glass-paper. Two years later Samuel Jones introduced the *Lucifer* which was tipped with chlorate, antimony sulphide and sulphur. In the following year the *Congreve* appeared, a match having a sulphur head with a phosphorus tip which would strike on any rough surface.[19]

[19] R. Miller Christy, *Bryant and May museum of firemaking appliances; catalogue of the exhibits,* Bryant and May, London, 1926, pp. 106, 108.

The discovery in 1864 of tetraphosphorus trisulphide provided match manufacturers with a more manageable material for tipping 'strike anywhere' matches, though it was another thirty years before the compound became widely used.

Children were employed in match factories, particularly for packing matches into boxes. This was described as 'bairns' work' by one Tyneside manufacturer who claimed that he could not afford to pay adults to do the work. The children were supervised by a man who suffered from 'phossy jaw' in an advanced state. He was described as resembling a death's head, the hole in the lower part of his face concealed by a dirty bandage, and exuding the loathsome smell of necrosis.

The basis of pyrotechny is gunpowder, that age-old mixture of charcoal, sulphur and saltpetre (potassium nitrate). The quality of the first two ingredients is of supreme importance but only potassium nitrate falls into the ambit of this chapter. Saltpetre was for a long time the bottle-neck in the manufacture of gunpowder.

In Tudor times saltpetre was made by the reaction between wood ashes (potassium carbonate) and calcium nitrate. The latter was found in lime-rich soils which had absorbed animal excreta. Likely sources were pigsties and hen houses, especially if built of brick when the mortar would furnish the necessary calcium. These places were sought out by the hated 'petre-men' who had power to invade private property and carry away suitable soil in whatever transport they cared to commandeer. Needless to say, complaints were rife. It was said that they 'dig in all places without distinction, as in parlours, bedchambers, threshing and malting floors, yea, God's own house they have not forborne'. The petre-men, it was claimed, disturbed livestock at lambing and calving time, and chased out sitting doves from dovecotes. They also undermined walls and left unfilled holes in the ground. Moreover they were accused of taking bribes and making extortionate demands on owners of horses and carts. In reply the petre-men pleaded 'the unwillingness of the King's subjects to do anything for this service'. Meanwhile saltpetre farms were set up in India, the product from Bengal being specially valued.[20]

From 1830 'Chile nitre' (sodium nitrate) came into this country to satisfy the demands of the sulphuric acid trade. The sodium salt was too hygroscopic for gunpowder manufacture, so it had to be boiled with potash to convert it into potassium nitrate, at the same time releasing useful sodium carbonate for the glass trade, but competing with the alum makers for potash. Only the opening out of the Stassfurt deposits of carnallite (potassium chloride) in 1865 relieved the pressure on potash. By that time however gunpowder was no longer the only explosive for civil engineering and military projects.

The next most important material in pyrotechnics was potassium chlorate, discovered by Berthollet in 1784. Though this was easily made by passing chlorine gas into hot potassium hydroxide solution; the equation,

$$3Cl_2 + 6KOH = KClO_3 + 5KCl + 3H_2O$$

[20] S.A. Gregory, *Chemicals and people*, Mills & Boon, London, 1961, p. 11.

indicates that potassium chloride is the major product. A cheaper alternative suggested by Thomas Graham involved saturating a slurry of slaked lime with chlorine:

$$6Cl_2 + 6Ca(OH)_2 = Ca(ClO_3)_2 + 5CaCl_2 + 6H_2O$$

The calcium chlorate was boiled with potash to form potassium chlorate, and no waste potassium salts were generated.

From the 1880s chlorates of sodium and potassium have been made by electrolysing hot solutions of the chlorides: a little potassium dichromate was added to catalyse the formation of chlorate. Fundamentally, electrolysis was no different from the earlier processes except that the chlorine and caustic alkali were generated in the same vessel.

Brilliant coloured flames were only possible after potassium chlorate became available; it was first used in pyrotechny by James Cutbush in 1823. Sodium chlorate found wide application as a weed killer, but the powerfully oxidising nature of the chlorates created hazards for the workers. In fireworks factories strict rules ensure that chlorate and sulphur pass through the works by separate routes.[21]

Coloured fireworks have long depended chiefly on barium and strontium nitrates. Barium was recognised by Scheele in 1774 and isolated by Davy in 1808. In 1602 a shoemaker in Bologna named Casciorolo observed that heavy spar (barium sulphate) became luminous when heated with charcoal. The impure barium sulphide so formed became known as Bologna phosphorus. Barium carbonate was discovered at Leadhills in South-West Scotland by a Birmingham physician named Withering; the mineral became known as witherite. The last witherite mine in Britain, at Settlingstones in Northumberland, closed some twenty years ago.

In 1790 a related mineral was found in the lead mines at Strontian in Argyll to which Crawford gave the name strontia (strontium carbonate). A mineral sulphate of strontium found near Bristol was named celestine. Barium and strontium nitrates were manufactured by dissolving the mineral carbonates in nitric acid. Celestine was formerly reduced to strontium sulphide by furnacing with carbon, the sulphide being treated with nitric acid. Table 3 shows the range of chemicals supplied, though not exclusively, to the pyrotechnic trade.

## 5  Copperas and Alum

### (a)  Copperas

Copperas contains no copper but is hepta-aquoiron(II) sulphate or green vitriol, $FeSO_4.7H_2O$. The manufacture of copperas depended on the weathering of iron pyrites $FeS_2$, found in coal measures as 'coal brasses' or 'fool's gold'. For both domestic and industrial purposes pyrites was an undesirable impurity in coal,

[21] A.St.H. Brock, 'The art and craft of firework making', *Chem. Ind.*, 1953, 250.

**Table 3** *Chemicals in the pyrotechnics industry ca. 1910*

| | |
|---|---|
| Sulphur | Aluminium powder |
| Charcoal | Magnesium powder |
| Saltpetre | Strontium nitrate |
| Potassium chlorate | Barium nitrate |
| Potassium perchlorate | Strontium chlorate |
| Iron filings | Barium chlorate |
| Antimony sulphide | Copper nitrate |
| Arsenic sulphide | Barium peroxide |
| Copper sulphide | Sodium peroxide |
| Sodium oxalate | |

and Faujas de Saint-Fond, a travelling geologist with an enquiring mind, saw the copperas trade as an example of the British concern to win profit from an otherwise intractable waste matter.[22]

A copperas field was a sloping area of about an acre, the topsoil having been removed and replaced by beaten clay. The heap of pyrites, perhaps six feet in height, was allowed to 'ripen' under the influence of wind and rain: care was taken in making the stack so that air and moisture could penetrate, otherwise the heap had to be turned periodically. The mass swelled and heated, and a strongly acid yellowish green solution of mixed iron sulphates ran down a channel in the clay and collected in a sump; scrap iron was thrown in to take up the excess acid and to reduce yellow iron(III) to green iron(II) sulphate. The sump liquor was concentrated by boiling, and a framework of twigs suspended in the cooling liquor; green crystals of copperas formed on the twigs and were lifted out to be dried.

The method of manufacture scarcely changed over three centuries. Agricola's description of a copperas works in *De Re Metallica* (1552) closely matches the account of a mid-19th century establishment. The first description in English was given by Matthew Falconer who made copperas from 'copperas stones' (pyrites) on the Isle of Sheppey in 1579. Sir William Brereton confirmed the details in 1624, and in 1678 the Royal Society was treated to an account of the Deptford Copperas Works.

The ripening process took from three to five years to complete, so the copperas maker usually engaged in some other trade, often connected with paint or varnish. Typical by-trades were wood distillation to yield solvents, the extraction of turpentine from resin, lampblack production, and the management of coke ovens. Since both pyrites and fireclay are associated with coal, the making of bricks and tiles was also commonly conducted at copperas works.

Copperas lay at the centre of an industrial network (Figure 1). The crystals were sold to dyers and ink manufacturers. In both cases a black colour was

---

[22] B. Faujas de Saint Fond, *Journey through England and Scotland (1784)*, tr. A. Geikie, Hugh Hopkins, Glasgow, 1907, vol. i, p. 142; J.U. Nef, *The rise of the British coal industry*, Routledge, London 1932, vol. i, p. 210.

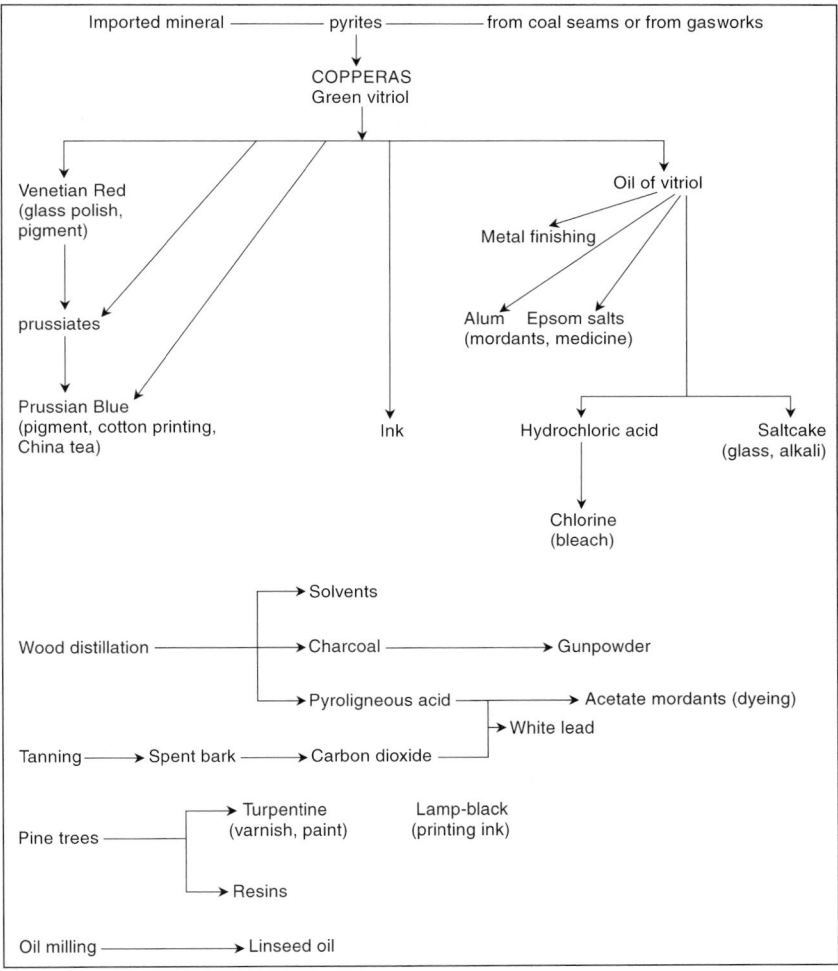

**Figure 1** *Industrial network related to copperas and its by-trades*

produced from an infusion of oak galls. This is the origin of Rudyard Kipling's advice to would-be writers:

> There may be silver in the blue-black, all
> I know of is the iron and the gall.

In the larger works copperas was roasted in a brick oven to yield oil of vitriol, a fuming sulphuric acid commonly known as Nordhausen acid. The residue in the furnace was red iron oxide, sold to paint works as the pigment Venetian red, or used for polishing plate glass under the name of jeweller's rouge; this work was usually done by women.

Copperas was also made from pyritous shale on the route to alum, and a good deal was made by dissolving scrap iron in sulphuric acid. The most recent source is the titanium works where iron is dissolved out of ilmenite, iron titanium oxide. Copperas from this source goes into the water purification industry.

## (b)    Alum

The ancient world recognised the importance of alum which lay in its capacity to fix natural dyes to woollen fabrics. It was also used in the Roman empire for tanning a durable leather for military sandals. Aluminium sulphate is extremely soluble in water and therefore difficult to crystallise. If however the salt is coupled with either potassium or ammonium sulphate an easily crystallised product of formula $KAl(SO_4)_2.12H_2O$ or $NH_4Al(SO_4)_2.12H_2O$ is formed.

Daniel Colwall reported to the Royal Society in 1678 that

> Alum is made of a stone digged out of a mine, of a seaweed, and of urine. The stone is found in most of the hills between Scarborough and the River Tees.

The stone was alum shale which contained pyrites and combustible matter. It was piled on a bed of brushwood, the stacks sometimes reaching 200 feet square and 100 feet high; fire was kindled in the brushwood and the heap left for some months to calcine throughout. The aluminium combined with the sulphur from the pyrites to form aluminium sulphate which was dissolved out in water. Sometimes the potash from the burnt brushwood was sufficient to convert the aluminium sulphate into alum, but usually some additional potassium salt was required. Suitable sources were potassium carbonate from wood ashes, kelp or burnt bracken, or potassium chloride from the conversion of salt into soda by boiling it with potash (p. 77).[23]

Ammonium alum was originally made by adding ammonium carbonate from stale urine. Later sources of ammonia were coke ovens and gas-works liquors.

Most of the alum went to the dyeing industry, but a smaller but significant portion was used in the manufacture of alumnus lakes. Small quantities were used in making white flour, and for pharmaceutical purposes, especially as a styptic.

The earliest alum works of considerable size was that started by Charles Macintosh at Hurlet near Paisley in Scotland in 1797. Under the name of Macintosh, Knox & Co., the works increased in size until 1805 when a second works was set up at Campsie near Glasgow. The combined operation, known as the Hurlet and Campsie Alum Co., was the largest alum works in the United Kingdom. They also made copperas, prussiates and Prussian Blue. After 1880 the firm concentrated on the manufacture of cyanides for the MacArthur gold extraction process, finally closing down during the First World War.[24]

The second major alum firm was founded by Peter Spence in 1846 at Pendleton, Salford. Spence came to alum by way of prussiates which he had made in London since 1834. Owing to complaints about pollution and nuisance,

[23] D. Colwell, *Phil. Trans. Roy. Soc.*, 1678, **12**, 1052.
[24] A. and N.L. Clow, *The chemical revolution*, Batchworth, London, 1952, p. 238.

Hurlet alum works *c*. 1835.

*Calcination of shale-heaps, the 'burnt shale' being conveyed by wheel-barrow to lixiviation tanks (foreground). From here the pumps convey the liquid to settling tanks*

*Settling tanks, under cover to prevent dilution by rain. The clear solution is then pumped to evaporators (Continued)*

Hurlet alum works *c.* 1835.

*Evaporators: the concentration of the alum solution was effected by furnaces, with hot gases drawn over the liquid surfaces. It was then crystallised in tanks*

*Washing crude alum crystals before the final recrystallisation*

the Pendleton works was transferred to Miles Platting, Manchester, in 1857. By this time a second alum works had been opened in Goole. In 1923 the Manchester operation was moved to Widnes, and the Goole works closed down in 1950. Ten years later the Peter Spence organisation was taken over by the Laporte Group.[25]

## 6   Oxygen and Hydrogen

For many decades after the discovery of oxygen in 1774 the gas was prepared in laboratory quantities by heating mercury(II) oxide, potassium nitrate, or 'oxygen mixture' (potassium chlorate with manganese dioxide). This last reaction was the indirect cause of numerous explosions when charcoal was accidentally mixed with chlorate in place of manganese dioxide; nevertheless it was carried out on a commercial scale to provide oxygen for theatrical limelights.

To produce the limelight, a coal gas/oxygen flame was directed on to a layer of quicklime in an iron tray. Both gases were contained in gas bags of leather or oiled silk fitted with brass nozzles and stopcocks. The pressure was controlled by pressing the bags between hinged boards, weights being placed on the upper board. The theatre manager would fill the coal-gas bag from the mains, but would send the oxygen bag to a nearby pharmacy for filling. Explosions sometimes occurred when the coal-gas bag was accidentally topped up with oxygen.

The first industrial process for oxygen was invented by the French chemist Boussingault in 1851 and exploited in 1880 by Leon and Arthur Brin. Barium oxide, prepared from the native carbonate, was heated in air to about 550 °C to convert it into barium peroxide. The temperature was then raised to 850 °C when the peroxide decomposed into oxygen and barium oxide. In a later development, the temperature was kept constant and the pressure alternately raised and lowered.

Brin's Oxygen Company began production in 1886 at Westminster, and soon there were oxygen plants in several major cities. The barium base to the process gave rise to a network of trades as shown in Figure 2.

Meanwhile Joule and Thomson had demonstrated that when a compressed gas is allowed to expand, the gas is cooled. Every gas has a critical temperature above which it cannot be liquefied; below that temperature, a suitable increase of pressure brings about liquefaction. The critical temperature for oxygen is −119 °C. Using the Joule–Thomson cooling effect and applying appropriate pressure, liquid air can be made and the oxygen separated by fractionation. Brin's Oxygen Company began to make oxygen by this method in 1906, changing its name to British Oxygen Company (now known as BOC). A partial Brin's process continued to operate as a route to hydrogen peroxide.

[25] Morris and Russell (ref. 14), p. 176.

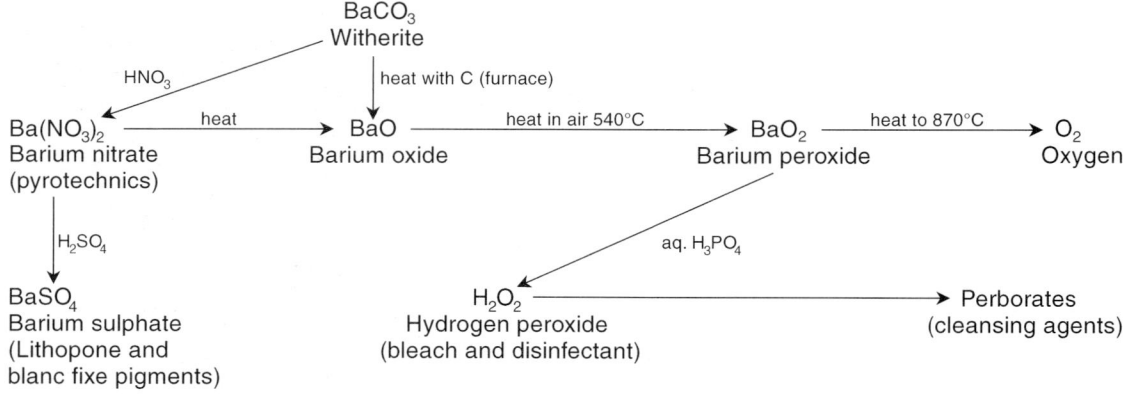

**Figure 2**   *Products based on barium chemistry*

## (a)   Hydrogen peroxide

Thenard and Gay-Lussac had discovered in 1818 that barium peroxide reacted with cold dilute sulphuric acid to form 'oxygenated water'. As Brin's process made barium peroxide available on a large scale, this former laboratory reaction became the basis of the industrial production of hydrogen peroxide. In 1888 Bernard Laporte opened a works in Shipley, Yorkshire, using this process to manufacture hydrogen peroxide as a bleaching agent for the wool industry. Ten years later he opened a factory at Luton, where the straw hat trade generated a demand for hydrogen peroxide bleach.

Hydrogen peroxide, along with other household and hospital bleaches, was also made by the Hedworth Barium Company at Jarrow Slake on the banks of the Tyne. For reasons of commercial security it was supposed that the workmen did not know what they were making.

The oxidising power of hydrogen peroxide led to an early recognition of its value as a general disinfectant; it had the merit of decomposing into water and oxygen, leaving no residue.[26] In this respect it was superior to that other popular household disinfectant, sodium permanganate, sold by the vinegar manufacturer H.B. Condy as Condy's Fluid. The permanganate was made by roasting together manganese dioxide, caustic soda and potassium nitrate.

Hydrogen peroxide was also used in the manufacture of sodium perborate, an important ingredient in dental preparations and such washing powders as Persil; this mixture of sodium perborate, sodium silicate, soda and soap was invented in Germany in 1906 and made by Crosfields at Warrington in 1910. Their perborate was made at Castner–Kellner in Runcorn by electrolysis of borax solutions, Crosfield taking the whole of the Runcorn output.

For pharmaceutical and household purposes, hydrogen peroxide is sold in strengths of '10 volume' (3%) and '20 volume' (6%), indicating that one litre of

[26] Thorpe's *Dictionary* (ref. 16), vol. iv, p. 19; vol. v, p. 304; vol. vi, p. 346.

*Laporte's new premises at Park Street, Luton*

# THE HEDWORTH BARIUM CO.,
## NEWCASTLE-ON-TYNE. LIMITED,

This Private Limited Liability Company was founded in 1891 to carry on the manufacture of **BARIUM COMPOUNDS**. It carries on a large trade in **Binoxide** or **Peroxide of Barium** of high strength, for the manufacture of **Peroxide of Hydrogen,** occupying the premier position in this particular industry.

Recently patents have been taken out all over the World for new and improved processes of manufacture.

The head of the Firm is J. C. ROLLIN, Esq., J.P., of Broomley Grange, Stocksfield-on-Tyne, The other Directors are Mr. CHARLES ROLLIN, B.Sc., F.G.S., of Bylton Hall, East Jarrow, and Mr. HUGH ROLLIN, of Overdene, Riding Mill.

*The Head Office is* **1, ST. NICHOLAS BUILDINGS, NEWCASTLE-ON-TYNE.**

*Advertisement by Hedworth Barium Company Ltd.*

the latter will decompose to give twenty litres of oxygen. This property led to the use of highly concentrated solutions of peroxide to drive submarines. Almost 100% pure hydrogen peroxide has been used in rocket propellants, the fuel being a suitable catalysed mixture of methanol and hydrazine hydrate.

### (b) Ozone

In 1785 van Marum noticed that the air around an electrical machine always developed a peculiar smell when the machine was working. This he attributed to 'electrified air', but in 1840 Schönbein recognised it as a new gas and named it ozone. Five years later it was revealed as an allotrope of oxygen with the formula $O_3$.

Several forms of laboratory ozoniser were invented, but all depended on passing a silent discharge through air or oxygen. Large scale production of ozone for industrial purposes had to wait upon the development of the electric power industry. The steam turbo-generator which made possible Merz's vision of 'electricity in bulk' did not come in until 1884, so the ozone industry dates from about 1890.

For a time, ozone was credited with almost magical powers ranging from improving the memory to increasing the output of silkworms. Two industrial uses have accounted for most of the ozone produced. One is the purification of air, as in the London Underground. The managers of the Paris Metro declined to follow London's example, pleading that there was already sufficient electric discharge from the trains to generate all the ozone needed.

The second is the disinfection of town water supplies, developed in Nice at the beginning of the 20th century. Smaller, but significant, quantities of ozone have been used in the organic chemicals industry, an example being the oxidation of isoeugenol to vanillin for food flavouring.[27]

### (c) Hydrogen

Hydrogen was discovered in 1766 by Cavendish who named it 'inflammable air', though the gas had been prepared a century earlier by Robert Boyle. The unparalleled lightness of the gas soon ensured its use in the ballooning craze which arose in the second half of the 18th century. For this purpose the gas was prepared from sulphuric acid and scrap iron. The most favoured vessel was a barrel coated internally with tar and fitted with a delivery tube, though sometimes the newly invented two-necked Woulfe's bottle was employed.[28]

Balloon ascents were social occasions, and rough grandstands were erected to accommodate the curious crowds, many of whom paid to sit for several hours to watch the inflation of the balloon. With so much acid in the vicinity there was an inevitable risk of droplets falling on to expensive silk dresses. Lavoisier had shown a less dangerous and more manageable way to prepare hydrogen in 1784,

---

[27] H. Caux, *L'Ozone et ses applications industrielles*, Dunod, Paris, 1904, pp. 360, 484, 490.
[28] A. and N.L. Clow (ref. 24), p. 151.

and one that was easily adapted to the industrial scale. He passed steam over red hot-iron in an iron tube when the following reaction occurred:

$$3Fe + 4H_2O = Fe_3O_4 + 4H_2$$

A century and a half later this reaction was used to fill the British airship *R101* which caught fire over France in October 1930. As early as 1836 Charles had demonstrated that coal gas would serve in place of hydrogen for filling balloons, though with less lifting power.

Lavoisier also knew of the reaction between steam and red-hot coke; this reaction was brought into limited commercial use in 1823. The resulting mixture of carbon monoxide and hydrogen burns with a blue flame and hence is known as blue water gas. In the presence of a suitable catalyst, a second reaction occurs between the carbon monoxide and steam:

$$H_2O + C = CO + H_2$$
$$H_2O + CO = CO_2 + H_2O$$

The carbon dioxide is easily washed out, leaving a reasonably pure hydrogen.

In the last decade of the 19th century hydrogen was produced as a by-product of the electrolytic alkali trade (p. 105). By the middle of the 20th century very large quantities of hydrogen began to be made in the course of 'cracking' petroleum to make motor spirit.

Two factors contributed to the rapid rise in the demand for hydrogen, the production of synthetic ammonia (Chapter 5) and the 'hardening' of oils. In 1897 two French chemists, Sabatier and Senderens, began to examine the catalytic hydrogenation of unsaturated organic compounds, working in the vapour phase. Four years later the German chemist Normann was able to convert oleic acid ($C_{17}H_{33}COOH$) into stearic acid ($C_{17}H_{35}COOH$) by addition of hydrogen using a nickel catalyst. This opened the way to the conversion of cheap vegetable oils to the more expensive animal fats, a boon to the soap industry and to the production of edible and cooking fats.

# 7  Fine Inorganic Chemicals

The difference between tonnage and fine chemicals lies chiefly in the purity of the latter, reflected in the price. Examples are sodium carbonate for the glass trade and for bath salts, zinc oxide for rubber filling and for cosmetics, and calcium phosphate for fertiliser and for tooth powder.

Although Paracelsus urged his readers to separate the pure from the impure, the concept of chemical purity was slow to develop. The Law of Constant Composition, on which tests for purity depend, was stated only in 1799, though some such idea must have underpinned the mineral analyses of the early 18th century. All methods of purification involve time, effort and energy, and so all are expensive.

One of the oldest methods available to chemists was sublimation described on

p. 157. In order to undergo sublimation (impure solid → vapour → pure solid) the solid–vapour equilibrium temperature of the compound at ordinary pressure must lie below its melting point, though industrial sublimations are sometimes conducted at reduced pressures. Among substances commonly purified by sublimation are iodine, arsenic(III) oxide and selenium dioxide.

In the laboratory, the traditional sublimation apparatus was an evaporating basin topped by an inverted filter funnel. In industry, great care is necessary in the design of sublimation plant, some form of continuous scraping being required to prevent clogging the condensers.

The other common method of purification was recrystallisation. Much space on the factory floor was taken up with deep vats for boiling the impure solution and shallow vats for crystallisation; the heat energy supplied to the former was dissipated in the latter, though in some kinds of factory it was possible to utilise waste heat from furnaces.

As a broad generalisation, fine inorganic chemicals were required for three purposes; laboratory chemicals for analysis, formerly known as 'tests' but since 1794 called reagents; photographic chemicals, and the wide range of substances, not necessarily medical, ordinarily sold by the pharmacist.

## (a)    Reagents

Reagents came into chemistry through the investigation of mineral waters brought about by the 18th century revival of interest in watering places with their strict codes of behaviour and dress. Torbern Bergman (1779) used barium chloride to detect sulphates, silver nitrate for chlorides, and tincture of galls for iron, though he was aware of yellow prussiate of potash, otherwise potassium ferrocyanide or potassium hexacyanoferrate(II).

This last reagent had been made by Macquer in 1759 by boiling Prussian Blue with potassium carbonate solution, but thirty years later C.W. Scheele made the pure salt and M.H. Klaproth described the preparation of a solution for analytical use. From that time onward its place as an analytical reagent was assured, so much so that W.T. Brande, professor of chemistry at the Royal Institution, regarded it as a universal reagent for metals (Table 4).

It was natural that such reagents should at first have been prepared by pharmacists. Allen and Hanbury, whose Plough Court pharmacy dated from 1715, began to supply reagents in commercial quantity in the first quarter of the 19th century, building up a stock of some two hundred.[29] Nevertheless, several later textbooks of chemical analysis gave instructions on the purification of reagents, together with tests for likely contaminants.[30] The great German pioneers of analysis, C.H. Pfaff and H. Rose, either prepared their own reagents in the 1820s or purchased them from pharmacists.

From 1850 Hopkin and Williams began to make laboratory fine chemicals at Wandsworth. In 1888 they merged with the old-established alkaloid manufac-

[29] Morris and Russell (ref. 14), p. 4.
[30] R. Galloway, *Manual of qualitative analysis*, 3rd ed., J. &. A. Churchill, London, 1870, p. 396; C.L. Bloxam, *Laboratory teaching*, 6th ed., J. Churchill & Sons, London, 1893, p. 293.

**Table 4** *Brande's table of metals detected by potassium hexacyanoferrate*

| Metal | Colour of precipitate |
| --- | --- |
| Manganese | White |
| Iron(II) | Pale blue |
| Iron(III) | Prussian Blue |
| Zinc | Yellowish white |
| Tin(II) | White, then yellow, then blue |
| Tin(IV) | Pale yellow |
| Copper(I) | Lilac |
| Copper(II) | Deep brown |
| Lead(II) | White |
| Cobalt | Pale green |
| Uranium | Deep brown |
| Nickel | Grey |
| Titanium | Deep blue |
| Mercury | Greenish white |
| Silver | Cream |
| Platinum | Yellow |

turers Howard and Sons who also made borax and pure sodium carbonate. In that same year, the firm of E. Merck of Darmstadt brought out a book of tests for purity embracing 150 reagents sold by the firm. One consequence of this was that the chemical world began to look to Germany for reliable reagents. Hopkin and Williams responded to the challenge in 1911 with their Analytical Reagents, standards and tests, covering 117 products.[31]

The outbreak of war in 1914 threw the British fine chemicals industry back on its own resources. A committee of interested chemical bodies, including representatives from Hopkin and Williams and BDH, recommended that the letters AR (denoting analytical reagent) should appear on labels for all chemicals meeting approved standards of purity. In 1926 BDH brought out its own book of standards for 158 chemicals. A similar range of reagents under the name of Judex was made by the Sudbury works of the General Chemical and Pharmaceutical Company.

It gradually became clear that the AR designation was being used too loosely; accordingly, Hopkin and Williams and BDH published a joint book in 1934, describing 200 products, in which the AR suffix was replaced by the word AnalaR.

Toxicological analysis, leading to results which must bear the scrutiny of the courts, presented a special case. Chemical analysis had been brought into disrepute by such spectacular trials as that of Marie Lafarge in Paris in 1840 at which it was said: 'Within two days the accused was declared innocent by the verdict of science, and now she is judged guilty by the verdict of that same science'. Orfila, commonly described as the father of toxicology, was believed to have used zinc contaminated with arsenic in Marsh's test for arsenic. If the test,

---

[31] T.F. McCombie, *The golden jubilee of AnalaR*, BDH, London, 1984.

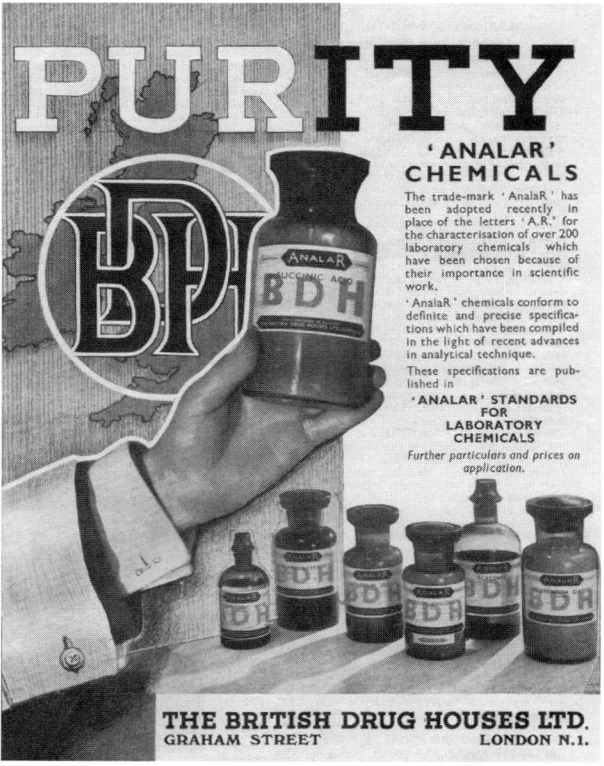

*BDH advertisement*

and its operators, were to regain any credibility then arsenic-free zinc and acid would have to be made available. Reinsch's test provided a useful alternative to that of Marsh, but failures occurred through the lack of arsenic-free copper. Had these pure reagents been available earlier, many a chemist would have been saved from humiliation at the hands of judges and counsel.[32]

As more alkaloids were extracted from plants, and as poisoners became more sophisticated, the need for pure salts of platinum and gold for use in alkaloid separation was perceived. These also found application as photographic chemicals.

Another special case was that of water analysis. Following several surveys of the health of towns, and the efforts of Edwin Chadwick on drains and John Snow on cholera, a mood of near-hysteria over the purity (whatever that might mean) of drinking water descended on the country. Microscopists led by A.H. Hassall and chemists led by Edward Frankland pressed their rival claims with ardent zeal. Among chemists a controversy sprang up between those like

---

[32] W.A. Campbell, 'Some landmarks in the history of arsenic testing', *Chem. Brit.*, 1965, **2**, 198.

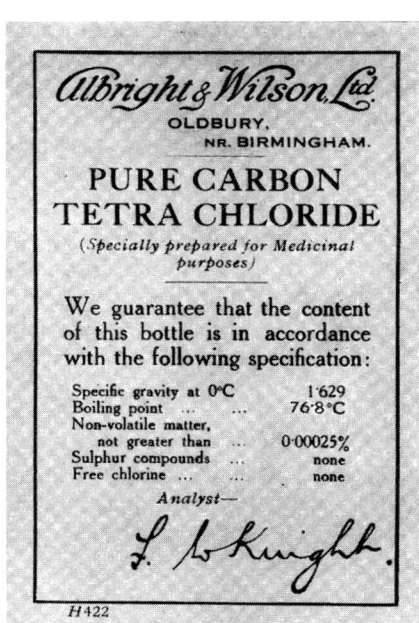

*Advertisement for Albright & Wilson's carbon tetrachloride (incredibly used for medicine!), guaranteeing its purity*

Frankland who saw the carbon content of water as the important indicator of impurity, and J.A. Wanklyn who believed that albuminoid nitrogen was more significant.[33]

Carbon was estimated by means of potassium permanganate, and nitrogen by Nessler's reagent for which pure potassium iodide and mercury iodide were required. Yet Normandy's *Commercial handbook of chemical analysis* (1850) revealed, among much else, that mercury iodide was frequently adulterated with red lead, which in turn was cheapened with brick dust!

## (b)   Photographic chemicals

While the water controversy was raging, photography was gaining daily in popularity, generating a demand for chemicals for developing, fixing, intensifying, reducing, toning and hardening. Some of these, especially the developers, were organic compounds but Table 5 lists twenty-six inorganic compounds sought both by amateurs and professionals. At first the need was supplied by local retail pharmacists who were quick to move into photography, soon undertaking the processing of plates.

---

[33] C. Hamlin, *A science of impurity; water analysis in 19th century Britain*, Adam Hilger, Bristol, 1990, p. 184; C.A. Russell, *Edward Frankland: Chemistry, Controversy and Conspiracy in Victorian England*, Cambridge University Press, Cambridge, 1996, p. 362.

**Table 5** *Photographic chemicals*

| | |
|---|---|
| Ammonium persulphate | Potassium ferricyanide |
| Borax | Potassium iodide |
| Boric acid | Potassium permanganate |
| Chrome alum | Potassium thiocyanate |
| Ferric chloride | Silver bromide |
| Gold chloride (chlorauric acid) | Silver chloride |
| Hydroxylamine hydrate | Silver iodide |
| Iodine | Silver nitrate |
| Mercury chloride | Sodium bicarbonate |
| Potash alum | Sodium carbonate |
| Potassium bromide | Sodium hydroxide |
| Potassium cyanide | Sodium metabisulphite |
| Potassium dichromate | Sodium thiosulphate (hypo) |

The firm of Johnson and Sons of Hendon, established as an assaying business in Maiden Lane, London, in 1743, and later well known for their books of test papers, was one of the first to enter the field. They made precious metal salts, especially gold chloride and silver nitrate, as well as the organic developers previously obtained from the Continent; later they also began to deal in cameras and other photographic equipment.[34] May and Baker, launched in 1834 to make such fine inorganic chemicals as bismuth salts, mercurials, and cyanides, entered the field while continuing to make general laboratory reagents. Until the 1920s the photographer who did his own processing had to be something of a chemist, carrying a stock of frequently dangerous chemicals, weighing and measuring with precision, and taking care to wash his apparatus. About that time, factory formulated developers began to appear and Burroughs Wellcome put out photographic chemicals in tablet form.

## 8  Uranium, Radium and Plutonium

Martin Heinrich Klaproth is remembered as a pioneer of mineral analysis in the 18th century. After analysing hundreds of mineral samples he collected the results into a massive work with the cumbersome title of *Analytical essays towards promoting the chemical knowledge of mineral substances*; the six volumes appeared between 1795 and 1815. Among his contributions to the art of quantitative analysis was his repudiation of the common practice of rounding up the analytical figures to 100%. Klaproth insisted that if the percentages did not add up, then either the figures were wrong or an unsuspected constituent had been missed.

Perhaps only a chemist with Klaproth's experience and outlook would have been attracted to the unlovely material called pitchblende on account of its appearance. In 1789 Klaproth extracted some pitchblende from the Joachimsthal mine with nitric acid, neutralised the solution with potash, and

[34] *Chem. Ind.*, 1933, 35.

obtained a precipitate which did not fit the description of any known metal. He was convinced that he had found a new element, and following the ancient custom of associating metals with heavenly bodies he called it uranium in honour of the new planet Uranus discovered by Herschel in 1781.

By heating the precipitate with a paste of oil and charcoal, Klaproth obtained a black powder which he believed to be the metal, but which was later shown to be the oxide $UO_2$. The new metal did not create a great stir. The gloomy philosopher Arthur Schopenhauer who attended Klaproth's chemistry lectures in Berlin in 1811 devoted only one line to uranium in his lecture notes: 'Easily converted to oxide. Reduces in the manner of molybdenum'. Even in the 1930s, a widely used textbook of inorganic chemistry running to more than a thousand pages contained only half a page on uranium.[35]

This lack of interest is not surprising, for uranium found few industrial uses. The oxide possessed the power of imparting to glass a range of tints from greenish yellow to ruby red; it also found a use in glazing pottery. Some attempts were made to turn uranium salts into useful pigments, but the cost was too high. In laboratory chemistry, uranium in the form of zinc uranyl acetate was used as a reagent for the detection of sodium.

## (a)    Radium

In 1896 Henri Becquerel was engaged in experiments on fluorescence when he observed that potassium uranyl sulphate would affect a photographic plate wrapped in paper and placed in a drawer. This was not fluorescence, and it was clear that a new kind of radiation from the uranium salt was at work. Soon the search was mounted for other materials displaying the same phenomenon.

The story of Pierre and Marie Curie in the ill-furnished shed which served as their laboratory has been told too often to need repeating here.[36] A few points, however, need to be noted. The raw material on which they worked consisted of pitchblende residues provided by the Austrian Government from the same Joachimsthal mine from which Klaproth had obtained his pitchblende; this mine had been worked continuously from 1512, originally for silver, then for bismuth and cobalt, and later for uranium. Without the early scaling up of the laboratory procedure for 'opening' a mineral, brought about by the needs of the small uranium industry, there would have been no residues for the Curies to investigate.

The Curies employed the classical methods of inorganic qualitative analysis, separating the constituents of pitchblende into the analytical groups, by successive precipitation and filtration. Pierre was an expert on piezo-electricity, and had developed specially sensitive electroscopes. These were used to monitor

---

[35] B. Engel (ed.), *Martin Heinrich Klaproth. Chemie nach der Abschrift von Arthur Schopenhauer (1811)*, Verlag für Wissenschafts- und Regionalgeschichte, Dr. Michael Engel, Berlin 1993, p. 116; J.R. Partington, *Textbook of inorganic chemistry for university students*, 4th ed., Macmillan, London, 1933, p. 941.

[36] P. and M. Curie, *Comptes Rendus*, 1898, **127**, p. 175; R. Reid, *Marie Curie*, Collins, London, 1974, p. 60; M. Curie, *Pierre Curie, with autobiographical notes*, tr. C.V. Kellog, 1923, Dover Reprints, New York, 1963, pp. 44, 91.

*Cartoon of Marie and Pierre Curie*

the direction in which the radioactive constituents were moving, into the filtrate or the residue, the crystal or the mother liquor. The work called for the highest degree of tenacity coupled with meticulous manipulative skill, for they had no coloured precipitates to guide them, only the discharge of the electroscope. In this way polonium (precipitated with bismuth) and radium (precipitated with barium) were discovered, and the term 'radioactivity' was coined.[37]

The production of radium from pitchblende took place in three broad stages; concentrating the ore from the gangue, precipitating barium/radium, and separating the radium from the barium. In addition to the Austrian source, pitchblende was found at Great Bear Lake in Canada and in what was then the Belgian Congo. Another ore, carnotite, occurred in Colorado and Utah. Very small amounts were found in the metal mines of Cornwall.

Radium was at first used for medical purposes, to give to malignant cells sufficient radiation to kill them with the minimum of damage to surrounding healthy tissue. This has largely been a matter for physics and production engineering rather than chemistry. The radium was often administered in the form of needles which were inserted into the affected tissue. The needles were thin tubes made from a platinum/iridium alloy, and contained a few milligrams of radium sulphate. Otherwise radiation was directed towards the tumour from

[37] E.R. Landa, 'The first nuclear industry', *Scientific American*, 1982, **247**, 154.

an external source. The industrial preparation of the needles and other sources brought problems which again were of an engineering rather than a chemical nature.

Both the Curies experienced bouts of extreme tiredness, and the ends of Marie's fingers became cracked and painful. Pierre deliberately burned his fore-arm with radium and recorded the progress of the lesion from the first reddening of the skin to the final dropping off of the scab.[38] When Pierre rose to deliver his (and Marie's) Nobel Prize lecture, several observers commented on how ill he looked.

Much radium was consumed in the manufacture of luminous paint for the dials of watches, clocks, and military equipment. Luminous paint consisted of a phosphor with a little radium to excite it. Earlier phosphors were made from ground oyster shells, charcoal and sulphur, stacked in layers in a crucible and heated at a high temperature. Later phosphors were essentially sulphides of calcium, barium or zinc with carefully controlled traces of such heavy metals as bismuth, lead and copper. While the chemical compositions were known, the preparative know-how was guarded with an almost alchemical secrecy.

Women were usually employed in painting the numbers on the dials. To get a fine point, the brush was sometimes passed between the lips, resulting in cancer of the lip or tongue. This practice was more prevalent in America than in Britain.[39]

Wild claims were made for the benefits of radium in products as varied as hair restorers, suppositories, bath salts, face powders, and mouth washes, and certain proprietors of mineral springs made much of the presence, real or supposed, of minute amounts of radium in their waters. One device, claimed to contain radium, was to be hung over the bed to dispel worry and bring contentment – surely an echo of the use of garlic to chase away vampires.[40]

A minor use of radium was in radiography to detect imperfections in castings and other metal structures, though X-rays soon displaced radium. The dispersal of static electric charge during the large-scale movement of textiles, paper, plastic sheet, and powders has important safety implications. and radium contained in thin gold foil has been used for this purpose. Fritz Pregl, the pioneer of microanalysis, recommended microanalysts to place inside the microbalance case a watch glass containing pitchblende to prevent dust adhering to polished glass objects through static charge. The proportion of radium going to these outlets was roughly 80% to medical work, 10% to luminous paints, and 5% to examining metals for fracture.

## (b) Nuclear fission

This picture of radium as the desired constituent of pitchblende with uranium the unwanted companion changed radically in 1938 when Hahn and Strassman announced the splitting of uranium into neutrons. That it happened on the eve

[38] J. Dunmore, *Anthology of French scientific prose*, Hutchinson, London, 1973, p. 114.
[39] D. Hunter, *Health in industry*, Penguin, Harmondsworth, 1959, p. 241.
[40] E.R. Landa (ref. 37), p. 162.

of a world war has largely determined its social and political consequences. In March 1939 Einstein and Szilard wrote to President Roosevelt suggesting that nuclear fission might lead to the construction of an atomic bomb. That possibility was enhanced by a further observation. The fissile isotope of uranium was $^{235}$U, but it was discovered that ordinary uranium $^{238}$U could capture slow moving neutrons, setting up a chain of events leading to the formation of a fissile plutonium isotope $^{239}$Pu.

The Department of Scientific and Industrial Research (DSIR) began organised research in October 1941; to disguise the real nature of the project, the name of Tube Alloys was invented. Work commenced at several British universities, particularly Birmingham and Oxford. At the former, the whole chemistry department was taken over for the war effort. From this time onward, the major thrust of the industrial chemistry of pitchblende was swung back from radium to uranium whence it started. For producing crude uranium, the old Joachimsthal methods sufficed. Beyond that there was very little knowledge of uranium chemistry in this country.[41]

There were two ways of arriving at quantities of fissile material. The first was to separate $^{235}$U from natural uranium, but industrial expertise on isotope separation was non-existent. The second was to form $^{239}$Pu and separate it chemically from the parent $^{238}$U. Here again, knowledge of uranium and plutonium chemistry was not up to the task. The most promising way to separate uranium isotopes was by gaseous diffusion of a volatile uranium compound, uranium hexafluoride being the obvious choice. Once more, there was very little knowledge and experience of fluorine chemistry in this country. It was clear that if the problem of bulk production of fissile substances was to be solved, pure scientific research had to go hand-in-hand with industrial development and a new kind of partnership had to be forged.

The General Chemicals Division of ICI was given the task of producing fluorine and converting it into $UF_6$. The only fluorine compound in general use, hydrofluoric acid, had been made in small quantities from fluorspar and sulphuric acid for etching glass. Now anhydrous hydrogen fluoride was to be electrolysed to provide fluorine, and ICI soon developed appropriate cells. By 1943 the production of uranium metal and its hexafluoride was well under way.

Originally transported in sealed glass ampoules, the increasing quantities of uranium hexafluoride called for larger metal containers which had to be conditioned by pre-treatment with fluorine. New lubricants and valve materials, resistant to uranium hexafluoride, were necessary, and the new science of fluorocarbons arose largely at Birmingham University. Much of the work on the diffusion method of separating uranium isotopes was carried out there also. The team at Oxford University gave support to work on suitable membranes through which diffusion might take place; the industrial side of the research was undertaken at Kynoch's, later Imperial Metal Industries, at Birmingham.

From 1943 to the end of the war, members of the British teams involved in the

[41] C.B. Amphlett, in P.J.T. Morris, W.A. Campbell and H.L. Roberts, eds., *Milestones in 150 years of the chemical industry*, Royal Society of Chemistry, Cambridge 1991, p. 223.

whole project moved to Canada where a joint research effort attacked the problems associated with the $^{239}$Pu route.

## (c)  Atomic energy

When the war ended, Britain was in possession of a great deal of knowledge and experience of the production and handling of fissile materials, and the decision was taken to erect nuclear power plants. These would supply much-needed electric power, by-pass the troubles associated with coal mining, and at the same time make plutonium for future military purposes. No longer protected by the compelling sanctions of the war effort, the industry had to work to far more stringent safety precautions.

In part, these depended on accurate analytical methods for plutonium. The chosen method involved coprecipitation with lanthanum fluoride followed by radiochemical assay using an alpha counter. Only by the strictest adherence to the prescribed procedure could reproducible results be obtained.

For separating plutonium from uranium, a solvent extraction process was chosen; $Pu^{IV}$ and $Pu^{VI}$ are soluble in organic solvents while $U^{III}$ is not. After much research, dibutylcarbinol (nonan-1-ol) was selected as the solvent and the whole process was tested at ICI's Widnes research laboratories. Chemists to work the solvent extraction and other kinds of radiochemical processes were being trained at the new Harwell laboratories.

When the uranium/plutonium mix was dissolved in nitric acid, brown fumes of nitrogen dioxide were evolved, to the annoyance of people living nearby, and exciting the curiosity of all who saw the sinister-looking cloud. Johnson and Co., the manufacturers of photographic chemicals, had developed a 'fumeless' method for dissolving silver and this technique was adapted for uranium. Even in 1948, little was known about the behaviour of such fission products as ruthenium during the process of solution. The precious metals firm of Johnson Matthey brought their experience to bear on the structure and properties of ruthenium/NO complexes – another example of the parallel progress of industrial and laboratory chemistry. The Windscale plant produced its first plutonium for weapon trials in 1952.[42]

Radioactive isotopes for medical and scientific purposes were marketed by a trading organisation at Amersham, Buckinghamshire, named the Radiochemical Centre. When our period closes, Amersham had become the sole distributor of radioactive products offered by the atomic energy industry.

The fission products of irradiated uranium could either be left in the fuel rods, or taken out in an operation called reprocessing. The latter course inevitably leads to stockpiling of radioactive material the storage of which carries health and safety implications. The Atomic Energy Authority settled for reprocessing and, finding that they had spare capacity, they offered the facility to other

[42] R. Spence, 'Chemical process development for the Windscale plutonium plant', *J. Roy. Inst. Chem.*, 1957, **81**, 363.

countries. This increased public concern and led to allegations that Britain was becoming the nuclear dust-bin of the world.

Several possibilities presented themselves to the Authority; radioactive effluent could be concentrated and stored in specially designed silos; effluent of low activity could be discharged into the sea; various kinds of waste could be buried in geologically acceptable sites. None of these is free from objections, especially from dwellers near the chosen sites.[43]

Absorbing liquid wastes into setting concrete in steel drums had been tried in Canada, but there were doubts about the long-term stability of such devices. In Britain experiments began on the incorporation of waste into glass-like materials designed to resist atmospheric or aqueous leaching. It was recognised that the best solution, though chemically extremely difficult, would be to purify such fission products as $^{137}$Cs and $^{90}$Sr for use in cancer treatment.

Apart from the chemistry of nuclear fission, a great deal of parallel research was generated into the production of moderators such as graphite. Much of this was done by Anglo Great Lakes on the site of a former German explosives factory at Lemington on Tyne, with co-operation from the Northern Carbon Research Laboratory at the University of Newcastle. Metallurgical research into such matters as boron steels for control rods was also a spin off from the atomic energy programme.

It is clear that the development of nuclear reactions for civil and military purposes has bred large amounts of 'pure' inorganic chemistry and has led chemists into corners of the Periodic Table into which they would not otherwise have penetrated.

---

[43] J.E. Littlechild, 'Products resulting from the irradiation of uranium', *J. Roy. Inst. Chem.*, 1957, **81**, 442; Sir John Cockcroft, 'The development of radiation chemistry and radiochemistry', *Royal Institute of Chemistry Lecture Reprints*, 1954, no. 1, p. 5.

Chapter 8 **_The Organic Chemicals Industry to the First World War_**

C.A. RUSSELL

## 1 Extraction of Organic Natural Products

Organic chemistry as a separate part of chemical science was a 19th century phenomenon. So it is hardly surprising that such manufacture of organic chemicals as did take place before 1800 was largely without scientific support and must have been what Archie and Nan Clow called an 'eotechnic' venture.[1]

Yet a number of products that _we_ should recognise as organic compounds were prepared and marketed on a comparatively large scale and it is worth glancing briefly at them. Here, in this earliest phase of the organic chemicals industry, may be recognised trends that have continued to this day: the strong pressures from society for rising standards of living, the equally robust responses by entrepreneurs to meet those demands, the transfer of technological skills from one area to another, the insistent tendency to improve, and the growing interconnectedness of different branches of industry to nourish one another.

Yet other things were different in those days. In the first place, the extent of a true chemical understanding of either materials or processes was extremely limited, as indeed it had to be in the days before structural theory, valency or even Daltonian atomism were accepted. Secondly, it was rarely the case that what we should call 'pure' products were obtained, partly because of inadequate techniques of separation but also because the criteria for purity did not exist as we have them. A third feature of this eotechnic technology was the great emphasis upon analysis rather than synthesis; the desired products were invariably extracted from more complex materials by various processes of destruction, not built up from simpler molecules as is the case today. A fourth difference resides in the simple fact that these more complex materials were always natural products, usually from the vegetable kingdom, and these were usually alive just before use (not fossilised as may be the case of primary feedstocks today). Finally there was an enormous wastage of natural resources, and environmental damage ranging from considerable air pollution to massive destruction of trees.

In the examples that follow it will be noticed that some of the techniques

---

[1] A. and N.L. Clow, _The chemical revolution_, Batchworth, London, 1952 [reprinted Gordon & Breach, Philadelphia, 1992], p. 91.

lasted long into the modern era, a fact that illustrates nicely the continuity of tradition in the chemical industry and also the folly of categorising too tightly any one period of time as eotechnic, palaeotechnic or neotechnic. We shall consider briefly three groups of products that were extracted from natural sources. In so doing we shall exclude several minor compounds, such as benzoic, succinic, malic and tartaric acids, as well as the many medicinal products of the older pharmaceutical industry.

## (a)  Sucrose

First, and in tonnage surely the greatest for any pure organic chemical, must be sugar (or sucrose). From that one substance the chemical industry has made a large quantity of important products by fermentation. Sugar itself has been known since antiquity as syrups, juices, honey *etc*. The crystalline substance may have been first brought to England by Hawkins in about 1560, from the Spanish colony of San Domingo. Sugar canes of various kinds were known in the Near East during the Middle Ages and were introduced into the West Indies in the 17th century. With the rapid development of sugar planting in the British colony of Barbados (from 1641) raw sugar began to be imported into England and later Scotland on an increasing scale. Its method of purification, a closely-guarded secret, involved addition of bullock's blood to a hot solution, when the coagulated albumen entrained floating and insoluble impurities, enabling them to be skimmed off. Subsequent filtration and evaporation yielded 'sugar-loaves'. From that day to this the principles of extraction are fundamentally unchanged: the juice expressed from cane sugar or extracted from sugar beet with hot water is filtered, decolorised with animal charcoal, re-filtered and then evaporated.[2] Crystallisation occurs on cooling, and the process is repeated several more times to yield crystals of variable purity and ultimately a thick solution known as molasses.

From the 18th century sugar-refining was big business in Scotland, at all the large cities but especially at Greenock.[3] However, the rising costs of fuel, and the abolition of slave labour from the West Indian sugar plantations, led to rising prices, and falling demands, for sugar. Nor were matters facilitated by a French embargo on imported British sugar during the Napoleonic wars. With the end of the war, and a lifting of the embargo, sugar refining in Britain began to prosper once again. It was further encouraged by the increasingly popular habit of tea-drinking amongst the working class, whose sweet tooth demanded the addition of sugar, and the appearance on their tables of a new confection known as *marmalade*. However a punitive tax on sugar inhibited domestic consumption until the tax was reduced in 1844. From then on there was a steep rise in importation of raw sugar for refining, from 206,472 tons in 1844 to 401,438 tons ten years later. Vast numbers of sugar sacks were sent to the paper mills, so reducing the latter's demand for fresh supplies of cellulose.

[2] N. Deerr and A. Brooks, 'Development of the practice of evaporation with special reference to the sugar industry', *Trans. Newcomen Soc.*, 1941/2, **22**, 1–19.
[3] *Ibid.*, pp. 515–535.

*Sugar refining in the early 19th century*

Two English firms may be specially mentioned. One was Finzell's at Bristol, with excellent access by ships from the West Indies and an unusually innovative proprietor. By the 1860s it had become one of the largest sugar-refiners in the world. The other firm, Severn, King and Company, were sugar-bakers at Whitechapel, London. In the small hours of 10 November 1819 they experienced a disastrous fire and, thereafter, a nearly equal disaster in the lawsuit with their insurers (the Imperial, Phoenix, and Globe companies). The court proceedings are of immense interest because of their revelation of the extent to which 'expert' chemical witnesses were sub-poenaed to testify. Sadly it was too early for the judiciary to take their witness as seriously as it should, but it was an important step in the progress towards the professionalisation of chemistry.

Of more immediate significance, perhaps, was the trial's disclosure of a large reservoir of technical knowledge about sugar-refining among those who could rightly be called 'chemists'.[4] Indeed by this time several major improvements in sugar technology had appeared. One was the use of vacuum evaporation which not only saved fuel but also minimised unwanted chemical changes at higher temperatures. Its patent by E.C. Howard in 1812/13 was immediately taken up by leading manufacturers in England and Scotland. Another was the use of animal charcoal as a decolourant, following a less successful bid to use wood charcoal. This process was patented by C. Derosne in 1830.

[4] J.Z. Fullmer, 'Technology, chemistry, and the law in early 19th-century England', *Tech. & Culture*, 1980, **21**, 1–28.

*Sugar beet works at Wissington, established in 1925*

Finally the question arose as to whether sugar itself could be grown in Britain. The German chemist A.S. Marggraf had discovered in 1747 that crystals identical to those obtainable from sugar-cane were extractable from the juice of beet roots. The British climate was too cool for sugar-cane, but in France and elsewhere in Europe sugar-beet had been successfully cultivated after its first introduction to Prussia in 1803. Napoleon introduced it to France in 1811 in response to the British blockade[5] which effectively cut off supplies of cane-sugar from the West Indies. Within a year France was producing 3 million lb (1.4 million kg) of beet sugar annually. An early experimental attempt to introduce it in Britain was made in Essex in 1832, but it did not succeed and many years elapsed before further trials were undertaken. In 1912 an attempt to grow the plant in Norfolk led to the creation of Britain's first processing factory at Norwich, others following at Ely and King's Lynn. However, the plain fact was that, because the yield of raw sugar is lower with beet than with cane, an unsubsidised production could never be economically secure. Matters improved in 1925 with the passing of the Government's Sugar Beet Subsidy Act, and a plant was established that year at Wissington (Norfolk) which, unlike the other three factories, had no directors from Holland. Ten years later the British Sugar Corporation was formed from the East Anglian companies. The Wissington plant became eventually the largest in Britain.

## (b)    Products of fermentation

The production of mildly intoxicating drinks from processes that we should now call fermentation was known in remote antiquity. The process and the

[5] J. Fournier, 'Sugar supply under the Continental blockade. Part I – Louis-Joseph Proust and grape sugar', *Actual. Chim.*, 1997 (6), 31–37.

product must on almost any definition be termed 'natural'. The active constituent was of course ethanol (or 'alcohol' or 'ethyl alcohol'), though it was present in considerable dilution with dozens of other organic substances at very low concentrations. It has been wryly remarked that, if we do not know the name of the first person to discover such a process, then neither do we know the identity of the one who originally conceived the bright idea of applying fiscal duties to the product. In the latter respect England was particularly afflicted – even more so than Scotland at one time – and this undoubtedly impeded the production of pure alcohol on an industrial scale. Relative fiscal burdens are indicated in Table 1 for the period up to equalisation; Irish rates tended to follow Scotland rather than England.[6]

**Table 1** *Duty on alcohol in England and Scotland, 1815–1860*

| Years | England: duty per gallon £. s. d. | Scotland: duty per gallon £. s. d. |
|---|---|---|
| 1815 | 0 10 2¾ | 0 9 4½ |
| 1817 | 0 10 2¾ | 0 6 2 |
| 1823 | 0 11 8¼ | 0 2 4¾ |
| 1826 | 0 7 0 | 0 2 10 |
| 1830 | 0 7 6 | 0 3 4 |
| 1840 | 0 7 10 | 0 3 8 |
| 1853 | 0 7 10 | 0 4 8 |
| 1854 | 0 7 10 | 0 6 0 |
| 1856 | 0 8 0 | 0 8 0 |
| 1860 | 0 10 0 | 0 10 0 |

The history of the brewing industries has been often told and will not be repeated here. Suffice it to say that the production of alcohol as an industrial chemical was from the 1840s to the 1940s exclusively accomplished by distillation ('rectification') of the products of fermentation processes. In the 19th century these products came mainly from the distilleries making whisky or gin (usually in Scotland). The invention in 1831 of a more effective distillation apparatus, the Coffey still, led to alcohol with higher degrees of purity. One of the agents used to achieve fermentation was yeast, and as early as 1680 Leeuwenhoek had showed that it was a mass of small cells, visible under a microscope. Its identification as a living organism, by Cagniard-Latour in 1838, was followed by recognition by Pasteur in 1857 that all fermentations required a living organism, and the demonstration of Buchner 40 years later that it is substances secreted by such organisms ('enzymes') that enable the reaction to take place. The analysis of alcohol was accomplished by Lavoisier in 1781 and its empirical and molecular formulae were established early in the next century, but its structure ($C_2H_5OH$) was not really proved until the work of A.W. Williamson in the 1850s.

[6] D. Bremner, *The industries of Scotland*, Black, Edinburgh, 1869, p. 450.

*The still house at Hazelburn Distillery at Campbelltown, Kintyre (1836)*

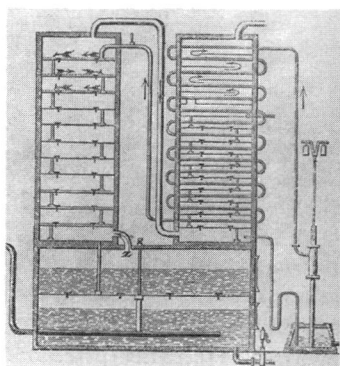

*A distillation apparatus invented in 1831 by Aeneas Coffey. Its system of perforated plates within the two columns enabled fractionation to be far more efficient*

At the beginning of the 20th century alcohol was chiefly manufactured in Britain from grain though an alternative source in molasses (the mother liquor from sugar refining) had been recognised in 1870. The process involved the enzyme-catalysed hydrolysis of maltose or sucrose to monosaccharides of which the glucose was then converted into ethanol. The three enzymes *invertase*, *maltase* and *zymase* all occur in yeast:

$$C_{12}H_{22}O_{11} + H_2O \xrightarrow{\text{invertase}} C_6H_{12}O_6 + C_6H_{12}O_6 ;$$

sucrose                        fructose     glucose

or

$$C_{12}H_{22}O_{11} + H_2O \xrightarrow{\text{maltase}} 2C_6H_{12}O_6$$
$$\text{maltose} \qquad\qquad\qquad\qquad \text{glucose}$$

Then:

$$C_6H_{12}O_6 \xrightarrow{\text{zymase}} 2C_2H_5OH + 2CO_2$$
$$\text{glucose} \qquad\qquad\quad \text{ethanol}$$

In 1877 six Scotch whisky distilleries amalgamated to form Distillers' Co. Ltd. In response to the growing demand for industrial alcohol (by the dyestuffs industry, for instance) they produced some of this product alongside the more conventional potable alcohol. They established a number of distilleries in England and, in the mid-1920s, established the Hull Distillery at Salt End, Hull.

Meanwhile it is appropriate to note that another kind of fermentation, involving the oxidation of alcohol to acetic acid, had been utilised for probably 2500 years in the manufacture of vinegar. The pure acid was first obtained by distillation of vinegar at the end of the 18th century. Another fermentation product of sucrose, lactic acid, was recognised as such in 1839 but was not prepared commercially until 1883 (in the USA), being first made in the UK by Bowmans of Warrington in the First World War.[7] The production at the same time of butanol and acetone by the Weizmann fermentation process is part of the history of that war.[8]

## (c) Natural dyestuffs

For several centuries before the modern period a number of natural colouring materials had been used as dyestuffs, especially in the processes of calico-printing (see also Chapter 3). Several important examples entered Britain as a result of the great age of exploration that accompanied the Renaissance. In 1662 William Petty gave a list of dyes then available in England,[9] including those shown in Table 2.

Most of these had appeared in Britain within the preceding hundred years or so, and other natural dyestuffs were added to their number over the next two centuries, such as archil (from lichens that could be grown indigenously), cudbear (from a lichen grown in Scotland), lac (from resins secreted by insects in trees growing in the Far East), and quercitron (from the bark of a North American tree). Use of such natural dyestuffs occurred in parts of Europe as

[7] H. Benninga, *A history of lactic acid making: a chapter in the history of biotechnology*, Kluwer, Dordrecht, 1990.
[8] D.T. Jones and D.R. Woods, 'Acetone-butanol fermentation revisited', *Microbiol. Rev.*, 1986, **50**, 484–524 (485–488).
[9] T. Birch, *History of the Royal Society of London*, London, 1756, vol. ii, p. 401.

**Table 2** *Dyes available in England, 1662*

| Dyestuff | Colour | Natural source | Countries of origin |
|---|---|---|---|
| Logwood | Blue | Wood of *Haemotoxylum campechianum* | Jamaica, Mexico |
| Indigo | Blue | Leaves of *Indigofera tinctoria* | Far East, W. Indies *etc.* |
| Woad | Blue | Whole plant *Isatis sativa* | England, France, Italy *etc.* |
| Kermes | Red | Leaf-eating insects, *Coccus ilicis* | Near and Middle East |
| Cochineal | Red | Cactus-eating insects, *Coccus cacti* | N. America |
| Sandalwood | Red | Wood of *Pterocarpus santalinus* | E. Indies |
| Brazilwood | Red | Wood of *Caesalpinia crista* | Brazil |
| Madder | Red | Roots of *Rubia tinctorum* | Holland, France |
| Redwood* | Red | Berries of *Rhamnus erythroxylon* | Widely in Europe |
| Fustic | Orange | Wood of *Morus tinctoria* | Cuba |
| Anotta | Orange | Seeds of *Bixa orellana* | S. America |
| Sumach† | Orange | Wood of *Rhus cotinus* | S. Europe |
| Saffron | Yellow | Flowers of *Carthamus tinctorius* | Egypt, China |
| Turmeric | Yellow | Roots of *Curcuma longa* | E. Indies, China, Java |
| Weld | Yellow | Whole plant *Reseda luteola* | England, Germany, France *etc.* |

\* Presumably Siberian redwood.
† Otherwise known as young fustic.

*Albert Mill, Keynsham, Somerset, the last place in Britain to manufacture logwood dyes (until 1964)*

well as Britain, being encouraged by the rise of calico-printing towards the end of the 18th century.[10] The production of many dyestuffs made prodigious demands on the environment. Thus as early as 1750 some 8000 tons of logwood were cut in Honduras, just to satisfy one year's demand in England.[11]

The problem of making these dyes 'fast', *i.e.* adhering strongly to cloth, demanded all the ingenuity and perseverance of which the early dyers were capable. The use of alum for this purpose was familiar in 17th century England as Petty and Robert Boyle both testify. Its manufacture, though for other purposes as well, has been styled 'The earliest chemical industry'.[12] In the 19th century it was made on a large scale by Peter Spence in England and Charles Macintosh[13] in Scotland. Since about 1650 tin salts have been used as occasional substitutes for alum and several other metal compounds appeared in the early 19th century. The use of mordants not only enabled dyes to adhere but also could yield different colours; thus Buckthorn berries yield a yellow juice that with alum gives a sap-green colour, while use of stannic oxide instead of alum enables madder to impart a 'fire-red' as opposed to a 'rose-red' colour. Other mordants used in the early period included iron(II) sulphate and the acetates of iron, lead and aluminium. The production of the much-prized 'Turkey red' colour from madder required an extremely complicated sequence of treatments by the dyer and was for centuries a closely-guarded secret until it appeared in Glasgow at the end of the 18th century. Nor must we forget purely inorganic pigments as Prussian Blue, made at Newcastle upon Tyne from 1770 and at Edinburgh fifteen years later (Chapter 7).

The chemistry of these substances was, of course, totally unknown until well into the 19th century, as was that of the recipes for applying the colours to fabrics. At the end of the 18th century William Henry remarked that 'few of our dyers are chemists, and few chemists dyers', though he did concede that 'the arts of dyeing and printing owe much of their recent progress to the improvements of men who have made chemistry their business'. One dye-manufacturer in 1824 discovered that 'the function of certain chemicals was determined only when the dyer forgot to add them to the vat on one occasion!'[14] Later, of course, the work of John Mercer and James Thomson, calico-printers in Lancashire, added substantially to the scientific understanding of the many processes then in use, though the detailed organic chemistry involved was not made clear until the

---

[10] A. Nieto-Galan, 'The use of natural dyestuffs in eighteenth-century Europe', *Strategies of chemical industrialization from Lavoisier to Bessemer*, Proceedings of a workshop at Liège in 1994 on 'The evolution of chemistry within Europe, 1789–1939', European Science Foundation, Strasbourg, 1997; reprinted from *Arch. Int. d'Hist. Sci.*, 1996, **46**, 23–38.

[11] J.A. Burdon, *Archives of British Honduras*, Sifton, Praed & Co., London, 1931, vol. i, pp. 77–78.

[12] C. Singer, *The earliest chemical industry*, Folio Society, London, 1948.

[13] On Macintosh see D.W.F. Hardie, 'Chemical pioneers – Macintosh', *Chem. Age*, 1957, **77**, 643; *idem*, 'The Macintoshes and the origins of the chemical industry', *Chem. Ind.*, 1952, 606–613; N.L. and A. Clow, 'George Macintosh (1739–1807) and Charles Macintosh (1766–1843)', *Chem. Ind.*, 1943, **62**, 104–106; H. Schurer, 'The macintosh: the paternity of an invention', *Trans. Newcomen Soc.*, 1951/3, **28**, 77–87.

[14] C.M. Mellor and D.S.L. Cardwell, 'Dyes and dyeing 1775–1860', *Brit. J. Hist. Sci.*, 1963, **1**, 265–279.

*Printing calico by blocks about 1800*

work of Perkin,[15] Hofmann[16] and a host of others after 1860. Meanwhile, however, the increasing mass of technical and empirical knowledge led to a sharp rise in patents for dyeing and calico-printing: about 5 p.a. until 1840; about 12 in the 1840s; and about 60 from 1850 to 1855.[17] The industry was certainly not static in the years before the introduction of coal-tar dyes; its long history has been related elsewhere.[18]

## (d)  Natural rubber

Many gummy products exuded from trees were known to the Victorians, gum arabic and gutta-percha among them. Of far greater lasting importance was gum elastic, otherwise known as caoutchouc or india rubber (because some came from the East Indies and most was used as a 'rubber' or eraser for black-lead writing on paper). Now it is simply called 'rubber'. It was obtained as a milky white latex from many species of trees, but chiefly from *Hevea brasiliensis*,

---

[15] On Perkin see especially J. Meldola, Perkin Memorial Lecture, *J. Chem. Soc.*, 1908, **103**, 2214–2257; D.H. Leaback, 'Perkin's pioneering enterprise', *Chem. Brit.*, 1988, **24**, 787–790.

[16] On Hofmann see L. Playfair, F.A. Abel, W.H. Perkin and H.E. Armstrong, Hofmann Memorial Lecture, *J. Chem. Soc.*, 1896, **69**, 575–732; C. Meinel and H. Scholz (eds.), *Die Alliance von Wissenschaft und Industrie: August Wilhelm Hofmann (1818–1892)*, VCH, Weinheim, 1992.

[17] Mellor and Cardwell (ref. 14), p. 274.

[18] A. and N.L. Clow (ref. 1), pp. 224–233; G. Turnbull (ed. J.G. Turnbull), *A history of the calico-printing industry of Great Britain*, Sherratt, Altrincham, 1951; P. Floud, 'The British calico-printing industry, 1676–1840', *CIBA Review*, 1961, (1), 2–7, and *idem*, 'The English contribution to the chemistry of calico-printing before Perkin', *ibid.*, pp. 8–14.

a native of the Amazon forests. By 1800 it had long been used to make bottles and was soon to find application as an adhesive (*e.g.* for book-binding).

Solubility of rubber in organic solvents was known at the end of the 18th century. Charles Macintosh of Edinburgh showed that naphtha from coal-tar was a suitable solvent, and in 1823 patented his water-proofing process for textiles. It fast became popular for outer garments, though as one Victorian commentator rather primly observed, 'it intercepts exudation, and causes it to be retained in the under garments'. Large quantities were manufactured for military use in the Crimean War, by George Spill and Co. in East London.

The great disadvantage of natural rubber, its loss of elasticity with temperature change and its tendency to self-adhesion when hot, was overcome by the process of vulcanisation, in which sulphur is added to the heated raw material. This was discovered in 1839 by Goodyear, and independently by Thomas Hancock of London in the UK in the early 1840s. However almost certainly 'the most important event in the history of the rubber industry' was the pneumatic tyre invented by J.B. Dunlop of Belfast in 1888.

Meanwhile demand for rubber was steadily increasing and the decision was taken by Government to create new plantations in British colonies in the Far East. After one or two abortive attempts to export trees grown in The Royal Botanic Gardens from Brazilian seeds, 2000 seedlings were transplanted from Kew to Ceylon in 1876. Eventually over 9 million acres were devoted to rubber production in Malay, Indonesia and Ceylon. In the first six years of the 20th century London prices soared by 135%.

## 2  Chemical Modification of Natural Products

So far the processes described involved little or no chemical modification of natural materials and a correspondingly small knowledge of chemical science. At the same time as some of these were being developed other techniques were being discovered that, with hindsight, we can see involved production of organic chemicals by chemical reaction. The first involved that most destructive of processes, thermal decomposition.

### (a)  Distillation of wood

By 1800, alongside sugar and alcohol, the products of wood distillation were beginning to appear in Britain. Wood was heated in cylindrical vessels and liquid products were collected. This was a different process from that known as charcoal-burning, though charcoal was a product of both operations. It may be summarised thus:

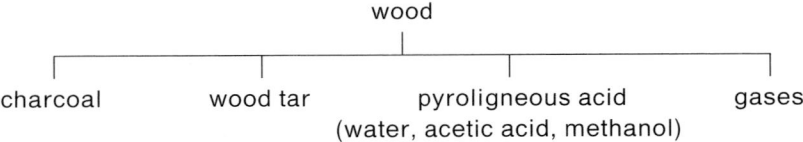

Wood distillation seems to have originated on the Continent (possibly Sweden with its product 'Stockholm tar'), and according to Richard Watson was introduced into Britain at Hythe in 1787, to produce exceptionally fine charcoal for gunpowder. There is evidence that it was carried on in Sussex also.[19] The first instance in Scotland appears to have been at the end of the 18th century, and by 1815 there were seven such distilleries north of the Border, four of them in Lanarkshire. Elsewhere in Britain wood distillation seems to have become concentrated in South Wales, Gloucestershire, Lancashire and Tyneside (sometimes using scrap wood from ship-yards).

Distilleries were established for a variety of reasons. In Scotland it was a way to use waste timber from ships or even from ship-wrecks. The Cannop Chemical Works was established in the Forest of Dean in about 1835 to dispose profitably of wood that was otherwise unsaleable. In 1791 pyroligneous acid was a by-product of the charcoal ovens established on the advice of Richard Watson at the Government gunpowder factory at Waltham Abbey. The large distillery of Hervey and Co. received the residues of trees like logwood from which dyers had extracted the colouring-matter they needed. It was a good way to clean up the environment and appears to have done little to harm it.

There was, of course, a demand for the products. Wood tar had been used for some time for caulking ships. The liquid by-products were at first a nuisance, though some acetic acid may have been converted into vinegar or lead acetate. In the early 19th century an extensive use by dyers of aluminium acetate created demand for acetic acid from this process which did not involve oxidation of the highly taxed alcohol. Later, distillation into lime separated the acid and left methanol (with some acetone) that in the mid-19th century was largely prepared by this method. However by the 1860s the industry began to recover from a temporary decline that had been partly due to better methods of vinegar production leading to cheaper acetic acid and partly to the displacement of wood tar by coal tar. The recovery was prompted by new demands for methanol by the aniline dyestuffs industry and by the need for acetone in the cellulose nitrate industry after about 1880. It was aided by two technical improvements in the treatment of pyroligneous acid; one was the gradual adoption of a more efficient three-vessel system for separating the components, and the other was a column still that enabled pure acetic acid and pure methanol to be obtained. For methanol this remained the only method of manufacture until the 1930s.

By the early 20th century the process was of minor importance in Britain and all its main products were obtained from overseas, where abundant supplies of wood served as raw materials (Russia, Germany, North America). However, in 1913 a modern plant was opened by the Office of Woods in the Forest of Dean, and another at Worksop (Shirley Aldred & Co.) was producing in the 1960s about one-third of the country's requirements for charcoal. Wood distillation

[19] H.W. Dickinson and E. Straker, 'Charcoal and pyroligneous acid making in Sussex', *Trans. Newcomen Soc.*, 1937, **18**, 61–66.

continued into the 1960s in Britain, though it was more important in Australia and survived until 1981.[20]

## (b)   Hydrolysis of fats

What is now called saponification, or hydrolysis of fats, was for centuries conducted with little understanding of the process but great regard for the prize at the end of it. As one Victorian writer sententiously put it:

> The article now to be considered is one of the utmost importance, as, with the progress of civilisation, it becomes an indispensable necessity to every human being; and as the art of its production is entirely of a chemical nature, the more it is studied under the guidance of chemical principles, the higher in quality and the better suited to its purpose will be the material.[21]

The material in question was of course soap. It was known in antiquity that wood ashes (containing potash) had a detergent effect, and crude forms of soap have been made from such materials and fats for many centuries, although usually in a highly impure form. It seems to have been manufactured in Bristol and London by the 16th century, though up to about 1800 it was made on a very small scale. By 1790 there were several hundred soap-makers in Britain (including the famous Pears firm in London), but their individual output was extremely small, averaging 16 tons each. However three events in the early 19th century conspired to cause a veritable explosion in the manufacture of this simple chemical. First was the phenomenal growth in textile production in the Industrial Revolution. Soap, whose previous uses had been primarily in cleaning clothes, now appeared as a necessity for both the raw materials (wool, cotton) and the finished product. Secondly, the researches of Chevreul in France led to a new understanding of the chemical nature of fats as esters of organic acids and of soap as their sodium salts (1823). Thirdly, the coming of the Leblanc process to Britain transformed the prospects of soap manufacture by providing for the first time almost unlimited quantities of the soda on which their industry depended. Sodium carbonate was converted into the hydroxide by treatment with lime ($Ca(OH)_2$), and this was then boiled in aqueous solution with fats from tallow, palm oil *etc.* Where $\sim$ is a glyceryl residue and R a long-chain alkyl group the reaction is:

$$RCOO\sim \ + \ NaOH \ = \ RCOONa \ + \ \sim OH$$
$$\text{fat} \qquad\qquad \text{caustic soda} \qquad \text{soap} \qquad \text{glycerol}$$

The detailed history of soap manufacture is not relevant here, being more concerned with competition than chemistry. However, a few key players may be briefly mentioned. The largest centre for soap production in the early 19th

---

[20] I.D. Rae, 'Wood distillation in Australia: adventures in Arcadian chemistry', *Hist. Rec. Australian Sci.*, 1987, **6**, 469–484.

[21] S. Muspratt, *Chemistry, theoretical, practical and analytical*, Mackenzie, London, n.d. [*ca.* 1860], vol. ii, p. 863.

century was Tyneside (the soap presumably made from waste tallow from candle making), being advertised as early as 1712. Although no major textile centre it had unlimited supplies of coal at its doorstep, and unrivalled deep water access for imported raw materials. Anthony Clapham, a chemist and druggist in 1801, had become thirty-five years later one of the largest soap makers in England and at one time employed Hugh Lee Pattinson. At almost the same time another firm, Doubleday and Easterby, were making soap on the other side of the town. Both firms were generating complaints and injunctions on account of their (mostly surreptitious) manufacture of soda to feed their soap-works, Clapham made the characteristic response of his day: to move his nuisance elsewhere, in this case to the south bank of the Tyne whence he was followed by many other manufacturers. Another Newcastle firm was that of Thomas Hedley,[22] who in 1839 became partner in a small firm of soap makers, John Greene & Sons, where he had been a junior clerk some years before. They were soon making candles (for the coal-mines) as well as soap and, in 1861, Hedley took over the business, which, nearly a century later, was acquired by Procter and Gamble.[23]

From 1809 to 1813 Anthony Clapham had an apprentice from Lancaster named Joseph Crosfield. By 1815 this young man had set up a soap-making business of his own at Bank Quay, Warrington. This was an ideal location, with alkalis readily available from Cheshire salt, local coal supplies, good access to the vast textile markets developing in Lancashire and even West Yorkshire, and a convenient place to import tallow *etc.* from the Mersey. His firm, eventually Joseph Crosfield & Sons Ltd., became one of the largest in Britain.[24]

Merseyside also attracted many other soap makers. These included William Gossage, a druggist-turned-alkali-manufacturer in Stoke Prior, Worcestershire (see Chapter 4). He diversified further into white lead manufacture and copper extraction, and then set up the highly successful firm of soap makers William Gossage & Sons Ltd. at Widnes. His discovery of the detergent-enhancing powers of sodium silicate led to the introduction of this substance to soap, an innovation adopted by many other manufacturers. This firm, and Crosfields, were acquired by Brunner Mond in 1911. In the same year as Gossage's arrival, 1855, a wholesale grocer named W.H. Lever[25] bought a soap-works at Warrington. His sales of 'Sunlight Soap' were so successful that in 1888 he was able to open a large new works on the Mersey, appropriately named Port Sunlight. With his brother J.D. Lever he formed the firm of Lever Brothers in 1894. They sold the carbolic soap known as

[22] W.A. Campbell, in D.J. Jeremy (ed.), *Dictionary of business biography*, Butterworths, London, vol. iii, 1985, pp. 151–152.

[23] *Idem, A century of chemistry on Tyneside 1868–1968*, Society of Chemical Industry, 1978, pp. 17–19.

[24] A.E. Musson, *Joseph Crosfield & Sons Limited, 1815–1965*, Manchester University Press, Manchester, 1963.

[25] D.J. Jeremy, 'The enlightened paternalist in action: William Hesketh Lever at Port Sunlight before 1914', *Business History*, 1991, **33**, 58–81; W.J. Reader, 'William Hesketh Lever' in D.J. Jeremy (ed.), *Dictionary of business biography*, Butterworths, London, vol. iii, 1985, pp. 745–752.

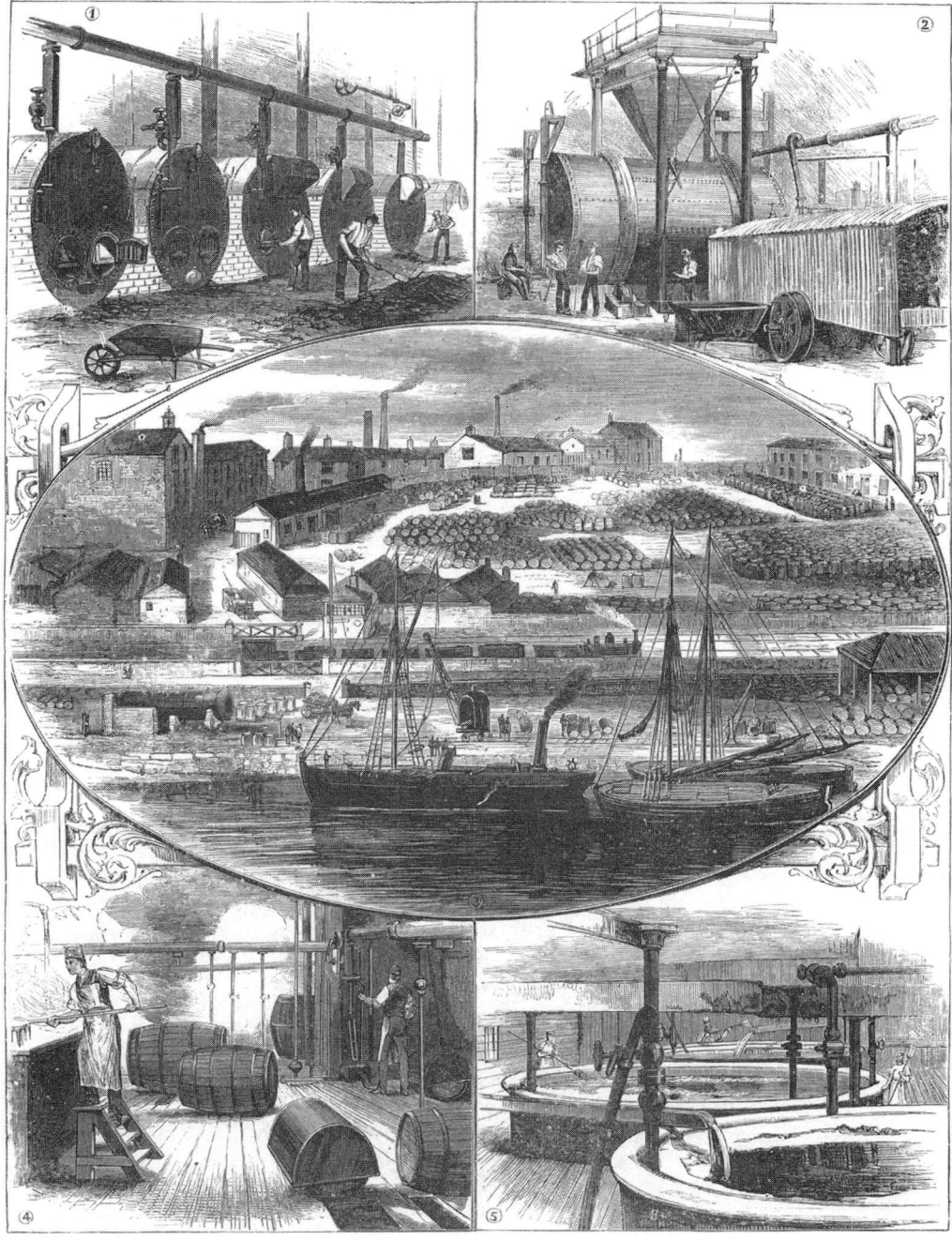

*Views of Crosfields' soap works from* Illustrated London News *for 13 November 1886;* (1) *the boiler shed,* (2) *the alkali plant,* (3) *the raw material as it arrives in the Mersey in casks,* (4) *the raw material being melted out of the casks,* (5) *some of the pans*

Lifebuoy (1894), Lux soap flakes (1899) and the abrasive Vim scouring powder (1904).[26]

In all these developments chemistry took a long time to play a major part in industrial practice. By mid-19th century Crosfields boasted a chemical laboratory eight feet long by five feet wide. Forty years later a rather bigger laboratory is shown on a map, created from converted cottages in Quay Fold, with each production department 'under the charge of a trained chemist'. Since the removal of the Soap Tax in 1857 Gossages were reported in 1863 as having 'been greatly advanced by the application of chemical science'.[27] On the other hand at Port Sunlight, where Lever had installed new laboratories, the prevailing emphasis was one of secrecy, several chemists having been sacked for apparent breaches in security though every effort was employed to retain the services of a disgruntled soap-boiler without scientific expertise, yet he and his family were the only people to have '*any idea* of soap-making'.[28] Yet, although the manufacture of this one organic chemical may seem to have long been independent of any real knowledge of chemistry, it was an important part of the chemical industry on account of its immense impact on textile processing, its dependence on production of that other major chemical product soda, and its production of a co-product with soap which chemistry was to transform into substances of great social importance. That substance was glycerol (or glycerine).

Glycerol was for many years discarded as an unwanted by-product of soap manufacture, adding its own toll to the deluge of pollution daily entering the streams and rivers of Britain. Its non-toxic nature and absence of odour may have not made its presence so obvious but it can be confidently asserted that in the first half of the 19th century at least 250,000 tons of unwanted glycerol were washed into the environment by this process.[29] Recovery of glycerol is not easy, partly because of its miscibility with water. However, it can be done by a process involving evaporation, crystallisation of solid material and distillation. This was first claimed as an industrial process by the west country firm of Christopher Thomas and Brothers, in the early 1870s. It was brought over to the north of England by a French inventor, V. Clolus, who erected a plant at Runcorn. By 1884 Crosfields had their own glycerol recovery plant that, twelve years later, was 'the most lucrative of any part of the business'.[30] By 1887 the Yorkshire firm of Joseph Watson & Sons also established a plant of their own by the River Aire. The latter was probably due to the influence of their new chemist J. Lewkovitsch who was to become an international authority through his book *Chemical technology and analysis of oils, fats and waxes* (1909).

The reason for this new attention to glycerol was the new demand for it in the production of high explosives. On treatment with nitric and sulphuric acids it

[26] C. Wilson, *The history of Unilever: a study in economic growth and social change*, Cassell, London, 1954.

[27] Musson (ref. 24), p. 68.

[28] Wilson (ref. 26), vol. i, pp. 31, 37.

[29] Calculated from taxable soap production figures 1801–1850 in B.R. Mitchell, *British historical statistics*, Cambridge University Press, Cambridge, 1988, p. 415.

[30] Musson (ref. 24), p. 78.

yields nitroglycerine, an explosive compound discovered by Sobrero in 1846, but too unstable to be of much practical value. In 1863 the Swedish inventor Alfred Nobel began experiments with it as a commercial explosive, developing a mercury fulminate detonator for the purpose.[31] It was still far too dangerous a material to be moved around but in 1867 he had the idea of absorbing it on a porous material. Using the siliceous earth kieselguhr he obtained a blasting material that could be safely handled and which he called dynamite. Following the British government's relaxation of restrictions on nitroglycerine, in 1871 Nobel established at Ardeer, in Ayrshire, the world's largest factory for its production (Nobel's Explosives Co. Ltd.). By 1875 Nobel had developed blasting gelatine, in which nitroglycerine was retained as a gel on collodion cotton. This too was made at Ardeer. Thus from the waste product of soap manufacture came products that seemed anything but beneficent in their application. Yet one must not forget that the major use of nitroglycerine for many decades was in blasting for mining and railway purposes, to say nothing of a minor role in medicine.

### (c)   Cellulose

One other natural material of great importance in the textile industry is cellulose, the main component of cotton fibres. Cellulose has long fascinated chemists, though without much practical effect. Thus in 1844 the Lancashire calico-printer and chemist John Mercer[32] discovered the effect of sodium hydroxide solutions on cotton, giving greater lustre as fibres swell in diameter and contract in length. This effect was applied in 1889 by H.A. Lowe to cotton under tension, giving silk-like threads of superior strength and a silk-like lustre. The product is still called 'mercerised cotton'.

Another variant can be obtained by precipitating cellulose from its solution in cuprammonium sulphate. This process was discovered in 1859 but never attained great popularity in Britain. However, more profound changes have been induced by forming cellulose esters.

*Cellulose nitrate* was discovered in 1838 by Pelouze but in the form of gun-cotton in 1846 by Schönbein; he treated cotton with a mixture of nitric and sulphuric acids and recognised its explosive tendencies. His patent was worked for a few months by John Hall & Co. at Faversham until terminated by a disastrous explosion in 1847. Following research by F. Abel at the Royal Military Academy at Woolwich it became clear that prolonged washing (to remove traces of sulphonates) rendered it a good deal less dangerous. It was then produced at the Royal Gunpowder Factory at Waltham Abbey, though a firm started by Thomas Prentice, later known as the New Explosives Company,

---

[31] On Nobel see, *e.g.*, N. Halasz, *Nobel, a biography*, Robert Hale, London, 1960; R. Trotter (ed.), *The history of Nobel's Explosives Company Limited and Nobel Industries Limited, 1876–1926*, ICI, London, 1938; and J.E. Dolan, 'The Nobel era', in J.E. Dolan and S.S. Langer (eds.), *Explosives in the service of man: the Nobel heritage*, Royal Society of Chemistry, Cambridge, 1997, pp. 26–32.

[32] E.A. Parnell, *The life and labours of John Mercer, the self-taught chemical philosopher*, Longmans, London, 1886; A. Nieto-Galan, 'Calico-printing and chemical knowledge in Lancashire in the early nineteenth century: the life and "colours" of John Mercer', *Ann. Sci.*, 1997, **54**, 1–28.

*Pans for nitrating cellulose at ICI's Nobel Division*

made Stowmarket the centre of gun-cotton manufacture in Britain. The manufacture of nitrocellulose for explosives was begun by Nobel at his Ardeer factory in 1881.

A related product is collodion, made by dissolving nitrocellulose in an alcohol/ether mixture, and its uses included treatment of minor wounds (it dries to a colourless elastic skin). A serendipitous discovery of great value arose one day when Nobel, nursing an injured finger covered in collodion, noticed that nitroglycerine in the laboratory combined with it in some way. In 1875 he discovered that a mixture of nitrocellulose and nitroglycerine in a ratio 8:92 gave a gelatinised explosive that did not need to contain kieselguhr. It was used for blasting and termed ballistite. A smokeless propellant with a ratio 65:30 was manufactured as cordite by the British government at Waltham Abbey from 1891. After mixing with 5% mineral jelly it was dissolved in acetone and then extruded.

Not all uses of cellulose nitrate depended on its power to explode. One major outlet was as photographic film, its major disadvantage being high inflammability. It was therefore not desirable to place it in a hot projector, for example. In 1862 Alexander Parkes, a Birmingham chemist with a strong inventive streak, discovered that a solution of nitrocellulose in an organic solvent when mixed with camphor gave a smooth solid material showing some similarity to tortoise-shell or ivory. His 'parkesine' was displayed at the International Exhibition in London in 1862 and was arguably the first plastic to be made. Commercially, however, it was unsuccessful, largely because of the large amounts of castor oil added as a plasticiser, causing it to soften in a few weeks. The problem was taken up by a Gloucestershire doctor, Daniel Spill, brother of George Spill, waterproofing manufacturer. Joined by Parkes, Spill resolved many of the technical and commercial difficulties confronting 'parkesine', and patented his own variation of nitrocellulose/camphor plastic under the name of Xylonite.

*An early advertisement extolling the water-proof qualities of celluloid*

Four years later an American inventor, J.W. Hyatt, made a similar material which he coated on to billiard balls. Because of the tendency to pyrotechnic properties of collodion it was reported by one Colorado saloon-owner that the sharp collision of the balls 'would produce a report like that of a percussion cap, causing every man in the establishment to draw his gun'. 'Celluloid' was the name coined by Hyatt's brother in 1870, and it was a pioneer plastic.[33]

In 1875 L.P. Merriam sought unsuccessfully to introduce the Hyatt process into Britain, so in the following year joined Spill to form the British Xylonite Company which became the only major producer in Britain. Less successful than its American rival in introducing cheap fancy goods (possibly because of British middle-class preferences) it achieved a breakthrough nine years later with celluloid collars and cuffs.

Celluloid film was marketed by Eastman Kodak from 1889. A commission appointed by the Home Secretary in 1912 recommended some safety precautions (as in cinemas) but recognised the widespread use of celluloid until such time as a non-inflammable substitute should become available. That was not to be until after the war.

The passage of collodion solution through holes as opposed to slits produced

---

[33] R. Friedel, *Pioneer plastic: the making and selling of celluloid*, University of Wisconsin Press, Madison, 1983.

a fibrous product, the first practical version being described by the French chemist de Chardonnet in the 1880s. This was one of the semi-synthetic fibres known as artificial silk, and by 1900 accounted for about 80% of the total. J.W. Swan had used carbonised nitrocellulose filaments in the production of his electric lamps. Two of his collaborators, C.H. Stearn and F. Topham later made them on a commercial scale.

*Cellulose acetate* was first prepared by French chemist P. Schützenberger in 1865. It is now generally made using acetic acid/acetic anhydride mixtures. Like the nitrate it can be dissolved in organic solvents (as methylene chloride) and extruded through holes or slits. From the former an 'artificial silk' can be obtained, and from the latter a non-inflammable film. A patent for artificial silk was taken out as early as 1883 by J.W. Swan, though in 1879 C.F. Cross and E.J. Bevan had acetylated cellulose (with acetyl chloride). They patented the first industrial process in 1894. Their product, the triacetate, was too insoluble in common organic solvents to be of much value, but a diacetate was shown by G.W. Miles in 1905 to dissolve readily in acetone. It was further developed in Germany, USA and Switzerland. Acetate safety film was introduced by Eastman Kodak in 1908, and acetate lacquer by the Dreyfus brothers at Basle in 1910. In Britain it was not until 1914, however, that its commercial potential was to be realised.

*Cellulose xanthate*: In 1892 Cross, Bevan and C. Beadle, consulting chemists, found that cellulose after treatment with caustic soda gave with carbon disulphide a solution of cellulose xanthate (ROCSSNa) which they called 'Viscose'. The hydroxyl groups in cellulose react thus:

$$\sim\!\!OH \xrightarrow{\text{NaOH}} \sim\!\!ONa \xrightarrow{\text{CS}_2} \sim\!\!O-\overset{\overset{\displaystyle S}{\|}}{C}-SNa \xrightarrow{\text{H}_2\text{SO}_4} \sim\!\!OH$$

Raw cellulose                                      Cellulose xanthate              Regenerated cellulose

This was patented in 1892, and was widely used despite the frequent leaks into the air of the obnoxious carbon disulphide. On extrusion of the xanthate into acid through a hole or a slit, fibres or films of cellulose were regenerated, the 'viscose rayon' being thus a further form of 'artificial silk', though simply consisting of pure cellulose. F. Topham (1900) patented a process for generating viscose thread by forcing viscose solution through a small orifice into a bath of sulphuric acid. The process was applied by Courtaulds, Coventry, from about 1906.

## (d)   Condensation of casein

Casein is a protein from skimmed milk. In 1899 a demand in German schools for white boards led to the patenting of formaldehyde-hardened casein plastics. In the UK a Syrolit Company was established to manufacture it from 1910 on the basis of a Latvian patent, initial trials being made in a disused cloth-mill at Stroud, Glos. Their failure led to a new company being formed in 1913, Erinoid

Ltd, whose modifications proved successful. The products were readily coloured and much used as replacement of tortoise-shell, for artificial horn and (especially) buttons, brushes and umbrella handles. In the UK casein plastics manufacture ceased in 1950.

## 3   The Supremacy of 'Old King Coal'

### (a)   Industrial progress and environmental hazard

If any single substance may be said to have had more profound effects on society and the environment than any other, that substance must surely be coal (at least until the advent of nuclear chemistry). The use of coal as a fuel in Britain goes back at least 400 years. The country was exceptionally rich in coal deposits and by 1800 was producing over 80% of the world's coal. At that time about half of the home consumption was for domestic heating, the rest being used by industry, including about 10% for iron making. From about 1830 the rate of coal production rose by a factor of nearly four, at least partly due to the growing railway network which made its conveyance to other parts of the country so much easier. Hitherto it had been mainly transported from the North East by sea (hence London's use of 'sea-coal'). By this time coal had found another use than merely as a fuel. Since about 1760 it had been in increasing demand as a source of coke, this now replacing the charcoal that had been the original material used for smelting iron ores (Chapter 10). At almost exactly 1830 the steeply rising curve of coke consumption for iron making crossed the falling curve of charcoal consumption.

The actual production of coke was soon conducted in ovens of various kinds, from all of which the volatile by-products were allowed to escape. As is well known the operation of any kind of coking plant produces three main products. From 1 ton of coal can be obtained coke (12 cwt.), tar (1 cwt.) and coal gas (15,000 cu. ft.). These are very rough estimates, the corresponding figures in metric terms being 1000 kg of coal yielding 600 kg of coke, 50 kg of tar, and 420 m$^3$ of gas. In addition smaller but significant quantities of ammonia and hydrogen sulphide are formed. The environmental pollution in the neighbourhood of an early 19th century coking-plant can hardly be believed. A few early attempts to remove the most offensive by-products by some kind of washing merely transferred contamination from the air to the earth and thence to the rivers and water supply.

However, many years before the huge increase in coking that was associated with the Industrial Revolution several observers had noted that heated coal would yield a gas that could be collected, and burnt with a highly luminous flame. These included the Rev. John Clayton of Wigan in 1684, George Dixon of County Durham in 1760, and in 1792 a Scottish engineer named William Murdock at his house in Redruth, Cornwall. An employee of the firm of Boulton and Watt, Murdock eventually persuaded them to erect an experimental gas apparatus at their factory at Soho, Birmingham in 1800. To celebrate the Peace of Amiens in 1802 the factory exterior was illuminated by

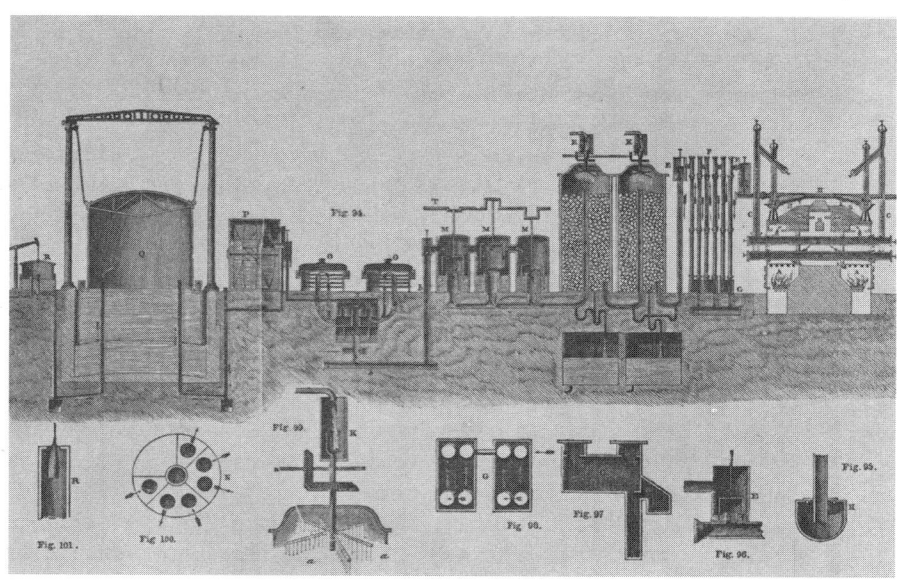

*Typical gas-works of the 19th century*

gas and from then on there was rising public interest. This was fanned into a fever of expectation through the uninhibited promotion of the projected Gas Light and Coke Company by a German entrepreneur F.A. Winzler [Winsor] who, in 1807, installed gas-lighting for the first time in London, at Carlton House Gardens. The rapid growth of public gas-lighting and a series of technical improvements have been well recorded and cannot detain us now.[34] Nor can the immense social consequences, such as the promotion of reading at home and school, the establishment of evening classes, a measure of crime prevention in the well-lit streets, the general lighting of theatres and music-halls, and a longer working-day in factories no longer feebly lit by candles. For the chemical industry the most important aspect of the meteoric rise of gas-lighting was an even greater demand for the pyrolysis of coal, now needed to make gas for its own sake or else as a by-product from coking plants. If coke-ovens had annoyed the neighbouring populace for their generally disgusting discharge of noxious by-products into the atmosphere, the installation of gas-works in the middle of large cities caused complaints to be multiplied, particularly when the offended parties were articulate and litigious. This was the case, for instance, in 1813 when one William Knight was indicted for 'divers noisome and offensive stinks and smells' at Whitefriars in the City of London or when the affluent citizens of Elswick, Newcastle, issued a stream of complaints after their local

[34] See D. Chandler, *Outline of history of lighting by gas*, South Metropolitan Gas Company, London, 1936; S. Everard, *The history of the Gas, Light and Coke Company, 1812–1849*, Benn, London, 1949; E.G. Steward, *Town gas, its manufacture and distribution*, HMSO, London, 1958.

gas-works was commissioned in 1857. And all this time the ever-growing mountains of tar presented almost insoluble problems of disposal.

In the initial stages of the gas-making industry no demand existed for the black, unsavoury material accumulating at an alarming rate every day. Attempts to carbonise tar to make more gas proved uniformly unsuccessful, as were efforts to reduce its production by passing the raw gas through red-hot pipes, since the illuminating power of the gas was much reduced (due to conversion of ethylene into alkanes). The magnitude of the environmental problem now being created may be gauged by the response of the producers. They attempted a number of fairly desperate measures.

First they would try to burn the tar in open pans, but the fire hazards created were too great for the process to be continued, at least in urban situations. Next they attempted to dispose of tar by dumping it in the nearest river; it is no accident that many gas-works were situated at the side of a canal or river, or that for some years works by the Thames had been permitted to dispose similarly of a foul product obtained by passing coal-gas through lime-water. However, many of the complaints from fishermen, brewers and other citizens overlooked pollution of the Thames by hospitals and slaughterhouses, and the immense volume of raw sewage entering the river from the rapidly growing population; one commentator has urged that they 'probably exaggerated the impact of industrial effluent'.[35] Some tried to follow the advice of Samuel Clegg: 'I decidedly think that the most economical use to which coal-tar can be applied, is to burn it beneath the retorts'. Finally, in desperation they would give it away to any who would take it, as was the case at Leith in 1822. By now a few potential uses had been discovered but there was nothing like enough demand to absorb all the available by-products for another quarter century at least. As late as 1885 the cry was still being heard 'What shall we do with our tar?'.[36]

The answer lay in one word: distillation. This oldest of alchemical techniques was brought into the service of a pressing modern problem. The first attempts were rough and ready. As early as 1815 F.C. Accum[37] suggested boiling down the tar in a still, so that the volatile fraction could be condensed and given (or sold) to the varnish maker as a substitute for turpentine. In Scotland Charles Macintosh began to exploit the coal-tar produced by Glasgow gas works. Seeking to extract the ammonia (for use in dyeing) he found that, on distillation, the tar produced alongside the pitch a volatile oil of low-boiling naphtha. Discovering that it was capable of dissolving caoutchouc (raw rubber) he found it could be used as a water-proofing agent for textiles, one of which has ever since borne his name. A firm of tar-distillers at Leith, Longstaff and Dalston, set up in business to provide Macintosh with his volatile 'spirit' (1822); this was the firm who obtained its tar *gratis*, paying only for its removal from the gas-works.

[35] W. Luckin, *Pollution and control: a social history of the Thames in the nineteenth century*, Hilger, Bristol, 1986, p. 12. A slightly safer and more complex design was patented by Read Holliday of Huddersfield in 1848. There was almost no end to the stream of ingenious responses to this problem.

[36] L.T. Wright, 'What shall we do with our tar?', *J. Soc. Chem. Ind.*, 1886, **5**, 558–565.

[37] On Accum see D.W.F. Hardie, 'Chemical pioneers – Fredrick Accum', *Chem. Age*, 1957, **77**, 1009–1010.

Twelve years later another tar-distiller in Manchester began to produce a low-boiling naphtha which was then used to dissolve some of the residual pitch in order to make a black varnish. At about the same time a lamp using naphtha as a fuel was seen by an American visitor, Joseph Henry, at Greenwich (though the associated dangers prevented its widespread adoption).[38]

The distillation of tar yielded five main products, generally as follows:

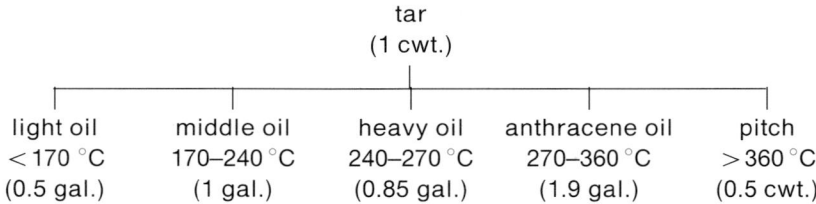

As it happened, experiments from earlier decades had already shown a way forward. For some centuries tar and oil from wood distillation had been indispensable in caulking the wooden ships of the Royal and merchant navies. In Britain the rapidly diminishing supplies of native timber led perforce to importation of nearly all this material, first from northern Europe and then from North American colonies. The idea that an effective substitute might be obtainable from coal was far from obvious despite a few earlier experiments in that direction. Thus in 1681 the German chemist J.J. Becher, then living in London, had taken out a patent for the making of tar from 'pit-coal', a product claimed to be superior to even the best production from Sweden. As was often the case nothing came of the idea and the inventor is said to have died in penury. In the next century many attempts were made to obtain an acceptable product from coal in both continental Europe and Britain (at Newcastle, Coalbrookdale and elsewhere) though to little effect. Success came first in Scotland.

Archibald Cochrane, ninth Earl of Dundonald, was a relatively impoverished peer who sought to restore his family's ailing fortunes by exploiting minerals on his own estate, most notably coal (p. 77). Having served in the Royal Navy he had already encountered the problem of making ships watertight and of finding an alternative to copper sheet for protecting them against worms. With this in mind he intentionally set about experiments to produce a new kind of tar from his own coal. By allowing some coal to burn in a restricted and regulated amount of air he was able to ensure that the rest underwent the kind of decomposition familiar to charcoal burners and considerable amounts of tar were collected. A tar kiln was set up on his estate at Culross and the procedure led, if not to his own financial relief, at least to the establishment of the British Tar Company (in which he was joined by entrepreneurs from Newcastle).[39] This time coke was the by-product, so they opened other tar kilns in the neighbour-

---

[38] N. Reingold (ed.), *The papers of Joseph Henry*, vol. iii, Smithsonian Institution, Washington, 1979, p. 273 [entry for 12 April 1837].

[39] D.W.F. Hardie, 'Chemical pioneers – Lord Dundonald and the Loshes', *Chem. Age*, 1957, **77**, 342.

hood of iron-smelting plants where the coke could be put to good use. Accidentally they became aware of the inflammable gas also produced though, unlike Murdock a few years later, they did not realise its potential as an illuminant. Another by-product was ammonia and this was collected, converted into ammonium chloride and sold to metal-workers as a flux for tinning. The main product was the tar and this was extremely well-received as an alternative to wood-tar, though not as a substitute for the copper cladding. Unfortunately, it did not enrich its noble discoverer to the extent he had hoped.

By the 1830s, however, circumstances had changed and with them had come an immense technical problem. This concerned the use and preservation of wooden sleepers on the railways. By 1830 these were beginning to replace the stone blocks that had been in favour despite their unwieldiness and weight. But the wooden sleepers tended to rot with such speed that their replacement required enormous quantities of wood at great expense. Treatment with various chemicals had been tried. including mercury(II) chloride (Ryan, 1832), copper sulphate (Lloyd, 1837) and zinc chloride (Burnett, 1838). None was very successful. Then, in 1838, came a patent by John Bethell for preserving timber by impregnation with the heavy oil from coal-tar. Within ten years the 'railway mania' was at its height with an escalating demand for treated wooden sleepers on which to lay the track. A timber merchant at Rotherhithe, H.P. Burt, saw the opportunity and began making his own coal-tar at Millwall and (later) at Silvertown and Southampton. By 1866 he had formed a partnership with a nephew (T.B. Haywood) and S.B. Boulton (later Sir Samuel Boulton), a former associate of the railway engineer Thomas Brassey. Sleepers were 'pickled' in horizontal iron cylinders in which occluded air was first removed under reduced pressure, giving timber that would last up to eight times that of unimpregnated wood in the same conditions. The firm Burt, Boulton and Haywood by the end of the century were producing creosoted sleepers 'on a colossal scale',[40] at which time creosoting of timber was still the largest outlet for tar oils, one gallon of which would be absorbed by one cubic foot of timber. It proved useful not only for railway sleepers but for telegraph poles, fencing posts and all wooden objects partly buried in the ground. Thus the threat to the environment posed by one large-scale chemical reaction (pyrolysis of coal) became largely neutralised by widespread application in the non-chemical sector of industry. In that sense potential disaster was turned into considerable social good. That became even more true of further developments within applied chemistry itself.

## (b)   Coal-tar acquires a chemistry

In all the work described above there was little understanding of the identity, let alone the chemical nature, of the constituents of coal-tar. The first one to be recognised was naphthalene, partly because it was a solid (it was to cause much trouble by settling in gas-pipes and blocking them) and partly because it was so relatively abundant (up to 43% of middle oil). In 1819 it had been isolated by

---

[40] R. Meldola, *Coal, and what we get from it*, SPCK, London, 1897, p. 70.

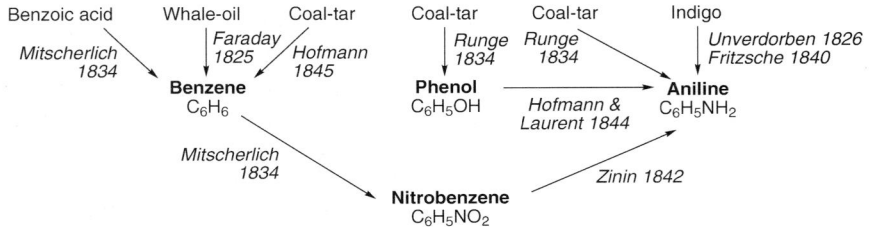

**Figure 1**  *Relationships between simple constituents of coal-tar*

Brande at the Royal Institution and also by Alexander Garden, an assistant of Accum. 'Naphthaline' was so-called by John Kidd, professor of chemistry at Oxford, who in the following year described it as 'a concrete substance as white as snow'. Its composition was determined by Faraday[41] in 1826. Anthracene was obtained by Dumas and Laurent in 1833 from the highest-boiling distillate. In 1834 the German chemist F.F. Runge obtained aniline (previously known from alkaline decomposition of indigo), phenol, pyrrole, quinoline and other bases in coal-tar, though he did not analyse them or give them these names. Benzene had been discovered by Faraday in whale oil in 1825 (and analysed by him), and later obtained by decarboxylation of benzoic acid with lime, but proof of its occurrence in coal-tar (chiefly the light oil) had to wait until A.W. Hofmann decisively showed its presence in 1845. Two years previously he had demonstrated the identity of products obtained from indigo, coal-tar and the reduction of nitrobenzene as what came to be called 'aniline'. As Figure 1 indicates, some of the relationships between the simplest constituents of coal-tar were by now beginning to become clear.

That was the situation when, in 1845, the already distinguished German chemist A.W. Hofmann arrived in London. Much of the research on simple aromatics had been conducted in Continental Europe during the 1830s and 1840s and the subsequent history of organic chemistry, and certainly the history of the British chemical industry, would have been very different if Hofmann had not been enticed to London by (amongst others) the Prince Consort in order to start up a new Royal College of Chemistry. Here, for the first time in England, a chemical research school was to be formed and, within a few years, was to be the scene of spectacular discoveries. And here, more than anywhere else in the world, were vast stocks of coal-tar awaiting exploitation.

Hofmann continued his studies of organic bases but within a few months of his arrival set an assistant to work on the distillation of coal-tar with a view to isolating its main constituents. This was Charles Blachford Mansfield whose great achievement was the isolation of nearly pure benzene and toluene (the latter having been previously prepared in 1841 by distillation of the fragrant tolu balsam). Mansfield went on to prepare large quantities of benzene,

---

[41] For recent biographies of Faraday see J. Meurig Thomas, *Michael Faraday and the Royal Institution*, IOP, Bristol, 1991; and C.A. Russell, *Michael Faraday*, Oxford University Press, New York, in press.

examining its properties and exploring its potential as a solvent. In so doing he discovered its ability to remove grease from clothing and (unlike turpentine previously used for such purposes) to leave no lingering odour afterwards. He was the father of the dry-cleaning industry. He joined the growing band of those who devised lamps that burned tar-oil distillate but were soon replaced by safer designs using paraffin. His own 'illuminating machine' just predated another by the Huddersfield manufacturer Read Holliday, with whom Mansfield was soon to collaborate, joining forces over lamp design and also passing on his unique expertise on distillation. That technique proved to be Mansfield's undoing. In February, 1855 he was purifying some crude hydrocarbon samples sent by Holliday in preparation for the Paris Exhibition when his apparatus exploded, bringing down the building in the former Agar Town, London (just north of the present St. Pancras station). Mansfield and his assistant were severely burned and unable to extinguish the flames on their clothing by jumping into the adjacent Regent Canal as that had frozen over. They died within a day or two.[42]

At this time several reactions of benzene and its homologues (as we might call them) were being examined, with varying degrees of success. One that proved of the greatest importance was nitration.[43] Faraday had apparently obtained a nitration product from the benzene he had prepared, and Laurent had used nitric acid to make picric acid (1841), a bright yellow substance capable of dyeing wool and the first coal-tar synthetic dye. Meanwhile Mitscherlich had used fuming nitric acid to prepare nitrobenzene (1834). Mansfield seems to have favoured this method and made large quantities of the substance. In a long patent issued in 1849 he refers also to the use of *mixed* acids (nitric and sulphuric), and is sometimes credited with the invention of that supremely important technique. However it had been claimed in 1842 (though not in a published report) by John Leigh, director of Manchester gas-works, and was certainly used by Hofmann and Muspratt in 1846 to obtain a dinitrobenzene using fuming nitric and concentrated sulphuric acids. Nitrobenzene was introduced into perfumery by Mansfield as 'oil of mirbane', becoming a common additive to soap. More important, it turned out to be a gateway into the vast area of aromatic chemistry to be explored in the next few decades.

The next stage in the story has been told countless times and is the discovery of the first synthetic dyestuff to be made on a commercial scale. Another assistant at the Royal College of Chemistry, William Henry Perkin, was set a task that combined two of Hofmann's priorities: one was an intense interest in organic bases related to benzene, and the other was the artificial formation of useful materials, in this case quinine. This fusion of the academic and the practically useful was to become a hallmark of the Royal College's ambitions in the years to come. By this time (1856) not only benzene but also toluene had been nitrated and their nitro-compounds reduced to amines. So the laboratory had access not only to aniline but also to the toluidines (not yet recognised as

[42] E.R. Ward, 'The death of Charles Blachford Mansfield (1819–1855)', *Ambix*, 1984, **31**, 68–69.

[43] E.R. Ward, 'Industrial mixed acid nitration', *Ambix*, 1976, **23**, 199–200; *idem, Incidents and personalities in the history of aromatic nitration and aromatic nitro-compounds*, privately published, Poole, nd.

*August Wilhelm Hofmann (1818–1892), German chemist, first director of the Royal College of Chemistry and pioneer of modern organic chemistry*

*William Henry Perkin (1838–1907), discoverer and manufacturer of the first synthetic organic dye, mauveine*

three separate compounds). Applying simple arithmetic to the known empirical formulae of factor and product Perkin assumed that addition of oxygen might produce quinine from the allyl derivative of toluidine:

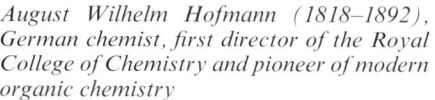

$$2C_{10}H_{13}N + 3[O] = C_{20}H_{24}N_2O_2 + H_2O$$
allyltoluidine                    quinine

Naive though such speculation may seem today, Perkin was unaware of the structural impossibilities involved for the simple reason that no structures were known and no structure theory then existed. With youthful optimism he treated the allyltoluidine with potassium dichromate. No quinine resulted, only 'a dirty reddish-brown precipitate' Undeterred, he tried something even simpler: oxidation of aniline itself (though this time even the arithmetic made little sense). The dichromate yielded an unpromising black sludge, but extraction with organic solvents disclosed a beautiful purple material that was soon shown to adhere well to fabrics and 'Mauve', as it became known, was clearly a substance of great commercial potential. In fact its appearance at all was entirely due to the presence in the aniline of considerable traces of toluidines (due to the contamination of the 'benzene' with toluene). Its formation and structure have recently been given[44] in revised form as shown:

R = H or CH₃

Almost overnight the supplies of aniline in Hofmann's laboratory were seen as woefully small. All the work on nitration and reduction now gained an urgent new relevance and with it came new problems of scale-up. These problems were gradually solved by Perkin in the next year or two, not in the cramped conditions of a research laboratory with the constant risk of fire, but in a factory that he set up at Greenford, with encouragement from the Perth dyers J. Pullar & Son and with financial support from his father. His subsequent success has been attributed to his approach to technical problem-solving which was 'as much in traditions of the great British craft industries as in the rigorous and exact methods of Hofmann's German style'.[45] Crude benzene was obtained relatively cheaply from George Miller & Co., tar-distillers in Glasgow, re-distilled and then nitrated. The nitric acid was prepared *in situ* from concentrated sulphuric acid and Chilean saltpetre. Nitrations like this are notoriously temperature-sensitive and, despite playing on the vessels with a hose or immersing them in water, explosions could and did occur. Moreover large amounts of the toxic nitrogen dioxide were released into the atmosphere so even on the initially small scale of operations the potential for nuisance was considerable. Complaints were rare, however, possibly because a large boiler-house chimney depicted in Perkin's own drawing of his works had probably been used also to release noxious fumes well above ground level. There was no mains drainage, but the Grand Junction Canal was conveniently and temptingly near.

Reduction to aniline, followed by dichromate oxidation of its sulphate, yielded the dyestuff, which, though good for many purposes, was rather more fugitive than some of those provided by nature. Its popularity – at first immense – gradually waned though it was still appropriately available for the period of national mourning following the death of Prince Albert in 1861. By then the British dyestuffs industry was in business.

[44] O. Meth-Cohn and M. Smith, 'What did W.H. Perkin actually make when he oxidised aniline to make mauveine?', *J. Chem. Soc., Perkin Trans. 1*, 1994, 5–7; O. Meth-Cohn and A.S. Travis, 'The mauveine mystery', *Chem. Brit.*, 1995, **31**, 547–549.

[45] A.S. Travis, *The rainbow makers: the origins of the synthetic dyestuffs industry in Western Europe*, Lehigh University Press, Bethlehem, 1993, p. 62.

*Perkin's sketch of his first factory at Greenford*

## (c)  The manufacture of synthetic dyestuffs

Much early work was carried out in the 2 or 3 years after Perkin's entry into the dyestuff business on other products from oxidations of aniline. Hofmann even found a red product ('Aniline Red') using carbon tetrachloride, as unlikely an oxidising agent as might be imagined.[46] In fact all were essentially oxidation products of aniline/toluidine mixtures, and they were the first of the triphenyl-methane dyes. First came rosaniline, pararosaniline or Basic Red 9, patented in France by Verguin in 1859 using tin(IV) chloride. Manufacture was commenced by CIBA at Basle. Soon after this a mixture of pararosaniline and its methyl homologues proved to be a dyestuff with enormous popular appeal, known variously as Rosaniline, Fuchsine and Magenta (the last being patriotically named after the town in Italy where Napoleon III defeated the Austrians in 1859). Magenta was first made by Renard Frères in France, and by 1865 was being manufactured in England by five companies. In 1861 a further patent was obtained for the production of Aniline Blue (or Bleu de Lyon), made by heating aniline red and aniline (1:1) at 165 °C.

Pararosaniline

Magenta

Aniline Blue

[46] A.S. Travis, 'Science's powerful companion: A.W. Hofmann's investigation of aniline red and its derivatives', *Brit. J. Hist. Sci.*, 1992, **25**, 27–44.

By 1863–1864 not more than five artificial dyes were available, namely Mauve, Aniline Blue, Magenta, Imperial Violet and Phosphine. Modest weights were produced at the beginning. Production often started in household equipment, as with Dan Dawson, who dried Magenta in a domestic oven *ca.* 1860 (specks of Magenta appearing on bread for weeks afterwards). Later many items of industrial equipment were developed that owed little to domestic or even laboratory practice, though may have been suggested in some cases by other technologies (as extraction of natural dyes). Few of them were protected by patent law. They included pressure vessels and filters, mechanical stirrers and the use of steam-jacketed apparatus.[47]

It is necessary to glance briefly at a few of the key businesses that constituted this important part of the British chemical industry. First of all there was of course the firm that became known as Perkin and Sons. Established at Greenford in 1857 as manufacturers of Mauve, the firm went on to major on another product of its founder's research, production of the 'natural' dyestuff alizarin by synthesis from anthracene (also in coal-tar). This process was patented in 1869 and became the kernel of the firm's business, Perkin being until 1872 the largest European manufacturer of synthetic alizarin.

*Production of alizarin*

Two important manufacturers of synthetic dyes began four years before Perkin, in 1853. One was the Manchester firm of Roberts, Dale & Co. At first they made general chemicals for dyers *etc.* and oxalic acid from sawdust. Encouraged by success of this process they diversified production and turned their attention to the new aniline dyes. In this they were aided by a number of young chemists who came across from Germany. One of them, Heinrich Caro, discovered an alternative route to something approximating to Perkin's Mauve, using copper salts in place of dichromates as oxidising agent. His

[47] W.J. Hornix, 'From process to plant: innovation in the early artificial dye industry', *Brit. J. Hist. Sci.*, 1992, **25**, 655–690.

process was patented in 1860. Further success came with Bismarck Brown, discovered in 1863 by another German chemist in their employ, C.A. Martius. Easily formed by diazotising 3-phenylenediamine, it was launched in 1865, and became the first commercially important azo-dye. At an early stage they began making picric acid as a colourant but in 1887 an explosion in a vessel of the material devastated the surrounding area and brought operations at that factory to an untimely end. Incredibly, only two lives were lost. At that time picric acid was not officially recognised as an explosive material! The other firm to be founded in 1853 was Simpson, Maule and Nicholson, a London partnership between George Simpson, a Kensington dealer in chemical apparatus, and George Maule who had been until recently an apprentice to the same druggist in Lancaster as his friend Edward Frankland. They were soon joined by Edward Nicholson, who (like the others) had studied under Hofmann at the Royal College of Chemistry.[48] They opened at Locksfields, Walworth, making ether, pyrogallol, collodion *etc.* and selling scientific apparatus. They soon became suppliers to Perkin of nitrobenzene and (later) aniline, for the Greenford works were now unable to produce enough of their own.

In 1859 they began making the new dye Magenta (Rosaniline hydrochloride) from aniline (containing some toluidine) and arsenate (patented independently by Medlock and Nicholson). There followed expensive litigation with the Yorkshire firm of Read Holliday, who also used the arsenate method. The court eventually concluded that the Medlock patent was null and void, largely on account of the ambiguity residing in the word 'anhydrous' employed in the patent. The firm's product was marketed as Roseine (the name previously given by their chief chemist D.S. Price to a short-lived product from oxidation of aniline by lead dioxide). Simpson, Maule and Nicholson used a French patent of 1861 to make Bleu de Lyon, by heating together Magenta and aniline (1:1) at 165 °C. To gain solubility they sulphonated it, to give Nicholson's Blue (the monosulphonic acid) and Soluble Blue (a trisulphonic acid).

The firm also sold Hofmann's Violets (arising from his discovery of alkylated rosanilines). They became a serious rival to Mauve. Other products included Regina Purple (from heating aniline acetate at 200 °C) and Phosphine, a chrysaniline, the first basic orange dye, a by-product from manufacture of Magenta.

Phosphine

[48] D.H. Leaback, 'Chemical enterprise from "The Elephant"', *Chem. Brit.*, 1992, **28**, 340–343.

It is said that here Nicholson introduced mechanical stirring into industrial chemistry, nitrating benzene in iron cylinders whose contents were agitated by steam-powered stirrers. A similar arrangement for the reduction of nitrobenzene to aniline elicited the following comment that stressed the indissoluble connection between environmental and economic gain:

> The condensation of the products, which at one time escaped and poisoned the atmosphere of the factory, now materially increases the yield of aniline; and, in fact, by this manner of working there is obtained a result which is generally observed on chemistry, *viz.*, each step made in the direction of healthfulness is an advance in economy.[49]

In 1865 Simpson, Maule and Nicholson built the Atlas Works at Hackney Wick. By that time it had the largest output of synthetic dyestuffs in Britain. George Simpson retired in 1866 and two years later Nicholson and Maule took early retirement, the former to life as a consultant, the latter to a leisured preoccupation with field-sports. They could well afford to do so. At this point the firm was absorbed into a new organisation, Brooke, Simpson and Spiller, when Edward Brooke, a manufacturing chemist from Manchester, together with his former chemist William Spiller, combined forces with Richard Simpson, brother of George.

The Yorkshire establishment of Read Holliday at Huddersfield entered the dyestuffs field in the 1860s. It was founded in 1830 to reclaim ammonia from gas-works, now in some demand as a substitute for the stale urine used since the Middle Ages to scour cloth (and last known to be so employed in the 1930s!). Once again a manufacturer sought to gain independence from other suppliers and make his own raw materials. Accordingly Holliday set up coal-tar distilleries in Yorkshire, Lancashire and Bromley-by-Bow. His firm became the largest producer in the north of England, if not the whole country. This operation continued until the early 1880s. Meanwhile Holliday further diversified into making the dyes themselves, beginning to make Magenta in 1860. It was this enterprise that initiated the litigation with Simpson, Maule and Nicholson. During all this activity a relatively small number of simple organic compounds was bring produced in enormous quantities from coal-tar: benzene, toluene, naphthalene, anthracene, phenol and their simple substitution products. Since they were nearly all destined for the production of dyes, the fortunes of the organic chemical industry of Britain were inevitably linked to those of the dyestuff manufacturers. At first it looked as though the prophecy of Hofmann might be fulfilled, and England would become

> at no distant date . . . the greatest colour producing country on the world; nay, by the strangest of revolutions, she may, ere long, send her coal-derived blues to indigo-growing India, her tar-distilled crimson to cochineal-producing Mexico, and her fossil substitutes for quercitron and safflower to China and Japan

[49] A.W. Hofmann, G. de Laire and C. Girard, 'Report on the colouring matters derived from coal tar shown at the French Exhibition, 1867', in M. Reimann, *On aniline and its derivatives: a treatise on the manufacture of aniline and aniline colours*, trans., rev. and ed., W. Crookes, Wiley, New York, 1868, Appendix, p. 111; cited in Travis (ref. 45), p. 93.

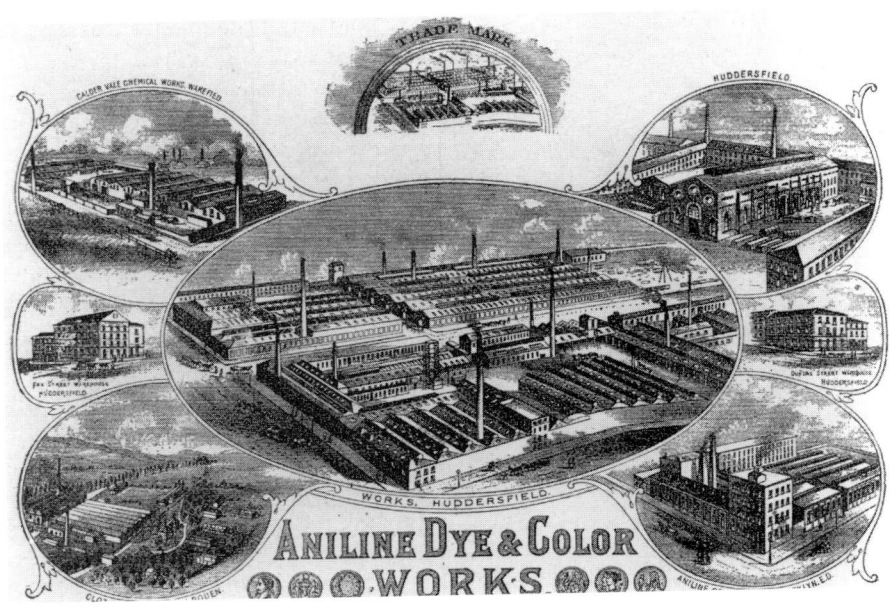

*Advertisement of 1887*

In fact the British supremacy in dyestuffs manufacture lasted a mere decade and by the late 1860s the writing was on the wall. It is sometimes hard to tell which were the causes and which were the effects, but by 1870 a great many significant indicators pointed in the same direction. The departure by 1868 of all the principals from Simpson, Maule and Nicholson was followed by a slow decline in prosperity of its successor and inventiveness seems to have peaked about then. One reason for these early retirements was the sheer aggravation of protracted litigation over patent law. When the cases of Simpson, Maule and Nicholson *vs.* Read Holliday and of Simpson, Maule and Nicholson and Renard Frères *vs.* Levinstein were finally settled in the courts in 1865 at least it was clear what could, and what could not, be claimed as patent infringement. Also emerging with startling clarity was the expense of such litigation and the difference in patent laws of different European countries. It might be truthfully said that the battle lines were now drawn up. Nor were matters rendered easier for the British by the collapse in 1868 of La Fuschine, successor to Renard Frères. The latter arose partly for reasons of bad financial management and partly on account of appalling pollution of the River Saône flowing through Lyon and receptacle of the firm's arsenic waste (with fatal consequences).[50] This opened the door to increased German imports, sometimes under false

[50] H. van den Belt, 'Why monopoly failed: the rise and fall of Société La Fuchsine', *Brit. J. Hist. Sci.*, 1992, **25**, 45–63.

names so as to avoid any little awkwardness at Customs. And the 1860s saw movement in the opposite direction as industrial chemists like Caro and Martius returned to their native land, depriving England of their services and yet bringing with them much invaluable information about the British dyestuff industry. Still more portentous was the eventual departure of Hofmann from the Royal College of Chemistry to complete his long career in Germany. It heralded the meteoric growth of organic research at German universities and industrial laboratories[51] and the rise of a new theoretical organic chemistry in that land. And all the while British organic chemistry was being systematically crippled by the punitive duty on alcohol which, according to Herbert Levinstein, eliminated 300–400 products that were made in Germany from even being considered in Britain.[52]

For the rest of the century the history of the British organic chemicals industry (which was effectively the dyestuff manufacturers) was not exactly one of unrelieved gloom but it did not augur well for the future. Brooke, Simpson and Spiller went on to acquire the Greenford Green works in 1874. They did not manage it well; the aniline dyes business was rapidly discontinued and the alizarin prices raised. Naphtha residues were for the first time allowed to accumulate, then sent in leaky casks by barge to the Phoenix Works at Hackney Wick (owned by W.C. Barnes, a partner in the 1874 take-over of Greenford). At the Atlas Works there was continued output but no new products. Sharp decline began in the mid-1880s. Several good chemists were employed there, but they were not encouraged by the firm and their discoveries were generally not patented. These included Meldola's Blue, the first oxazine blue (from base-catalysed condensation between 4-nitrosodimethylaniline and 2-naphthol) and A.G. Green's dehydrothiotoluidine (from 4-toluidine and sulphur) and its co-product Primuline. The latter could be diazotised and coupled *in situ* on cloth, giving rise to the immensely important Ingrain Process.

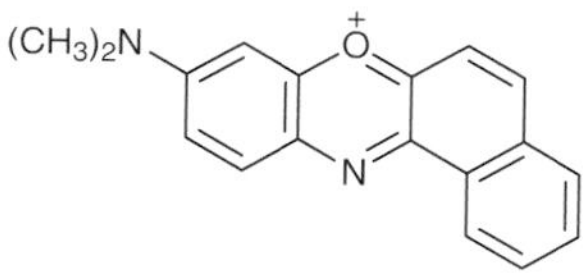

Meldola's Blue

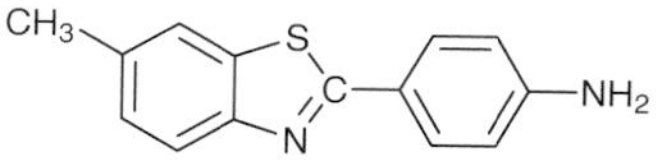

Dehydrothiotoluidine

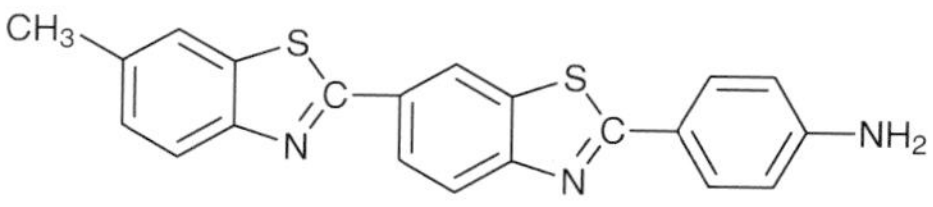

Primuline base

[51] E. Homburg, 'The emergence of research laboratories in the dyestuffs industry, 1870–1900', *Brit. J. Hist. Sci.*, 1992, **25**, 91–111.
[52] M. Fox, *Dye-makers of Great Britain, 1856–1976*, ICI, London, 1987, p. 43.

As a side-line to their sleeper-creosoting business the firm of Burt, Boulton and Haywood sold off naphtha and benzene. This was so profitable that they expanded their tar-distillation activities in general, and in so doing began to accumulate considerable quantities of anthracene. Presumably this encouraged them to buy out the run down Greenford establishment of Brooke, Simpson and Spiller in 1876 and shortly afterwards to extend their production of alizarin at a new works which they established next to their distillery at Silvertown. In 1882 they became the sole British member of the new Alizarin Convention. By 1914, now as the British Alizarine Co. Ltd., they were supplying 80% of British consumption of the dyestuff.

Another firm that was later to become involved in important amalgamations was Levinstein Ltd. It was founded by Ivan Levinstein, a chemically-trained German who set up a dyestuffs factory at Blackley, Manchester, in 1864. There he made Magenta and Aniline Blue. In 1881 the firm was sued by BASF for infringement of its 1878 patent concerning azo-dyes from naphthylamine. Levinstein lost and transferred his process to Holland where national patent law did not threaten manufacturers in the way that it did in England. Levinstein employed foreign chemists, often for a short time, and actually provided a cultural centre for immigrants. The firm survived with some difficulty until its fortunes improved with the First World War which brought the opportunity to supply dyes for military uniforms. Its rivals Read Holliday had manufactured picric acid from 1899 (*via* 2,4-dinitrochlorobenzene) and they became the chief constituent of British Dyes Ltd. in 1915.[53]

The Yorkshire firm of Hickson & Welch sprang from an earlier enterprise of Ernest Hickson who had previously worked for Brooke, Simpson and Spiller. In about 1894 he started making sulphur blacks at Shipley, Bradford, and soon afterwards began to make lactic acid required by the textile industry for mordanting purposes. His sulphide black shed was twice destroyed by fire as open-fired vessels were in use. The firm of Clayton Aniline was founded in 1876 by Charles Dreyfus at Clayton, Manchester, initially to make aniline. It diversified into such a wide range of coal-tar products and their nitro- and amino-derivatives that its output has been described as 'a motley range of products'. The firm was joined by A.G. Green from Brooke, Simpson and Spiller as manager of the colour department from 1894 until his appointment as professor of colour chemistry at Leeds in 1901. A takeover by CIBA of Basle in 1911 gave that company its first plant in the UK.[54]

An unlikely producer of dyestuff intermediates was the firm of Alexander Morton & Co., weavers, in Carlisle. James, the son of the founder, had been demoralised by the fugitive character of many commercially available dyes and determined to enquire into the causes (*i.e.* the chemistry). With the help of Rudolf Hübner, the head dyer, Morton had discovered how to make 2-aminoanthraquinone, needed for the preparation of the self colours Indanthrene Blue R and Yellow G (by treating it with, respectively, caustic potash

[53] The firm L.B. Holliday was subsequently formed as an independent entity.

[54] E.N. Abrahart, *The Clayton Aniline Company Limited, 1876–1976*, Clayton Aniline, Manchester, 1976.

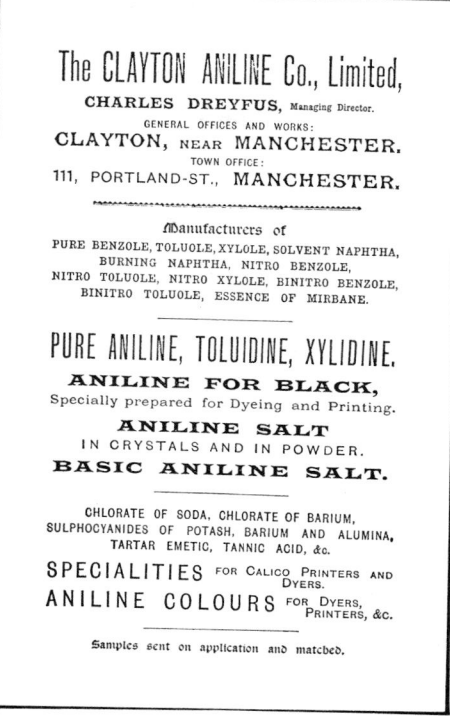

*The Clayton Aniline Company near Manchester, founded in 1876 by Charles Dreyfus*

fusion and antimony pentachloride in nitrobenzene). The prized intermediate was obtained by heating anthraquinone-2-sulphonic acid with ammonia at 180–200 °C and about 50 atmospheres pressure. Here also they invented the important wool dye Solway Blue, and a separate company was established to make it, the Solway Dyes Co.

The supremacy of German dye-production was never more vividly illustrated than by the establishment before the First World War of several outposts of the German chemical industry in Britain itself. This followed the Patent Act of 1907 which required foreign patentees to work British patents in the UK. Two German consortia established bases on Merseyside, in the very heartland of the British chemical industry. One was BASF (founded in Mannheim in 1865 as the Badische Anilin- und Soda-Fabrik), Agfa and Bayer. It had been the destination of Caro on his departure from Manchester in 1868 and now in 1907 acquired 24 acres at Bromborough in order to make azo-dyes and intermediates. Another was Meister, Lucius and Brunning who in 1908 at Ellesmere Port began to produce the synthetic indigo that was at last to replace the natural dye,[55] even

---

[55] M. Byrne, 'Indigo-dyeing: past and present', *J. Consumer Stud. Home Economics*, 1981, **5**, 219–227.

though it used phenylglycine imported from Germany.[56] The dye had been manufactured in Germany since 1897, its first successful synthesis having been effected by Baeyer in 1880, after research costing nearly a million pounds.[57] The phenylglycine process, due to Heumann (1890), improved on the original Baeyer method which effected ring closure on the very expensive 2-disubstituted benzenes (as 2-nitrophenylacetic acid).

*The phenylglycine process*

Owing to a subsequent legal interpretation of the Patent Act (the disastrous 'Parker ruling') manufacturers were relieved from the necessity to *make* all the material needed by the British market (though they could import all they wanted), the Act thereby losing much of its force. So the two Merseyside establishments produced very little. The Ellesmere Port factory was taken over by Levinsteins at the outbreak of war.

## (d)    Other products from coal

It is easy to imagine that all coal-tar products were directly or indirectly converted into dyes. Their role as dyestuff intermediates can hardly be exaggerated, yet the fact remains that other realms of human activity have gained from the new chemistry that has emerged. Most significant in the years up to 1914 were the drugs and artificial flavourings that began to be synthesised. This was specially the case in Germany, but not quite everything was imported. To take a single example the firm Bush Boake & Co., founded 1851 to make flavours, in 1886 absorbed Potter & Moore, a company dedicated to the distillation of essential oils. They made salicylic acid from phenol by the Kolbe process (using $CO_2$) but their synthesis of vanillin (1910) began with eugenol obtainable not from coal but cloves.

---

[56] P. Reed, 'The British chemical industry and the indigo trade', *Brit. J. Hist. Sci.*, 1992, **25**, 113–125.

[57] For this and later German developments associated with BASF see H. Schmidt, 'Indigo – the king of dyes turns 100', *Melliand Textilber.*, 1997, **76** (6), E88–89; *idem*, 'Indigo, 100 years of industrial synthesis', *Chemie in unserer Zeit*, 1997, **31** (3), 121–128.

A further use for chemicals derived from coal-tar was in the field of explosives, though before 1914 they were less important than those derived from glycerol or cellulose (pp. 242–245). Picric acid (2,4,6-trinitrophenol) had been known since 1771, and was manufactured as a yellow dyestuff in Manchester from 1849, both by John Dale and by F. Crace-Calvert. Its explosive properties were first employed by the French army in 1885, when, as Mélinite, it found use as a high explosive shell-filling. Great Britain followed suit in 1888, and after trials at Lydd in Kent, used it as Lyddite in the South African War. Just before 1914 it was being produced at 35 tons per month by the nitration of phenol. It was manufactured by Read Holliday at Huddersfield who, distressed by the rising price of coal-tar phenol, devised the first route to the synthetic material:

$$C_6H_6 \xrightarrow{H_2SO_4} C_6H_5SO_2OH \xrightarrow{\text{alkali fusion}} C_6H_5OH$$

Trinitrotoluene was soon to displace picric acid, though before 1914 only small amounts were used, and most was imported by the government.

## (e)  Back to the environment

We began this survey of organic chemicals from coal by reference to the enormous environmental problems created by accumulation of vast amounts of crude coal-tar. We conclude by observing that, with the utilisation of that tar for chemical purposes, one environmental problem has been replaced by another. Yet, as will become obvious, the chemical industry was far from being unaware of the difficulties and, on many occasions, took drastic steps to deal with them.

In the first place there was the question of working conditions, and it has to be admitted that as in other branches of the chemical industry these were far from ideal and at times appalling. First, there were *dangers from toxic substances*. In Levinstein's works at Blackley several fatalities arose from inhaling hydrogen sulphide; in this case genuine attempts were made to remove the gas by 'iron sludge' or, more effectively, by caustic soda solution. An apparent case of intoxication after a night-shift at Clayton Aniline Company was attributed to the same cause; elimination of leaks from the plant improved both the product and the operatives' health. For six years men were exposed to considerable risk of lead poisoning at Read Holliday through the use of Seidler's route to sodium nitrite:

$$NaNO_3 + Pb = NaNO_2 + PbO$$

Men working for Perkin and Son were liable to mercury poisoning from the use of mercuric nitrate in the Magenta process. Precisely for that reason the humanitarian decision was taken to abandon that process. In some cases a management intention to eliminate danger to personnel was frustrated by external advice. Levinstein objected to dangers imposed upon his workforce by fuming sulphuric acid, though he was contradicted by Professor Roscoe whose experiments were conducted on very small quantities, on a flat roof and with a strong wind blowing offending vapours away! In general workers were often in close contact with chemicals, whether or not known at the time to be harmful. The case of 2-naphthylamine is perhaps the most notorious, though such carcinogens were banned only since 1952 or even 1960. In quite recent times Maurice Fox could recall the sight of faces so coloured by dyes at L.B. Holliday & Co. that it was possible to identify their working unit at a glance.[58]

A second hazard resided in the ever present possibility of *fire or explosion*. This was exaggerated by the use of open pans (*e.g.* those used for converting naphthalenesulphonic acids into naphthols), and open-fired vessels, as was the case when a shed was twice destroyed by fire at Hickson & Welch. Explosions could arise from a multiplicity of causes. One at Read Holliday's in 1900 seemed to have been caused by ignition of picric acid from a spark from a chisel. More dangerous was the use of high pressure vessels as autoclaves. An earlier explosion at Read Holliday in 1871 led to decapitation of one worker standing near. The hazards were increased still further by the kind of foolish practice recorded at Scottish Dyes Ltd. in 1915. Production of 2-aminoanthraquinone required a pressure of 50 atmospheres; at one stage this was accomplished in 1915 in an improvised autoclave with the high pressure maintained by the head dyer sitting on the safety valve!

Thirdly the sheer *crudity and chaos of the working environment* provided its own share of hazards. While making sulphide blacks at Read Holliday workmen had to stand for 30 hours, agitating by hand. The fatigue and consequent carelessness can be imagined. At Blackley a workman was scratched on the leg by a sack containing 2,4-dinitrochlorobenzene, and this led to blood poisoning and death. The roads between buildings were unpaved and unsurfaced, a quagmire in winter from clogs, hooves, wheels or the rolling of casks. Washing/cleaning facilities were rare, though at Levinsteins washing was made compulsory for workers dealing with naphthylamines (more probably on account of their smell than of any suspected carcinogenic effect). Protective clothing was rarely provided, and even then was frequently resisted by piece-workers as inhibiting their social movement. Food was regularly consumed on site.

The wider effects of pollution have already been touched upon. It is possible to construct a general scenario of the cases for prosecution and defence.

*Evidence of damaging pollution:*
1. Undeniable damage to vegetation or property: this mainly consisted in statements about adjacent crops which withered away or else gave low yields of produce. The cases were not as clear-cut as, for instance, those concerned with alkali works
2. Undeniable damage to health: here medical evidence was sometimes given as to the illness caused by the effluents.

[58] M. Fox (ref. 52), p. 147.

*New washing facilities for staff at Clayton Aniline*

3. Unacceptable smells: a bad case was that of Levinstein's 1-naphthylamine, called 'monkey' from its smell and produced at a rate of 100 tons per month. There was a stream of actions from Manchester Corporation over a 20-year period. Read Holliday was similarly threatened with closure of his Turnbridge works because of 'noxious gases'.
4. Explosions: these could hardly be denied, especially when one occurred at the Turnbridge works during a visit by the Inspectors!
5. Removal of proprietors and family away from the works: many proprietors followed the example of Read Holliday who before one indictment had already moved his family and home to the country. This did not impress the magistrates.

*General defence:*
1. Other pollution was much worse: Read Holliday pleaded that more significant pollution came from dye-houses, slaughterhouses, offal-boiling establishments, gas-works and (especially) lack of drainage from houses. He might well have been right.
2. Chemical fumes were not noxious: a brazen defence plea on behalf of Read Holliday alleged that naphtha and creosote oils were used in the treatment of, respectively, consumption and bronchitis, and so cannot be harmful! This plea was entered by John Leigh (having by then made the curious career change from gas-works director to Medical Officer for Manchester Corporation[59]). The Huddersfield magistrates were unimpressed and on this occasion imposed a fine of £1 and costs. Even less convincing was evidence given on behalf of Levinstein by one 16-stone butcher who attested that 1-naphthylamine actually increased his appetite.

[59] R.H. Kargon, *Science in Victorian Manchester: enterprise and expertise*, Johns Hopkins University Press, Baltimore, 1977.

*Pollution over Widnes: once seen as a famous indicator of chemical pollution in fact it reveals that much of the smoke comes from factory chimneys and seems to be chiefly due to burning of coal*

3. Chemical operations necessary to meet social demands: the consumer benefits of industrial chemistry were frequently paraded, though rarely did this kind of evidence stand alone.
4. Chemical operations necessary to local economy: the frequent cry was heard that to close the works would throw hundreds of men out of work.
5. Special case: in some instances it was genuinely possible to point out that others were not persecuted for similar offences, and that was common when (as at Huddersfield) the local authority was specially diligent.
6. Efforts had been made to limit effluent: many cases have been encountered already and they formed a good basis of defence. John Robinson & Co. (founded in Huddersfield in 1853) patented a multi-tiered furnace as a 'smoke preventer' to reduce atmospheric pollution and economise on fuel (1855).

Thus well before the end of the Victorian period questions of pollution had become issues of major importance for the organic chemical industry. The pre-eminence of coal was still unchallenged as the 20th century dawned. The industry might have continued its gradual decline and, as coal stocks eventually dwindled, ground to an eventual halt. In fact it needed a major cataclysm to alter its direction and change its fortunes. That was provided in 1914, and still more in 1939, by the outbreak of war.

Chapter 9    *The Age of Polymers and Petrochemicals (Industrial Organic Chemistry from 1914)*

C.A. RUSSELL

## 1   Organic Chemistry and the First World War

It was not merely production of poison gases as chlorine that led to the ascription of the First World War as 'the chemists' war'. The organic chemicals industry had a crucial part to play. Yet the contrast between Britain and Germany in 1914 was vast, with the former's organic chemical industry limited to coal-tar based dyestuffs and a few simple aliphatics from natural sources. They were made in response to urgent contemporary demands.

### (a)   Explosives

Explosives were clearly essential in war, and the chemical industry immediately became heavily involved in their production.[1] The manufacture of nitro-glycerine and nitrocellulose had been conducted in Britain for over 40 years when war was declared. The former was widely used as a sensitiser for high explosives, the latter (as gun-cotton, cordite *etc.*) as a shell propellant. Alternatives derived from coal-tar were picric acid and TNT; their importance and their relation to the nitrogen industry has already been discussed (Chapter 5), but they also constituted a problem for the organic chemists since the raw materials from which they were made were simply not available in sufficient quantities in coal-tar.

With the declaration of war immediate steps were taken to increase the supply of the traditional aromatic-based explosives, especially picric acid. L.B. Holliday was recalled from military service to supervise the erection of new factories. The standard method of manufacture was the sulphonation of phenol, followed by nitration of the resultant sulphonic acid. Such was the shortage of phenol (only about 1% of coal-tar) that alternative modes of production had to be devised, one using the sulphonation of benzene, followed by hydrolysis; another depended upon alkaline hydrolysis of 2,4-

---

[1] S. Miall, *A history of the British chemical industry*, Benn, London, 1931, pp. 39–60 (based on lecture by W. McNab, *J. Soc. Chem. Ind.*, 1922, p. 353).

dinitrochlorobenzene. At Brookes Chemicals, near Halifax, a continuous method was devised for the nitration of phenolsulphonic acid.

*Picric acid from the hydrolysis of 2,4-dinitrochlorobenzene*

Picric acid has a corrosive effect on metals and other disadvantages. From 1904 it tended to be replaced by TNT (2,4,6-trinitrotoluene) though ten years later it was still the commonest explosive used in shells. TNT was first used in Germany in 1890.[2] In 1914 the weekly output of TNT in Britain was minuscule, though by the end of the war 238,000 tons had been manufactured.[3] The first government factory for TNT was by Chance and Hunt at Oldbury (1915). Such was the shortage of toluene derived from coal-tar that another plant was erected on an adjacent site by the Asiatic Petroleum Co. to conduct nitrations of toluene derived from their Borneo crude (which was unusual for a petroleum in containing substantial amounts of aromatics). A much larger TNT factory was later erected at Queensferry, at Chester, producing up to 100 tons a day. Even this was not enough, and by 1915 Lord Moulton, president of a new government Committee for Explosives Services, had persuaded reluctant service chiefs to 'dilute' the TNT with a quarter of its weight of ammonium nitrate, so giving a new explosive mixture 'amatol'.

The cellulose-based explosives were often derived from cotton waste, and that presented no serious problems of supply. The same was not true of some of the other chemicals needed, most notably acetone. For cordite, a mixture of cellulose nitrate and acetate with 5% mineral jelly, acetone was needed as a solvent for the mixing process and then the extrusion into short rods. Although some acetone was recovered and recycled it was at one stage in critically short supply and of dubious quality. Thus the quality of acetone in loading charges was blamed by an Admiralty Board of Inquiry for the fact that naval shells dropped into the sea far short of their supposed range of 5000 yards. The chemical had traditionally been obtained in small quantities from wood, and mostly imported from the USA. Now, plants were erected to make acetone from acetic acid, passed over heated barium carbonate or lime. This was supplemented from 1916 by the new Weizmann fermentation process, for which the Admiralty requisitioned six whisky distilleries.

By the end of the war explosives manufacture had vastly increased, most of it in private as opposed to government installations. Many of the smaller explosives firms and gunpowder factories came together with Nobel's Explo-

[2] W. Taylor, *Modern explosives*, RIC Lectures, Monographs and Reports, 1959, no. 5, p. 22.
[3] D.W.F. Hardie and J.D. Pratt, *A history of the modern British chemical industry*, Pergamon, Oxford, 1966, p. 100.

sives Co. to form, in 1918, Nobel Industries Ltd. Environmentally the consequences of explosive manufacture were all too obvious: enormous loss of life, devastation of buildings and even towns, and despoliation of the country-side (especially in France) with wholesale destruction of trees and blight of agricultural land with unexploded shells *etc.* Less noticeable, but of perhaps longer term importance, was the liberation into the atmosphere of millions of tons of $CO_2$ and nitrogen oxides.

A further environmental hazard associated with this development of the British chemical industry was that of accidental explosion. On 19 January 1917 a munitions factory at Silvertown exploded, causing a shock wave of such proportions that a large gasometer across the Thames at Greenwich caught fire, the conflagration consuming 8 million cubic feet of gas. It was the biggest disaster ever to strike the British gas industry. Some years before, in 1914, Hickson had established a factory at Castleford with the primary purpose of producing TNT for the government. In 1930 an enormous explosion and fire caused 13 fatalities and left several hundred people homeless.

The actual process of manufacture also presented considerable environmental problems. The discarded 'nitre cake' (largely $Na_2SO_4$), from the production of nitric acid from sodium nitrate, was discharged into the Solway Firth at Gretna and into the Irish Sea at Queensferry. The original method of making picric acid was accompanied by copious evolution of brown nitrogen dioxide fumes, 'causing an almost intolerable nuisance': the operations were therefore conducted 'under cloak of darkness'. However, as production greatly increased the industry responded to both the environmental and economic disadvantages of such waste by using absorption towers. And prolonged exposure of workers to the adverse effects of TNT vapour led to considerable, if selective, distress.[4]

Finally one cannot ignore the immense amount of sea-pollution as explosives were successfully employed in sinking such ships as those carrying supplies of Chilean nitrate, vast amounts of which went to the bottom of the Atlantic Ocean. This was desperately needed for the production of yet more explosives, though it was chiefly the German rather than the British chemical industry that caused such naval losses, particularly through the activities of the German U-boats.

## (b) Aircraft construction

In 1914 the army had altogether less than two dozen aircraft (perhaps 0.5% of the world's aeroplanes). Their potential for military purposes was slowly realised, first for surveillance, then for aerial attack on ground troops (by machine-gunning or bombing) and later in the destruction of enemy Zeppelins over Britain. Within a year aircraft could not be assembled quickly enough for the Royal Flying Corps and the Royal Naval Air Service. They were largely constructed from wooden frames covered in fabric. To tauten the fabric, and also to make it water-tight, it was treated with 'dope'. At first the latter was a

[4] Miall (ref. 1), p. 49.

*The aircraft used in 1919 by Alcock and Brown to cross the Atlantic, with wooden wings covered with fabric water-proofed by cellulose acetate 'dope'*

solution of cellulose nitrate but its high inflammability was an obvious disadvantage. From 1916, a solution of cellulose acetate in acetone began to replace it (though dense industrial nitrocelluloses were still being employed to some extent as an 'aircraft dope' after the Second World War). As aerial combat developed the pilots of biplanes were at a great disadvantage in that the upper wing shielded from view any approaching enemy aircraft until it was too late to take avoiding action. Accordingly sheets of cellulose acetate were inserted in those upper wings, the transparent panels thus enabling pilots to observe hostile manoeuvres in the sky above them. Later the same material was employed for wind-screens and gun-turrets, being replaced by Perspex only just before the Second World War.

For both cellulose acetate and its solvent acetone there came ever-increasing demands. Only a small works of the Safety Celluloid Company near Ealing produced British cellulose acetate before the First World War. Their contributions were augmented by those from the Dreyfus brothers who formed the British Cellulose and Chemical Manufacturing Company at Spondon in Derbyshire. After the war the firm became the British Celanese Company, introducing acetate fibres in 1925.

The acetic acid used in the acetylation of cellulose was manufactured at another commandeered works, Crosfield's at Warrington, by a liquid phase oxidation of acetaldehyde. The latter was obtained either by hydration of acetylene or by oxidation of ethanol. And of course the provision of acetone continued to depend on acetic acid also.

$$C_2H_5OH \xrightarrow{-2[H]} CH_3CHO \xrightarrow{+[O]} CH_3COOH$$

$$\uparrow {\scriptstyle +H_2O}$$

$$C_2H_2 \longrightarrow$$

## (c)  Dyestuffs

As soon as the war was declared it became obvious that the textile industry was about to face a dye-famine as the staple raw materials for dyestuffs (organic and inorganic) were being diverted into munitions production. At government behest a step was taken towards some kind of national dye industry, with British Dyes Ltd. launched in 1915, though its sole acquisition was the Huddersfield firm of Read Holliday & Co. In addition to making various grades of khaki dye for the army, Hollidays were soon diverted into high explosives production. Its new owner soon began to construct a vast factory near the River Colne in Huddersfield, not far from the existing works. The extent of environmental loss to the valley may be gauged from the statistics available: 450 acres, 40 million bricks, 250,000 tons of cement, 3000 tons of cast iron and a work-force at one stage approaching 2000. In fact the building process was so slow that by the end of the war small-scale production had only just started. It is interesting to note that during the First World War Read Holliday was less successful than Levinstein, its output being based on benzene and toluene. Levinstein on the other hand was also dependent on naphthalene (which is more abundant), and in addition had a better research team. Both firms set up their own desperately needed plants for producing oleum. In 1919 Levinstein merged with British Dyes to form the British Dyestuffs Corporation, later to be subsumed in ICI.

## (d)  Rubber tyres

Rubber tyres were in demand for both military vehicles and also for aircraft. By the end of the war the demand for tyres had almost doubled since 1900. After 1914 the UK Government tried to restrict exports of Malayan rubber, but Germany and the USA replied by reclamation of rubber, one of the first large-scale cases of industrial recycling. On both sides of the English Channel there was talk of providing some synthetic alternative. Little was known of the theory of polymers in general, rival opinions being the micelle theory suggested in 1853 by Nägeli, and a chain theory proposed at the BAAS in 1910 by S.S. Pickles. In 1889 J.H. Gladstone and W. Hibbert had added bromine to rubber, showing a high degree of unsaturation, and later estimated its molecular weight to be about 6500.

In the UK, the hydrocarbon isoprene had been obtained from natural rubber in 1860 by Greville Williams. In 1882 William Tilden reported polymerisation of isoprene (from turpentine) to a yellowish rubbery solid.[5] The idea of a *synthetic* rubber was first proposed by W.R. Dunstan of the Imperial Institute in *ca.* 1910, an idea first taken up by Fritz Hofmann at Bayer, in Germany. A polyisoprene was obtained but research was halted in 1909 as progress seemed uneconomic.

---

[5] D.M. Bate, R.S. Lehrle, E.J. Place, S.L. Willis, D.S. Campbell and C.D. Hull, 'The first sample of synthetic rubber made by William Tilden in 1882 – modern work reveals a mystery', *Polymer*, 1997, **38** (21), 5261–5366.

*Sir William A. Tilden (1842–1926)*

For similar reasons a later collaboration with BASF also came to nothing, being abandoned in 1919.

By 1919 the British chemical industry had become transformed by the pressures of war. It has been suggested that there were three principal effects: a great increase in the scale of production of some common chemicals, modification of some existing processes and introduction of some new ones, and the entry of the State into chemical manufacture.[6] The increase in scale may be judged, for example, from the rise in annual output of TNT at Nobel's from 9000 to 50,000 tons between 1914 and 1918; the total production of British dyes was 4000 tons in 1913 but by 1925 it had risen to 14,592 tons.[7]

With the sudden cessation of demand for aircraft by the new RAF the cellulose acetate industry found itself with enormous spare capacity, and turned its attention to alternative outlets, most notably the production of artificial silk. High explosives continued to be made for civil as well as military use, but on a much restricted scale. The expertise in nitration (such as it was) was turned to good use in the more peaceful area of dyestuffs chemistry. With the return of those chemical workers who had survived warfare in the trenches came a welcome accession of old skills, never more necessary than in the manufacture of organic chemicals. It may not be possible to agree with the second part of the following quotation, but there is much truth in the first:

> The only thing either Britain or America got out of the World War [I] was a synthetic organic chemical industry; but it was worth all it cost.[8]

[6] Hardie and Platt (ref. 3), p. 98.
[7] W. Haynes, *This chemical age: the miracle of man-made materials*, Secker & Warburg, London, 1945, p. 62.
[8] *Ibid.*, p. 60.

Some of the consequences of these tumultuous changes will now be considered for the inter-war period. As a preliminary general point it is worth noting that industrial research had now acquired a new urgency. The government's Department of Scientific and Industrial Research (DSIR) was set up during the war, though its contributions to the chemical industry (organic or otherwise) were for long hampered by the industry's wish to conduct its own research. Similarly new links forged between that industry and the academic world were weakened by the differences in time-scale and probably ideology as well. A classic case was the disappointing collaboration between ICI and University College, London, to produce a new detergent for use in hard water.[9] Despite the presence of leading chemists as N.K. Adam and F.G. Donnan the project was inconclusive and was terminated in 1936.

## 2   Synthetic Polymers

The inter-war years were marked by a great efflorescence of an industry concerned with synthetic polymers, some of which were new. By 1939 they had come to be of considerable importance in the manufacture of items for the domestic market. What was new was their variety and extent of use. Because they are nearly all excellent insulators they additionally found extensive use in electrical appliances, from the 1930s 'wireless set' to insulation for wiring in communications and (eventually) radar systems. The age of plastics had arrived.[10]

Much development in these areas of industry took place overseas, especially in the USA and Germany. But the British chemical industry became gradually more involved in monomer and polymer production as the 1930s decade drew to its climactic close.

### (a)   Phenol–formaldehyde resins

L.H. Baekeland was a Belgian by birth who, from 1889, had settled in New York as chemist to a photographic manufacturer. He set up the Nepera Chemical Company, manufacturing photographic papers capable of being handled in subdued light (still remembered by older photographers as 'Velox' papers). Baekeland claimed to have invented this immensely useful product, and in 1898 sold his firm to George Eastman for $1m. Free now to follow his whims for research he turned first to electrochemistry and then to the production of a substitute for shellac. By 1905 he was examining the products from the reaction between formaldehyde and phenol in a search for such a new material. His product instead proved useful as a resinous binder for asbestos and other inert

---

[9] G.K. Roberts, 'Dealing with issues at the academic-industrial interface in interwar Britain: University College London and Imperial Chemical Industries', *Science & Public Policy*, 1997, **24** (1), 29–35.

[10] See M. Kaufman, *The first century of plastics: celluloid and its sequel*, Plastics Institute, London, 1963; P.J.T. Morris, *Polymer Pioneers: a popular history of the science and technology of large molecules*, Center for the History of Chemistry, Philadelphia, USA, 1986.

powders. He called it Bakelite. His combination of chemical and engineering skills enabled him to use temperature and pressure changes to control the hardening process and thus produce a resin that could be moulded. He began to manufacture the new substance on a large scale from 1910, setting up the General Bakelite Co. in New Jersey. The firm was eventually taken over by Union Carbide (1939). Bakelite was the first major industrial synthetic plastic.

Meanwhile, in England, new electrical insulation materials were sought by James Swinburne, a former associate of Joseph Swan. Using a method patented by A. Luft (1902) he too developed a commercially successful phenol–formaldehyde plastic. Business was slow, so Swinburne switched to phenol–formaldehyde lacquers, in demand in Birmingham since the invention there of japanning in 1740. His London firm, the curiously named Fireproof Celluloid Syndicate (the products were neither fireproof nor based on celluloid), was thus replaced in 1910 by another of equally ingenious name: The Damard Lacquer Co. at Birmingham (their products being supposedly 'damn 'ard'!). The names reflect two essential fears: that resins, being organic, would be inflammable, and that they would be too soft (another company rejoiced in the title Rockhard Resins Ltd.).

In 1916 Swinburne acquired rights to exploit Baekeland's patents and his Damard company took over a factory that Baekeland had recently established at Cowley in Middlesex. Until the end of the First World War it concentrated exclusively on Government contracts for electrical equipment materials. After the war its operations, now dealing also with luxury and domestic items, were transferred to a new factory near Birmingham.

By the 1920s new demands arose for insulating materials from both the motor-car and electrical industries; the length of cables was often determined by the quality of insulators. Gramophone records (78 r.p.m.) constituted the first important outlet for phenol-formaldehyde resins; by 1925 such resins were established in paint formulations.[11]

In 1926–1927 two rival firms, Mouldensite of Darley Dale and Redmanol of London, fused with Damard to form Bakelite Ltd. (reflecting similar moves by their parent companies in the USA). In the UK cresols were also employed, like phenol being obtained from coal-tar. Subsequent take-overs and amalgamations brought British Bakelite production eventually under the control of BXL Plastics Ltd, a subsidiary of BP Chemicals Ltd.

By the 1930s bakelite was used chiefly as moulding materials for extruded tubes, telephones, radio cabinets, for oil varnishes, and (as laminates) for decorative panelling. In 1923 UK production was 500 tons, and in 1936 it had risen to 15,000 tons.

## (b)    Urea–formaldehyde resins

Owing to the dark colour of bakelite colourless substitutes were sought to which bright colours could perhaps be added. In 1926 condensation products from

---

[11] Hardie and Platt (ref. 3), p. 189.

formaldehyde and a mixture of urea and thiourea (obtained from sulphur recovery products in coal-gas manufacture) were made by an English chemist Edmund Rossiter. They were marketed as 'Beetle' products, first by British Cyanides Co. Ltd. of Oldbury (a subsidiary of Chance and Hunt), and then in 1936 by British Industrial Plastics Ltd.

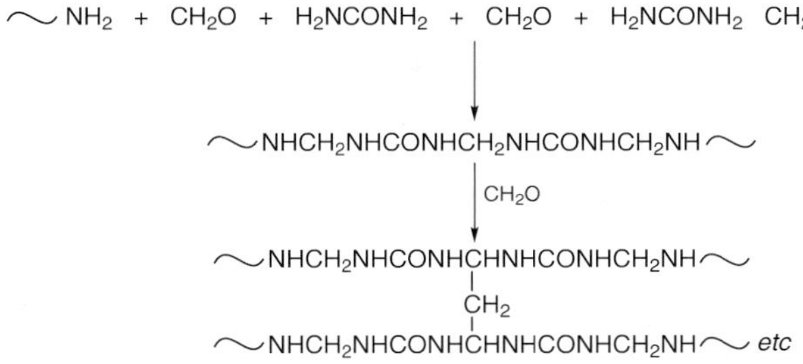

$$\sim NH_2 + CH_2O + H_2NCONH_2 + CH_2O + H_2NCONH_2 \; CH_2O \; H_2N \sim$$

$$\downarrow$$

$$\sim NHCH_2NHCONHCH_2NHCONHCH_2NH \sim$$

$$\downarrow CH_2O$$

$$\sim NHCH_2NHCONHCHNHCONHCH_2NH \sim$$

$$\underset{|}{CH_2}$$

$$\sim NHCH_2NHCONHCHNHCONHCH_2NH \sim etc$$

*Condensation of formaldehyde and urea*

It was discovered in 1926 by researchers at Tootal, Broadhurst Lee & Co. that formation of urea–formaldehyde resins in between cotton fibres imparted impressive crease-resistance. This was 'the first link between the textile and plastics industries'.[12]

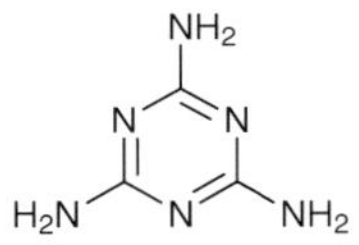

Melamine

A product obtained by heating urea alone is melamine, though it was more often made by heating cyanamide ($H_2NC\equiv N$). Like urea it condenses with formaldehyde, as shown in 1939 by Henkel. The resultant resins have had widespread use as decorative laminates (Formica *etc.*).

## (c)  Alkyd resins

Berzelius in 1847 had observed a resinous product from the action of glycerol and tartaric acid. Such products from polyhydric alcohols and dicarboxylic acids are known now as the alkyd class of polymers (from *alc*ohol-ac*id*). In 1901 Watson Smith, editor of *Journal of the Society of Chemical Industry,* discovered that glycerol and phthalic acid yield, not a phthalein as expected, but a resinous, transparent solid. It was suggested as a good cement for glass and earthenware. Known as a 'glyptal' resin (*gly*cerol + ph*thal*ic acid), it was not at first developed industrially as its usefulness was not generally realised and as raw

---

[12] *Ibid.,* p. 190.

materials were not readily available until after the First World War (phthalic acid by oxidation of naphthalene). In the late 1920s glyptal resins were examined by GEC in the USA, and by British Thomson-Houston and ICI in UK, and manufactured in large quantities.

## (d) Poly(methyl methacrylates)

The methyl ester of methacrylic acid, $CH_2 = CH(CH_3)COOCH_3$, yields with a peroxide initiator a polymeric material that is strong, highly transparent and of high refractive index. This was 'Perspex'[13] which, together with polythene, led to the rise of ICI as a major world producer of industrial polymers in the 1930s (marketing Perspex under the name of Mouldrite Ltd.). It was extensively used in the Second World War for military aircraft, being made by ICI at Billingham, and has since been found extensive application in safety-glass, contact lenses *etc*. The monomer was made as shown:

$$CH_3COCH_3 \xrightarrow{+HCN} CH_3\underset{\underset{OH}{|}}{\overset{\overset{CN}{|}}{C}}CH_3 \xrightarrow{-H_2O} CH_3\overset{\overset{CN}{|}}{C}{=}CH_2 \xrightarrow{hydrolyse} CH_3\overset{\overset{COOH}{|}}{C}{=}CH_2 \xrightarrow[esterify]{MeOH} CH_3\overset{\overset{COOMe}{|}}{C}{=}CH_2$$

Methyl methacrylate

## (e) Nylon

In 1928 W.H. Carothers at du Pont in the USA produced a polymer from adipic acid and hexamethylenediamine.

$$\sim OC(CH_2)_4COOH \ + \ H_2N(CH_2)_6NH_2 \ + \ HOOC(CH_2)_4COOH \ + \ H_2N(CH_2)_6NH \sim$$

$$\downarrow$$

$$\sim OC(CH_2)_4COHN(CH_2)_6NHOC(CH_2)_4COHN(CH_2)_6NH \sim$$

Nylon-66

It was manufactured in the USA on an experimental scale in 1937 (mainly for toothbrushes). By 1939 the first nylon stockings went on sale and proved a spectacular success: 64 million pairs were bought in the first year! Large scale production commenced in Delaware from 1940. In the UK manufacture began in 1941, after ICI had acquired the rights from du Pont. Together with Courtaulds it set up British Nylon Spinners, where nylon thread was manufactured for tow-ropes, parachutes and tyre-cord. Nylon mouldings were also made for gear-wheels which needed no lubrication, were almost silent in operation, did not rust and were cheaper to produce than their metal counterparts. Bearings made of the same material were also found effective against other moving nylon parts or even against steel, lubricated only by water. Nylon thus became another polymer of great military use.

[13] M. Chisholm, 'Plastic fantastic', *Chem. Brit.*, 1998, **34** (4), 33–36.

## (f)  Polyethylene

The high pressure polymerisation of ethylene was accidentally discovered in ICI laboratories at Winnington in 1933 by R.O. Gibson and E.W. Fawcett.[14] A small scale operation in 1937 was followed by the first commercial production two years later. In the ensuing war one of its first important applications was as a cable insulator for radar equipment.

## (g)  Polystyrene

Since the 1930s the monomer styrene has been made by a Friedel Crafts reaction between ethylene and benzene, followed by catalytic dehydrogenation of the resultant ethylbenzene:

$$C_6H_6 + C_2H_4 \longrightarrow C_6H_5C_2H_5 \longrightarrow C_6H_5CH{=}CH_2 + H_2$$

The styrene was then used either for production of the GRS (Government Rubber Styrene) co-polymer or to be polymerised on its own to give polystyrene. The resultant polymer was used for electrical equipment. Later, it became a component of furniture (as foam), a ubiquitous packaging material and a thermal insulator for refrigeration equipment. More recently polystyrene has been criticised for its brittleness in certain fabricated objects as toys, and much more for its extremely smoky flame when burnt (having a fairly high C:H ratio of 1:1).

At first the chief producers were Germany, the USA and Canada. In Britain the only polystyrene produced before 1950 was from a small-scale production by Boake, Roberts & Co., whose wartime process manufactured styrene by dehydrating the essential oil phenylethanol. Later developments in the UK belong, however, to the petrochemical period, which began when styrene was produced by Forth Chemicals at Grangemouth in 1953 or by Shell at Carrington from 1955.

## (h)  Poly(vinyl chloride) (PVC)

PVC[15] was first observed as long ago as 1838 by Regnault. It was first made industrially by the German firm of IG Farben, in 1931, and two years later in the USA. Before the Second World War most research on PVC, and all the European manufacture, was in Germany. In the UK there was no incentive to replace natural rubber, and strong pressures from interested parties not to do so. There were genuine problems with the poor stability and dielectric properties of some imported specimens. However, in the late 1930s some investigations were mounted by ICI at a 'New Resin' section at Billingham and at the

[14] R.O. Gibson, *The discovery of polythene*, Royal Institute of Chemistry, London, Lecture Series 1964, no. 1.

[15] M. Kaufman, *The history of PVC: the chemical and industrial production of polyvinyl chloride*, Maclaren & Sons, London, 1969.

Dyestuffs Division at Blackley. With the outbreak of war decisions were taken to manufacture PVC, chiefly for purposes of cable insulation, in a plant to be established at Runcorn. Full production began in 1942.

The original method was from acetylene, with a catalyst of mercury(II) chloride and charcoal:

$$HCl + CH{\equiv}CH \longrightarrow CH_2{=}CHCl$$

the acetylene being generated from calcium carbide. At Runcorn the early production of PVC for cable insulation used lead acetate as a coagulant; the lead also acted as a stabiliser and brought about a significant reduction in water absorption. The large-scale addition of lead to the working environment was even then seen as problematic, and it was never used for 'leathercloth'. In wartime Germany, however, lead was not surprisingly *not* used in PVC, partly because of the more stringent factory legislation in that country regarding lead compounds. Despite the availability of numerous alternatives even in the late 1960s most British PVC was still being made with lead salt stabilisers. ICI at Runcorn had facilities for manufacturing both HCl and acetylene. Later it switched to petroleum as a cheaper source of acetylene, and later still to processes based on ethylene, such as the liquid phase chlorination to 1,2-dichloroethane, which is subsequently cracked at about 500 °C over a pumice or kaolin catalyst:

$$CH_2{=}CH_2 + Cl_2 \longrightarrow CH_2ClCH_2Cl \longrightarrow CH_2{=}CHCl + HCl$$

The dichloroethane was first produced by BHC, in 1961, from a plant at Grangemouth but shortly afterwards in South Wales, near the PVC plant owned by British Geon.

By the end of the war UK production of PVC was a mere 5000 tons or so, but twenty years later (1965) it had risen to 200,000 tons. PVC had the advantages of relative cheapness, low flammability and the ability to be fabricated into a huge variety of shapes, being used for record sleeves, lightweight mackintoshes, imitation leathercloth, piping, insulation and several billion bottles. It made possible the introduction in 1950 of unbreak-able, long-playing gramophone records. PVC has, however, the great envir-onmental disadvantage of many organohalogen compounds, namely of releasing into the atmosphere free radicals that do damage to the ozone layer. Indeed, even on standing PVC will slowly release HCl, though this can be inhibited by addition of suitable plasticisers. And, although it is fairly non-inflammable at low temperatures, on strong heating it pours into the atmo-sphere not only organo-chlorine compounds but large amounts of the suffocating HCl, together with dense clouds of carbon. Given the immense quantities of PVC still in our cities a major catastrophe like a nuclear explosion would yield such a blanket of carbon that a 'nuclear night' becomes a real possibility, with the sun's rays obscured for months at a time. Such a damaging prospect is a function of the uses we make of the

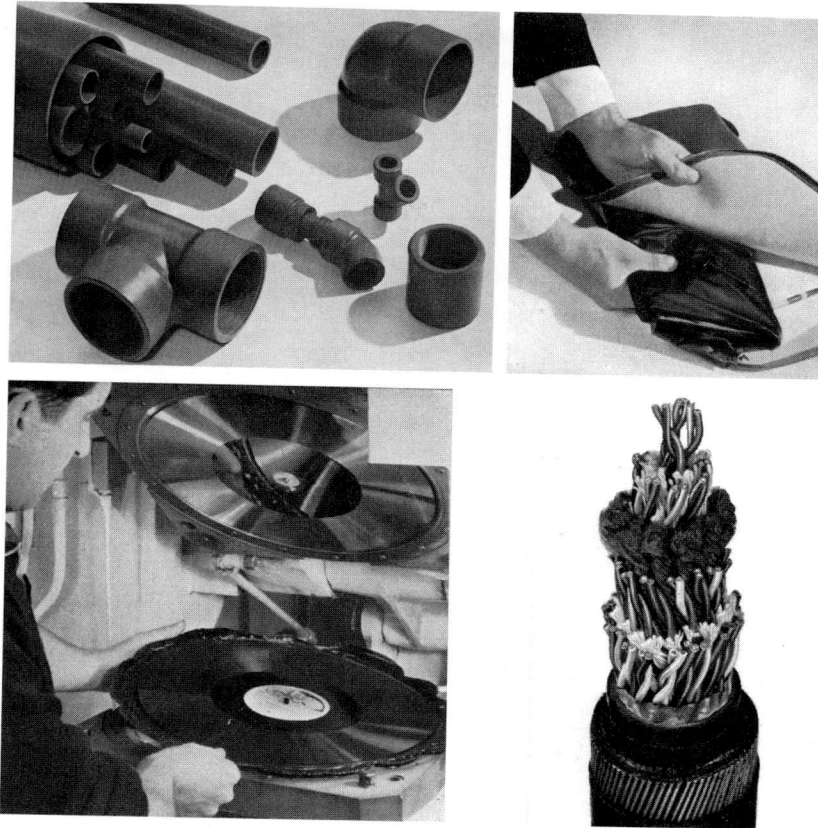

*Numerous uses for PVC in the 1960s: piping; LP gramophone records; lightweight macs; insulation in a multi-core power cable in the coal-mining industry*

material, and our insistent demands for it, not of the manufacturing process of the chemical industry.

## (i)   Poly(vinyl acetate) (PVA)

This has been made since 1928 in the USA and Germany from polymerising vinyl acetate, itself produced from acetylene and acetic acid in the presence of a catalyst:

$$CH_3COOH + CH{\equiv}CH \longrightarrow CH_3COOCH{=}CH_2$$

It did not appear in Britain until after the war (1949), alternative routes then being used from acetaldehyde and acetic acid (British Celanese) and from ethylene (ICI). The polymer does not have the objectionable combustion characteristics associated with PVC or polystyrene, and has the advantage of readily forming emulsions. As such it has been particularly used in adhesives, paints *etc.*

## (j)    Synthetic rubber

As we have seen, wartime efforts to produce a synthetic rubber substitute were unsuccessful. In Germany IG Farben embarked on synthetic rubber research from 1926. By 1929 they were producing Buna rubber, obtained from the action of sodium on butadiene, discovered independently by F.E. Matthews of the English consultancy firm Strange and Graham and by C. Weizmann. It had the great merit of being comparatively oil-resistant, though was soon replaced by copolymers of butadiene with styrene (Buna S) or acrylonitrile (Buna N). The latter was specially valuable for tank linings *etc.* With the onset of the Second World War Buna S production was massively increased, and by 1941 Germany was a leading producer, its output peaking in 1943.

In the USA the loss of Malayan rubber plantations to the Japanese in 1941, following Pearl Harbor, led to an immense increase in GRS production over the next four years. Meanwhile neoprene (from chlorobutadiene), discovered at du Pont, came into production in 1933, though eventually deemed inferior to Buna S for tyres.

However, in Britain synthetic rubber had no place in industrial production until after the war and the advent of an indigenous petrochemical industry. It had been argued by several chemists before the last war that 'artificial rubber must be better than natural rubber', and that it was in the national interest to take it seriously, not least because output could be 'expanded rapidly in case of national emergency'.[16] In Britain those words fell largely on deaf ears.

## (k)    Polypropylene

This polymer really belongs to the petrochemical era, but investigations on the polymerisation of propylene dated back long before that. The failure to produce anything other than viscous liquids or rubbery semi-solids was difficult to square with the different behaviour of the simpler ethylene. In 1954 the Italian chemist G. Natta discovered that the catalyst needs to be stereospecific to ensure regular arrangements of the methyl groups in the polymer. Thereafter it was successfully manufactured in Italy by Montecatini, with uses generally similar to those of polythene.

In Britain polypropylene was made by ICI at Wilton from 1960 ('propathene'), and at Belfast from 1963; production by Shell began in 1962 at Carrington.

---

[16] W.J.S. Naughton, *Synthetic rubber*, Macmillan, London, 1937, pp. 19, 27.

### (l) **Polyester resins**

Of these the most important is polyethylene terephthalate, from ethylene glycol and terephthalic acid (benzene-1,4-dicarboxylic acid). The latter was obtained by oxidation of 4-xylene, mainly obtained from petroleum refining. Following work by J.R. Whinfield at the Calico Printers' Association in 1941 the first full-scale British production was at ICI's Wilton plant from 1955 (Terylene). Its American counterpart, made by du Pont, was known as Dacron. Despite early difficulties with adherence of dyestuffs, later overcome, its crease-resistance

*Whinfield demonstrates his discovery of terylene, drawing a thin thread from a viscous solution in a test tube*

*Interior of ICI's terylene plant at Wilton (1955)*

made it extremely popular for fabrics and its rot-resistance led to widespread use in hoses, sails, ropes and tents.

To summarise: the British polymer industry was well established before the large-scale application of petrochemicals. Its later development will be touched on in the final section of this chapter. The situation in the inter-war years was thus characterised by continuing manufacture of old-established plastics as Bakelite, alkyds and cellulose derivatives, and by the production of four new materials: polyethylene, nylon, PVC and polystyrene. By 1962 the UK output of plastics was increasing at about 14%, thanks largely to the availability of new raw materials derived from petroleum.[17]

## 3  Non-petroleum Based Primary Organic Chemicals

The advent of new dyes, polymers and plastics generated demands that were no doubt themselves increased by traditional advertising, promotion through events like the British Empire Exhibition at Wembley in 1924/5, frequent references on the radio, and a general increase in the standard of living in the years after the Great Depression. Thus by 1928 there were 20 firms in the UK making cellulose acetate, the industry having quite recovered from an immediate post-war slump in the demand for aircraft 'dope'. Socially, the change was from 'bargain basement to *haute couture*'.[18]

The chemical industry was therefore forced to expand its production, and at the same time look for more and better sources of the primary intermediates. Three other features of the inter-war years helped to shape the development of British manufacture of such vital organic intermediates. The first was protective legislation in the form of the Safeguarding of Industries Act of 1921, whereby a Key Industry Duty (33%) was imposed on a variety of industrial imports, including most organic chemicals. Then, throughout the period, a second feature was a large number off take-overs and mergers that overall improved efficiency and of which the formation of ICI in 1926 was by far the most important. This effect was not of course limited to organic chemicals. Thirdly, the Finance Act of 1928 re-introduced a 'Petrol Tax' (replacing one withdrawn in 1921). This was a customs duty of 4d a gallon on imported light hydrocarbon oils, ostensibly to protect the Scottish shale oil industry but doubtless also to bring in needed extra revenue to the Baldwin government from the growing use of the motor-car. Successive increases to that duty not only infuriated motorists; they also put off by many years the day when Britain should catch up with overseas rivals and have its own petrochemical industry. Until then the key intermediates were obtained in more traditional ways.

### (a)  Acetone and related chemicals

In the early post-war years acetic acid was crucial in two main ways. First, it was

[17] Hardie and Platt (ref. 3), p. 11.
[18] Haynes (note 7), p. 202.

used, as we have seen, to make acetone, the most useful solvent for cordite and cellulose acetate. Secondly, the latter depended on acetic acid or its simple derivative acetic anhydride as the acetylating agent for the cellulose. In the 1920s British Celanese tried the acid-catalysed direct carbonylation of methanol at 310 and 200 atm.:

$$CH_3OH + CO \longrightarrow CH_3COOH$$

but the method was unsuccessful, with considerable problems of corrosion. They reverted to the direct dehydrogenation of ethanol, but quickly replaced that by catalysed vapour phase oxidation (1931). The Distillers' Co. also made acetaldehyde this way from the early 1930s.[19]

The ethanol for these processes came at first from fermentation. The Hull Distillery at Salt End was joined in 1928 by another subsidiary, British Industrial Solvents, and the two Hull factories became Britain's leading installation for making ethanol by fermentation (of molasses). DCL's experience was relevant in establishing deep fermentation processes for acetone and butanol (at King's Lynn and Bromborough) and, from 1946, of penicillin at Speke, Liverpool. By now biotechnology had long left its eotechnic phase and, as we shall see, was to play an important role in later developments. Later, of course, petrochemical sources of ethanol came into being, though the first one in Britain was not until 1942 with the British Celanese plant for producing ethylene for this specific purpose.

## (b)    Methanol and formaldehyde

Known since 1868 (Hofmann), formaldehyde has gained an importance for the polymer industry that can hardly be exaggerated, as in the production of alkyd, melamine, phenolic and urea-based resins. It has also been used as an intermediate for many non-polymeric materials, including several widely-used explosives. Cyclonite, or RDX, developed between the wars at Woolwich, is half as powerful again as TNT and was much used in under-water warfare in the operations against U-boats. It was obtained by nitration of hexamethylene-tetramine ('hexamine'), itself made from ammonia and formaldehyde.

Hexamine                RDX

[19] See F.J. Weymouth and A.F. Millidge, 'The manufacture and uses of acetic acid', *Chem. Ind.*, 1966, 887–893.

Another explosive that rose in importance in the Second World War was PETN (pentaerythritol tetranitrate), obtained from the alcohol pentaerythritol which could be synthesised from acetaldehyde and formaldehyde: $C(CH_2ONO_2)_4$. It was extensively used as a component of bursting charges. Formaldehyde can be easily prepared by oxidation of methanol, and it is this substance that now acquired a new importance. Methanol was first made by the destructive distillation of wood, and this continued in the UK until the 1930s (p. 207). However, in 1925 a complex series of political manoeuvrings involving Brunner, Mond & Co., the British Dyestuffs Corporation and (among others) the German firm of IG led, ultimately, to the formation of ICI, but more immediately to the acquisition by Brunner Mond of German know-how regarding the production of oil from coal. More specifically it led to the establishment at Billingham of Britain's first plant to use 'synthesis gas', a mixture of carbon monoxide and hydrogen. From this methanol could be obtained in the presence of zinc oxide catalyst and at high temperatures and pressures:

$$CO + 2H_2 \longrightarrow CH_3OH$$

Since then ICI have opened a second plant at Heysham and another has been established at Grangemouth by British Hydrocarbon Chemicals.


### (c)  Calcium carbide and acetylene

The production of acetylene[20] from calcium carbide (simply by the action of water) has been known for many years, but the invention of a satisfactory electric furnace did not come until 1892 (in the USA). Although the raw materials (limestone and coke) are readily available the huge amount of electrical energy consumed raises questions of environmental priorities as well as financial viability. A plant built by British Celanese at Spondon during the First World War was closed down soon after the Armistice, as acetylene was supplanted by the cheaper ethanol as a source of acetic acid. Acetylene was also used as a welding gas and (later) in some chemical operations, but all through the inter-war years it had to be imported. The lack of hydroelectric power in most of the UK was the chief reason (though tentative plans existed for an establishment in the Highlands of Scotland).

Once again it was war, or the threat of war, that goaded the British government into some kind of action. In 1939 a carbide works was opened at Kenfig in South Wales, near to plentiful supplies of raw materials. It was operated by British Industrial Solvents and, after 1953, leased to Distillers Co.

[20] P.J.T. Morris, 'The industrial history of acetylene: the rise and fall of a chemical feedstock', *Chem. Ind.*, 1983, 710–715.

Ltd. By now acetylene was chiefly required for synthesis of vinyl chloride, the precursor of PVC:

$$HC{\equiv}CH + HCl \longrightarrow H_2C{=}CHCl$$

and this was made by British Geon at nearby Barry. The gas was also used in the manufacture of the solvent dichloroethane:

$$HC{\equiv}CH + 2HCl \Longrightarrow CH_3CHCl_2$$

and calcium carbide was employed in the production of calcium cyanamide, used as an intermediate in the formation of melamine:

$$CaC_2 + N_2 \xrightarrow{1000\,°C} CaCN_2 + C$$

In 1943 ICI opened a carbide plant at Runcorn and other establishments followed until about 1965 when supply roughly equalled demand in the UK.

## (d)   Coal tar aromatics

Right through the inter-war period coal-tar remained the chief source of aromatic chemicals in the UK. Unfortunately the proportions of products were rarely those required by industry, so there was often a glut of some chemicals and an acute shortage of others. This is well illustrated by the case of phthalic anhydride.

By the end of the First World War the Solway Dyes Co. spawned another firm with extensive plant in Scotland, at Grangemouth. It was called Scottish Dyes Ltd. and was established in 1918. One of the finest cotton dyes ever produced was made at their Carlisle works in 1920, phthalic anhydride being an essential starting-point: Caledon Jade Green.

Phthalic anhydride had been traditionally made by oxidation of 2-xylene, a very minor constituent of coal tar. It can, however, be prepared by the mercury-catalysed oxidation of naphthalene, a relatively abundant constituent of the tar. At Grangemouth they started to make phthalic anhydride under licence in 1921, in order to give an alternative route to anthraquinone not using coal-tar anthracene, a dyestuffs intermediate not present in any large quantities in coal-tar ($<0.5\%$ of the tar).

| Naphthalene | Phthalic anhydride | Anthraquinone |

*Preparation of anthraquinone*

Scottish Dyes was acquired by the British Dyestuffs Corporation in 1925. Shortly afterwards an entirely new kind of pigment came to light when, in 1928, a blue residue was found in an iron pot that had been used for the conversion of phthalic anhydride into phthalimide. The new material was called Monastral Blue, the first of the phthalocyanine dyes, large ring compounds like porphyrins with the metal atom at the centre. Their structure was elucidated at Imperial College in 1934 by Linstead *et al.* Much needed for glyptal resins, and for the new phthalocyanine dyes, phthalic anhydride was thus also in demand for new syntheses of anthracene derivatives.

Caledon Jade Green                              Phthalocyanine

Traditional coal-tar products continued to be available through the Second World War. Thus TNT continued to be manufactured from coal-tar based toluene, though, as we have seen, some had been derived from petroleum even in the previous war. Moreover it was beginning to be displaced by newer explosives such as RDX and PETN. Another newcomer was tetryl (2,4,6-trinitrophenylmethylnitramine) made by nitration of dimethylaniline, itself derived from coal-tar benzene. Specially useful in detonators, it was manufactured by ICI Dyestuffs Group, and also at a government factory at Deighton. The replacement of these conventional explosives, though presaged by Hiroshima and Nagasaki, is still in the future.

## 4    The Early British Petrochemical Industry

### (a)    Petrochemicals come to Britain

Figure 1 indicates the steep rise in UK consumption of petroleum products from 1900 to 1955.[21] About one-third of this consumption was by petrol-driven motor vehicles whose numbers on British roads in that period increased from a few thousand in 1900 to nearly 6.5 million in 1955. Derv, kerosene, lubricants, asphalt *etc.* accounted for the remainder consumed. Before the Second World War almost all these petroleum products had to be imported, since Britain had no recognisable deposits of oil other than that from shale. Moreover the

---

[21] Data from *Oil for Britain*, Esso Petroleum Co. Ltd., London, n.d. and *Statistics relating to consumption and refining production*, UK Petroleum Bureau, 1949–1955.

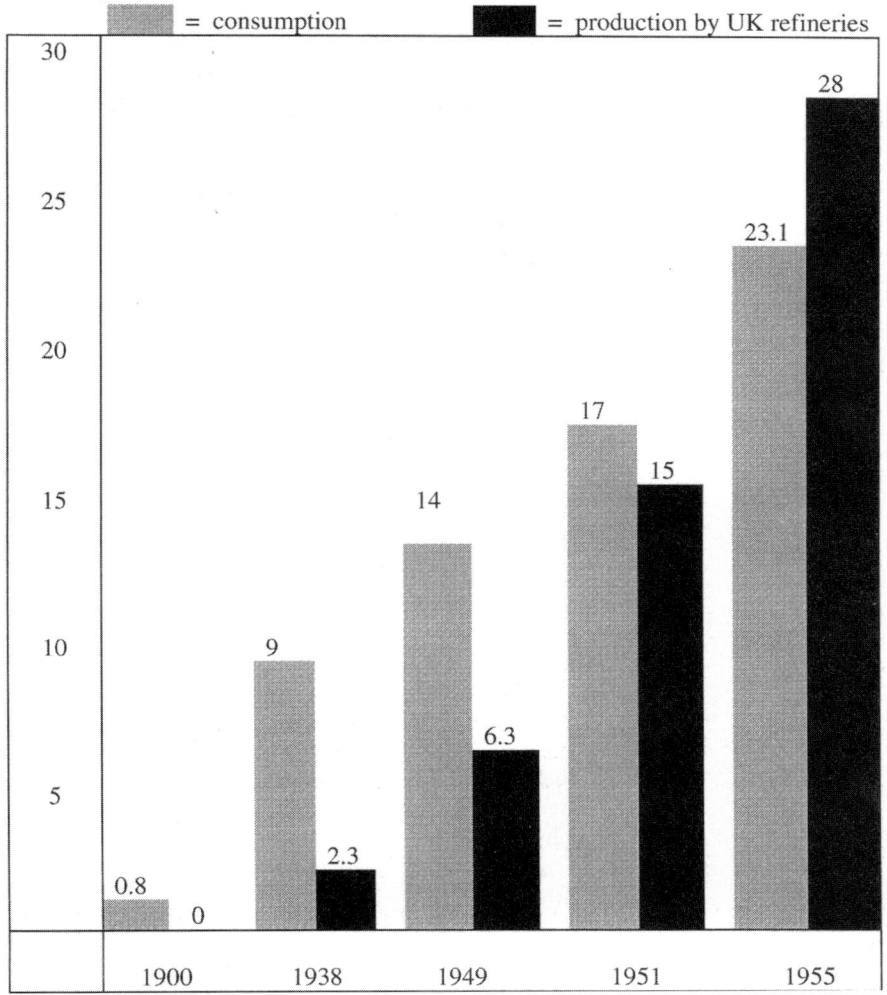

**Figure 1** *UK consumption of petroleum fuel 1900–1955 (millions of tons)*

petroleum had to be refined at or near the point of origin since there were few processing facilities in the UK. Only six small refineries existed with a capacity of over 100,000 tons: at Fawley (1921, Esso), at Pumpherston (1884), Llandarcy (1921) and Grangemouth (1924) (all BP), and at Ellesmere Port (1934) and near Manchester (1938) (both Burmah Oil). After hostilities ended the battered British economy had to save massively on foreign exchange. The very hard winter of 1946–1947 was so severe that much of the country's fuel stocks were consumed, and £200m was lost on exports. In 1949 the crisis was deepened by a devaluation of the pound from $4.03 to $2.80. At the same time there were rising volumes of crude oil in the Middle East.

*Petroleum transport (clockwise from top left): turn of the century; 1910; 1923; 1935*

Clearly it made sense to import the crude and refine it within the UK. So, with Government encouragement, a programme of construction was begun in 1947 that within the next 26 years was to cost over £750m. Within three years six British refineries were on stream, each producing over a million tons per annum. They were at Fawley (Esso), Stanlow, Shell Haven and Heysham (Shell), and Llandarcy and Grangemouth (BP). They have since been joined by many others.

Nor was petrol shortage the only disadvantage facing the recently victorious UK. Its chemical industry was in some disarray, and apparently between five and ten years behind Germany and the USA. This was a bitter pill to swallow for a country that had seen the birth of industrial chemistry and in Victorian times had boasted that it was built on iron, coal and limestone and surrounded by salt water. Yet the fact was that, with dwindling coal supplies, none of these commodities was of much value to that part of an industry concerned with organic chemistry.

There were two obvious explanations for Britain's decline in this sector. The first was a lack of indigenous calcium carbide that could yield acetylene and with it the chemical products so assiduously exploited by war-time Germany. And the reason for this lack was clearly the prohibitive cost of electricity to run the carbide furnaces; other than in Scotland the UK had no hydroelectric power, and so could not compete with countries like Canada, the USA, or even Norway. This deprived Britain of an astonishing range of chemical intermediates.

*A tanker unloads crude oil from the Middle East at Shell's refinery at Shell Haven on the Thames. This refinery closed in December 1999*

*Distillation units at Shell's Stanlow refinery*

Of course the second constraint was the absence of native petroleum. So, unlike the USA, Britain had to obtain phenol from coal-tar, urea and formaldehyde from coke, and ethylene than ethanol. Given an adequate supply of crude petroleum the carbide shortage hardly mattered. The solution to the problem was therefore to create in these islands large new petrochemical installations. Where better than in the immediate neighbourhood of the refineries that were still being built? So, at the six major centres already mentioned, and at others later, Britain at last began to share in the world's burgeoning petrochemical industry.

Petrochemical developments in the UK are mainly post-1960, but such was the speed of change that by 1962 Britain had become the largest petrochemical producer in Europe and the third in the world, with an annual output of 1.4 m tons. All this began with a small plant erected at Spondon to 'crack' imported petroleum wax to make alkenes for the production of a much needed synthetic detergent, Teepol (a secondary alkyl sulphate).

Ethylene and propylene were both made at Grangemouth, in Stirlingshire, by British Petroleum Chemicals Ltd., founded in 1947 by Distillers and BP, and soon renamed as British Hydrocarbon Chemicals. Ethylene has been made from petroleum gases since 1951 by ICI at Wilton, where it replaced the time-honoured method of dehydrating ethanol. A new ethylene plant at Grangemouth for BHC opened in 1960, the largest of its kind outside the USA.

### (b)    The shape of the new industry

Although a thriving petrochemical industry had existed in the USA for nearly thirty years before Britain was able to follow suit, many of its features would have presented themselves as a startling novelty to inhabitants of these islands. First there was its appearance. Instead of a traditional chemical plant, with individual parts instantly recognisable, a petrochemical installation was all gleaming pipes and vertical vessels. Gone were the old smoke-stacks, belching carbon and even more unmentionable waste products into the atmosphere. In their place was a slim chimney with a wisp of burning methane at its mouth, just visible against an unpolluted blue sky. Absent also were vast piles of solid by-products awaiting transport to a suitable burial site for, in truth, there virtually was no waste from the new processes. And, as we shall see, there were other signs of an industry that not only cared for the environment but was prepared to sink substantial sums of money to do so.

However, it was what went on inside these mysterious and futuristic towers that would have been most challenging to anyone trained in classical organic chemistry. In fact a whole new chemistry had emerged with the advent of new oil-processing techniques. Many, though by no means all, reactions were in the gas, not the liquid, phase, and many depended on high-temperature transformations that had hitherto been confined to laboratory study. Much of the new chemistry was homolytic as opposed to heterolytic, with free radicals taking the place of the familiar ions and molecules in much conventional solution chemistry. It was also a new age for catalysis, though this was less of a novelty than it might seem. Catalytic reactions had held an honourable place in industry at least since the days of the Haber and contact processes. Yet an unreconstructed classical organic chemist might have been forgiven for supposing that *anything* was possible in a petrochemical plant if only the right catalyst might miraculously appear.

To summarise the new technology even up to 1960 would require not a few sentences but a whole book, and excellent accounts are readily available

*BHC petrochemical complex at Grangemouth*

*Ethylene production plant at Grangemouth*

elsewhere.[22] However, it may be stated that the whole industry depends upon highly efficient separation of hydrocarbon fractions from the raw crude, usually by fractional distillation. The majority of these hydrocarbons are alkanes (paraffins), regarded by classical organic chemistry as the ultimate in unreactivity and stability. Moreover most of them are straight-chain alkanes, whereas a modern petrol engine works most efficiently with those whose chains are branched. Much work was done by manufacturers of motor fuel to bring about a catalytic transformation of straight- to branched-chain hydrocarbons.

A further achievement of the petroleum industry was the development of cracking, both thermal and catalytic, whereby long chains could be broken into shorter ones more suitable for their purposes. Such transformations, however, inevitably lead to at least one molecule being unsaturated (*i.e.* an alkene), as in the following case:

$$C_{16}H_{34} \longrightarrow \underset{\text{Octane}}{C_8H_{18}} + \underset{\text{Octene}}{C_8H_{16}}$$

From this work and much else came the ability to produce large quantities of desired alkenes, notably ethylene and propylene. The former has been made from petroleum gases since 1951 by ICI at Wilton, and from 1960 by BHC. It was then required for manufacture of polythene, though uses mentioned earlier included also production of PVC and styrene. Since the 1950s ethylene has been used to make ethanol by hydration, so reversing the familiar reaction. Another important use has been the catalytic oxidation by air to ethylene oxide, from which may be obtained ethylene glycol ('anti-freeze'), the higher polyglycols used as solvents, hydraulic fluids, plasticizers *etc.*, and a number of important detergents (as 'Stergene') containing a repeating unit $\sim(CH_2CH_2O)\sim$.

*Production of glycols*

*Production of 'Stergene'*

[22] One of the best accounts of this early period, though not restricted to petrochemicals, is A.J. Gait, *Heavy organic chemicals*, Pergamon, Oxford, 1967; publications of all the major oil companies in the UK are often very helpful, especially *The petroleum handbook*, Shell International Petroleum Co. Ltd., London, 5th edition, 1966. For the earliest history see R.J. Forbes, 'Oil from the earth', *Ciba Rev.*, 1967 (3), 3–40. A large number of booklets from the oil companies – too numerous to mention here – can be valuable source material for details of their operations and, especially, the post-war refineries in Britain. There is also relevant material for the lay-person in C.A. Russell's revision of F.W. Gibbs, *Organic chemistry today*, Penguin, Harmondsworth, rev. ed., 1970.

Another alkene, propylene, has also been used to generate the parent alcohol (propan-2-ol, or *iso*-propyl alcohol) as well as being the starting point for polypropylene and other ethylene analogues like polypropylene glycol. Another alkene, butadiene, has of course been used in the production of synthetic rubber, and in Britain has been manufactured from the same cracking process on light petroleum that also yields ethylene. By the late 1960s all four major producers of ethylene were also recovering butadiene. The Fawley plant was the first (1956), the product being fed into a plant owned by Dunlop's International Synthetic Rubber Co. on adjacent ground.

The cracking of petroleum does not merely produce alkenes, however. Acetylene has been made by high-temperature pyrolysis of methane, though this method did not appear in the early days of the UK industry. More surprising, perhaps, is the appearance of aromatics as petrochemical products. Traditionally regarded as coal-tar chemicals, they do occur in small quantities in some kinds of natural petroleum (as that from Borneo), but they can also be extracted from distillates resulting from cracking and other processes. Large-scale production in the UK began with a take-over in 1955 of a small plant owned by Petrochemicals Ltd. (which had produced aromatics since 1949). Benzene was used to make styrene, and 2- and 4-xylenes to produce respectively phthalic anhydride and terephthalic acid (for terylene). Although it had been cheaper to import aromatics in the early 1960s a large plant at Wilton for ICI was only one of several to appear in the UK for petrochemical production of aromatics.

One remarkable use of such materials was the production of the much-needed phenol.[23] A method was discovered independently in 1950 by the Hercules Powder Co. in the USA and Distillers' Co. in Britain. The Distillers–Hercules process involved the akylation of benzene by propylene to yield cumene (*iso*-propylbenzene), which could be then oxidised in the liquid phase at about 120 °C to yield a hydroperoxide which, on hydrolysis at 45–60 °C, produces phenol, together with acetone. The latter was a bonus; traditionally made from wood-spirit or by the Weizmann process, it was also being made by oxidation of propan-2-ol. At one stage it looked as though its production from cumene would cause a severe glut in the market. But the process was adopted by BHC and a large plant opened at Grangemouth in 1960.

## (c) Some environmental considerations

There remains the question of the environmental impact of the new petrochem-

[23] F.E. Salt, 'Synthetic phenol manufacture', *Chem. Ind.*, 1953, S46–S49.

ical industry that has come to the UK. This can be seen as a threefold problem, relating to their *supplies*, their *plant* and their *products*.

Taking the matter of supplying raw materials to the petrochemical sites in Britain it is obvious that this has been almost exclusively by tanker. It was ironic that just as the British industry was being established the Suez Canal should be closed in 1956. One effect of this was the necessity to transport oil round the Cape, so necessitating even larger super-tankers. When, eventually, the Canal reopened they were too large to use it. The appalling pollution caused by tanker accidents, from the *Torrey Canyon* onwards, is too well known to need repetition. In more recent years spectacular cases of sea pollution have occurred as far apart as South Wales and Shetland. However, it may justifiably be remarked that the chemical industry can hardly accept responsibility for maritime disasters by its suppliers. However, many tankers belong to petro-chemical companies and most UK oil companies now belong to TOVALOP (Tanker Owners' Voluntary Agreement concerning Liability for Oil Pollution), of which several (including Esso) were founding members. According to this agreement members would pay up to £4m as compensation to countries injured by their oil pollution. Highly sensitive to environmental criticism they also imposed strict guidelines for procedures of unloading cargoes.

A similar sensitivity has been displayed by petrochemical producers to the design and administration of their plant. All major installations had their own fire stations and health centres, so were ready to cope with sudden emergencies. But all went much further than that.

A classic case is the Esso refinery/petrochemical installation at Fawley on Southampton Water, by 1955 the largest in Britain and still in that position twenty years later. According to the chairman of the company by 1956 this Esso refinery was already saving the nation an estimated £150m in foreign exchange, to which another £100m would be added by expansions that year.[24] Such an immense project could hardly be invisible, though it was adjacent to the New Forest as well as a highly popular stretch of the coast-line. The designers sought to minimise any visual pollution by planting (eventually) 27,000 trees, thus screening it from view as far as possible. Noise pollution was at first a problem and consultants were called in from Southampton University to help with minimising the noise level of refining plant and furnaces. But of course it was chemical pollution that posed the biggest threat. To eliminate as far as possible the nuisance of chemical smell a deodorising plant was specially installed, and chimney stacks were built 100 ft (30 m) higher than the height required by law. Special measures were taken to prevent pollution of Southampton Water, with oil–water separators and other equipment. In the early years of Fawley the cost of anti-pollution measures amounted to over £1m. So confident was the Company that an early brochure pictured a shepherd with his dog and flock apparently at ease in the middle of a refinery (Fawley?). The picture bears the optimistic title 'sheep may safely graze'! There is no doubt that strenuous efforts have been made by the industry to clean up its operations as well as its image.

[24] *The Times*, 8 July 1956.

*Sound-absorbing baffles around furnaces reduce noise levels near the plant*

Yet there are still problems. Visually a petrochemical site will endear itself to a very small number of people. If the chimneys are unusually high they will be unusually visible. At the Shell site at Stanlow a water-cooling tower was 341 ft (104 m) high, which the company described with more truth than perhaps they intended as 'a well-known landmark in the county'! An environmental problem common to all petrochemical sites is the need for effective water-cooling, and enormous quantities of water are required. Thus each hour Shell Haven took 3 million gallons of water from the Thames, the Kent Oil Refinery required 3½–4 million gallons from the Medway, while Fawley abstracted 6 million gallons from Southampton Water. This must therefore mean a discernible raising in temperature of a very large volume of water. The long-term effects of this warming-up will need monitoring for years to come. Additionally there is always the possibility of inadvertent pollution as the natural waters pass through a chemical plant. Effluent separators (as at the Kent refinery) were installed both to counteract this problem and to cope with tanker ballast also. However, it would be quite unfair not to point out that by 1970 about 80% of the oil refining industry in Western Europe belonged to the Conservation of Clean Air and Water in Western Europe (CONCAWE) based at the Hague.

Finally what of the products of petrochemical operations in post-war Britain? It must be said that many of the objections arose well after the mid-1960s when we take our leave of the industry. As they applied to products in circulation well before that date, however, it is pertinent to glance briefly at them here. The first observation is that, by necessity, there is virtually no waste. All products, unwanted at first, have been subject to further treatment to yield

*A far cry from 19th century industrial housing – a village for Esso employees near Fawley*

*'Sheep may safely graze': romanticised statement by Esso of its environmental awarenes*

saleable materials. Wax residues have been thermally cracked to produce intermediates of lower molecular weight; sulphur extracted from effluents has been oxidised to sulphuric acid; even the most intractable organic product may be utilised as a fuel. The cumene process demonstrates nicely the immense range of chemicals obtainable from petroleum feed-stock, and also the avoidance of waste products.

Although some petrochemical products have been criticised on various environmental grounds, their replacement may be costly in terms of energy. Thus fuels that have been desulphurised will have a loss of up to 8% efficiency, while packaging with metal or glass instead of plastics adds significantly to the weight and thus may lead to increased energy consumption. Two other objections to the use of synthetic polymers have had more substance.

Many petrochemical products are not biodegradable and therefore cause short-term but serious hazards to animals that might ingest them (as when plastic bags are found on agricultural land). Research on this problem continues to this day and some biodegradable plastics, and many biodegradable detergents, have come on the market. In the latter case it was soon recognised that alkyl chains are more susceptible to such degradation if they are unbranched than if they have numerous branchings.

The discovery that the stratospheric ozone layer is threatened by organic chemicals containing halogens, especially CFCs, has led to their withdrawal from use as refrigerants. Other manufactured organic compounds have been shown to have a variety of undesirable effects on the human body, such as tetraethyllead, benzene, carbon tetrachloride and so on. Again, action has been taken to withdraw them from common circulation. One additional advantage of the cumene process was that it replaced an alternative route to phenol which involved chlorination of benzene and subsequent removal of chlorine. Such a process was frowned upon by some environmentalists because it may introduce unwanted chlorine into the biosphere.

Other criticisms of petrochemical products have been made and the industry has at times been forced on the defensive. Justifiable criticisms have been levelled at the problems caused by the flammability of plastics, particularly the undesirable combustion characteristics of PVC and polystyrene. These have been greatly minimised by a more judicious use of such materials, and by the addition of inert fillers. However the industry could hardly be held responsible for the deliberate and gross abuse of its products, as in the craze of solvent sniffing or in the wilful combustion of old tyres. The indictments levelled against other products as pesticides and fertilisers have been shown often to be one-sided and ill-conceived.[25]

Yet when all this is said the benefits conferred by the organic chemical industry in general have been immense and the admitted disadvantages must be carefully weighed against these. In an imperfect world there is always room for improvement and long before 1950 those who manufactured organic chemicals in the UK often took great pains to minimise the undesirable effects of either process or product. It remains true today that the organic chemical industry has still not lost its sense of neighbourly responsibility.

---

[25] See Chapter 11; also C.A. Russell, *The Earth, humanity and God*, University College London Press, 1994, Chapter 5, and references therein.

# Chapter 10 *Metal Extraction and Refining*

## C.A. RUSSELL AND S.A.H. WILMOT

## 1 Introduction

The chemical process that has most profoundly altered the course of history and changed man's environment for good is that of extracting metals from their ores. Only in those rare cases where metals occur in the free state (as gold, silver and sometimes meteoric iron) is chemistry not involved in their production. Even then it is often concerned in subsequent purification. So profound is this truth that two of the most formative ages in early human history are named after the metals that dominated them. Apart from the most noble metals gold and silver, used for decoration or adornment, the first to be liberated from a combined state was probably copper, followed shortly afterwards by tin. The alloy from these two metals gave its name to a period from about 3000 to 1000 BC: the Bronze Age. In Britain the dates are about 2000–450 BC. Useful for making drinking vessels and other containers, bronze lacked the strength of the next metal to dominate the civilisations round the Mediterranean basin, and that was iron. Once this newcomer was obtainable from the immense deposits of iron ore so widely available, it could clearly replace bronze for most purposes (except the decorative) and could be of inestimable value for a whole range of totally new uses. The literature of the ancient world has many references to metallurgical techniques and their remarkable effects, not least in respect of iron. One need look no further than the Old Testament.[1] The Iron Age had arrived. It is generally said to have come to Britain in about 450 BC.

Generally speaking it appears that the order in which free metals appeared in our environment was determined by the ease with which they could be prised out of their compounds (or conversely the difficulty with which they could be persuaded to combine with other elements). More precisely it was the order of increasing electrode potentials, with gold and the alkali metals isolated in the early 19th century at the two extremes. There were, of course, other considerations, not least the availability of ores rich enough in the desired metal to be exploited. Yet even in the ancient world one fact is starkly clear, and that is the dominance of military considerations. It has been said that, without any weapons at all, men would beat each other to death with their bare fists, and that is probably and sadly true. But given iron to make swords and spears the process is considerably simpler. To be sure, swords can be accompanied if not replaced by ploughshares, but the clear impression remains that the driving

---

[1] *E.g.* Genesis 4:22; Deuteronomy 3:11 and 8:9; Joshua 17:16; Judges 1:19 and 4:3 *etc.*

force behind much early metallurgy was the desire to make strong and effective weapons of war. This environmental effect will be encountered more than once in this chapter as will others of a more pacific but still detrimental nature.

In Britain the extraction of metals goes back several thousand years. The Romans sought tin in Cornwall and lead in the Somerset Mendips, though almost no traces survive of the industry in which these metals were extracted from their ores. Iron was also made at that period and as the centuries passed a few other metals were extracted, notably copper from Cornwall, lead from Derbyshire, and (much later) zinc from the Mendips, used to make the alloy brass. In recent times metals extracted in the UK have varied from the rare strontium to the almost ubiquitous aluminium. None, however, has so far had the colossal social and environmental effects of the one that, after aluminium, is the commonest metal on Earth: iron.

## 2   Iron Before Steel

The element iron constitutes just over 4% of the Earth's crust. As we know well today it almost always occurs in combination with other elements and, in Britain, as some kind of oxide. The discovery that, like copper, it could be extracted by heating with charcoal is lost in the mists of antiquity but news would have spread over all the civilised world, eventually to its most north westerly edge in Britain. But it must have become quickly obvious that iron-making had two problems not present with the less reactive metals like copper and lead. The first was its higher melting temperature (Fe 1535, Cu 1084 and Pb 327 °C), requiring a very efficient furnace if the metal were to be run off as a liquid as free as possible of slag (which would otherwise have to be laboriously hammered out into a 'bloom'). The second problem was related to the first in that, if the red-hot iron did not run out quickly and remained for a long time in contact with charcoal, it would absorb some of the latter, becoming in the process hard and brittle, what we call 'cast iron' with up to 4.5% of carbon.[2] Therefore the earliest iron-extraction furnaces were on wind-swept sites, often on hill tops, to enable the forced draught to raise the temperature of the fire even a little. They were manually operated, the man in charge introducing fresh fuel and ore as required and, when he was not doing that, using a primitive bellows to augment the natural draught to the fire. Perhaps a few pounds of iron would be produced from each 'furnace' in one day. It was hardly mass production, and the product had little effect on the lives of ordinary people, unless they were in the army or navy. Environmentally its impact was negligible.

Thus things might have remained for an indefinite period had not rumours reached Britain in the late 15th century of a remarkable development that had just arrived in the west of Europe. Where it had originated no one knows. It was the first of a series of technical innovations that were to transform British industry, and with it the course of world history. It was the blast furnace.

---

[2] Only in the 15th century was it discovered how to convert cast iron into wrought iron by burning away the unwanted carbon.

Exactly when and where it first appeared in Britain is not certain but it is likely to have been in or around 1500.

## (a) The blast furnace

In this epochal development an ordinary iron-making furnace was modified in essentially two ways. The most basic feature was that air was forced into the interior by means of a *mechanical* blast. For a century and more this was only possible through pumps driven by water power (wind-mills could not be relied on to give the necessary constancy of output). The air blast was introduced towards the base of the furnace through blast-holes (or tuyères), while molten slag and iron were withdrawn through separate holes at the bottom. As before the ore and charcoal were fed in from the top. As we now know the reactions were essentially the formation of the gaseous carbon monoxide from the coke and air, followed by its reduction of the ore to iron in a much more effective way than could be achieved by simply heating two solids together. In simplified form the process can be put:

$$2C + O_2 = 2CO$$
$$3CO + Fe_2O_3 = 2Fe + 3CO_2$$

The second modification depended on the first: furnaces could now be made much larger as they no longer depended upon human muscular power for air-injection. Indeed the output could be increased by factors up to nearly a hundred.

To operate a blast furnace one therefore needed iron ore, a plentiful supply of charcoal (usually made by slow combustion of coppice wood), freely running streams or rivers, and accessibility to markets. In practice all four conditions were not always met in one place, in which event it was usually the iron ore that was brought in from elsewhere, by cart or by ship. The conditions applied in the Forest of Dean and the wooded areas of the Scottish Lowlands, but especially in the well-forested parts of Kent and Sussex known as the Weald.[3] There iron beds were often found in association with Wadhurst clay, and were excavated from shallow pits. Rivers such as the Rother, the Sussex Ouse and the Arun provided an adequate if not great source of water power, and all around were vast tracts of ancient forest once devoted to deer-hunting and, through the Middle Ages, more than sufficient for supplying the tiny iron industry. Evidence for iron-making in Kent and Sussex includes an abundance of hammer-ponds, together with furnace remains at Coleman's Hatch (the oldest), Ashburnham, Cowden, Wych Cross *etc.* It is supported by local names such as Furner's [furnace] Green and by visible traces of red iron oxide in water-courses. Yet today only isolated pockets of woodland survive, as Ashdown Forest in Sussex.

[3] E. Straker, *Wealden iron*, G. Bell & Sons, London, 1931.

*Charcoal burning in the Forest of Dean*

*Ruins of the broken bay of a Wealden iron furnace at Warren near Lingfield (1887)*

Wealden iron production was stimulated by the new blast furnace and by the arrival of new entrepreneurial landowners following the dissolution of the monasteries in 1538. In 1543 the first iron cannon in England was cast, at Buxted. During Tudor times the Wealden forest supplied most of the iron for the Royal Navy, mainly as cannon and cannon-balls. Such provision was extremely timely for the succession of wars and semi-piratical enterprises that marked the reigns of Henry VIII and Elizabeth I. Manufacture of cannons had taken place on a large scale at Calais, using imported Wealden timber. Following the loss of Calais in 1558 yet more gun-casting establishments were started in the Weald. In peacetime domestic items were produced, such as fire-side irons, firebacks and farm implements.

During the 16th century two associated pieces of equipment improved the procedures at an iron works. The *finery* was a charcoal-fired hearth, assisted by an air-blast, in which excess carbon in cast iron was burned off to make wrought iron, more suitable for most purposes except casting. The *tilt-hammer* was a mechanical device in which the hot soft iron was subject to successive blows from a hammer dropping under gravity and then being raised mechanically. In each case water-power provided the energy.

Capital investment was fairly high, and the cost of manufacture was greatly reduced if all or most commodities were available in the estates of one owner. Consequently iron-making tended to become the prerogative of the landed class who, indeed, saw it as a means of 'improving' their estates. Economically it was wasteful in that excess carbon monoxide was burnt at the furnace mouth and the slag was dumped. Partly for these reasons progress was slow, more markedly in Scotland, with one blast furnace in Britain in about 1500 and 85 a century later. Yet during most of the 16th century England enjoyed a virtual monopoly of cannon production, though the lead passed to Sweden after the onset of the Thirty Years War (1618). The production of iron decreased after the death of Elizabeth I largely due, it seems, to the rising cost of charcoal resulting from a growing shortage as wood itself became scarcer. In this case environmental despoliation led to economic loss, not to its gain as happened so often in later years.

The environmental pollution to air and rivers was not yet on a large enough scale to attract much adverse comment, but that is certainly not true of the massive changes, visual and social, brought about by the wholesale destruction of forests in the cause of charcoal production. A commission was set up in 1549 'to be directed to certain men of Sussex concerning the hurt done by mills and furnaces made for the same'. Its disclosures reveal the extent of popular resentment at the environmental problems generated by the blast furnace. For example, in the area round Lewes and Hastings there were 50 iron-making establishments, each of which consumed 'at the least yearly 1500 loads of great wood made into coals, besides the great and noisome spoil of the said woods which is engendered for lack of cherishing of the same so felled to the use of the iron mills'. However, near Lewes it was reckoned that 3000 loads of wood were consumed by one 'hammer and furnace'. If the present trend continued there would soon be insufficient timber to build and repair houses, windmills or (most dangerously) sea-defences; the poor would suffer for lack of fuel;[4] and there was always the question of waste 'which we be not able to answer unto'. The daunting prospect would have to be faced of a mass migration away from the area, leaving decaying towns and villages 'not to be inhabited for lack of timber and fuel'. The doom-laden jurymen of Lewes could only bring themselves to admit that 'what harms and hurts hath been done by occasion of the mills and furnaces we cannot express, it is so great, and what will follow hereafter we cannot say'.[5]

Despite relocation of their gun-foundries from Calais to Sussex the government did take some ameliorative action. A general restriction on timber use was imposed in the time of Henry VIII, while in the first Elizabethan era Acts of Parliament in 1558, 1581 and 1585 prohibited the use of trees for iron-smelting if grown within fourteen miles of a navigable river or coast (excepting the Weald), and restricted new furnace construction near London or in the South East generally. An Act of 1611 forbade the use of wood fuel by glass-makers. Not all shared the apocalyptic expectations of the Sussex juries in 1549 and it is true that for nearly another two centuries their part of Britain continued iron production on a grand scale. Partly this was because the larger trees as oak and birch were no longer used but rather coppice wood grown for that purpose. So, in 1724, Daniel Defoe could write as if there were no tomorrow, the three south eastern counties being 'one inexhaustible store-house of timber never to be destroy'd but by a general Conflagration'. However, by the end of the 18th century the Weald had lost its blast furnaces. Responsibility for the decline lay with the sluggishness of the rivers (not improved by exceptionally dry weather for the first half of the century), competition from foreign iron imported into London, the rise of alternative coke-fired furnaces and above all the escalating costs of charcoal.

It was a similar story elsewhere. In Scotland iron production in the Lowlands was notably less than in England, but deforestation was so significant that an Act by the Scottish Parliament forbade the making of iron from wood. In the Furness area of Lancashire (now Cumbria) there were rich supplies of haematite but, by about 1737, the local timber supplies were exhausted and, for a while, ore was shipped to the Sussex ports. As supplies from the Lowlands forests diminished iron-masters in Furness and West Cumberland turned their attention to the Highlands, one firm building a blast furnace at Bonawe which ran from 1752 to 1862, the iron ore being transported from Lancashire by sea. Other furnaces were at Invergarry (1729), and at a site near Inverary now called Furnace (1754). Despite great distances these projects were successful, though Furness 'charcoal bar' was overtaken by Bessemer steel by the middle of the 19th century.

What has justly been called 'the timber famine'[6] was not confined to Britain

---

[4] That consideration would not apply to the poor in the North East where coal was their usual fuel; those in the South East, who relied on wood in any case, did not suffer such general poverty.

[5] Straker (ref. 3), pp. 117–121.

[6] A. and N.L. Clow, 'The timber famine and the development of technology', *Ann. Sci.*, 1956, **12**, 85–102.

*The 18th century charcoal-burning blast furnace at Bonawe, West Highlands*

and by the mid-17th century it was felt all over Europe, where the quality of cast iron produced seems to have been determined by the cost of charcoal.[7] In noting the environmental disaster of deforestation in Britain, however, we should not attribute all of it to the extraction of iron. During the Middle Ages many wooded areas had been cleared for pasture or cultivation. Timber was used (by those who could afford it) for building construction and as a fuel. By the 16th century the demise of the Wealden forests was hastened by huge demands for timber for construction of ships. Other industries that were to take their toll of wood supplies in the next century included the making of gunpowder, soap and glass, the clothing industry with its charcoal-heated dye-vats, and even the manufacture of beer (for oast-houses needed charcoal fuel and hop-fields required wooden poles). Alongside the production of diminishing amounts of charcoal went of course the formation of wood-ash from which alkalis could be extracted. The escalating cost of such alkali helps to explain the urgent need for a synthetic product (see Chapter 4).

Yet there can be no doubt that iron extraction played a major part in deforestation. One ton of bar-iron requires charcoal from almost 200 cubic yards of timber (half the cost of production). In 1720 (the first year for reliable data) it is estimated that British production was 27,000 tons, a figure that stayed fairly constant for the next three decades. That would need over 150,000 trees. It was said that one worker alone near Durham had felled over 30,000 oaks in the cause of iron production, and he had not finished his life's work yet.[8] However, on the credit side one may note the close connection between the iron industry and agriculture, facilitated by the agricultural interests of most furnace-owners.

---

[7] B.G. Awty, 'Early cast irons and the impact of fuel availability on their production', *J. Hist. Met. Soc.*, 1996, **30** (1), 17–22.
[8] W.G. Armstrong *et al.* (eds.), *The industrial resources of the Tyne, Wear and Tees*, 2nd ed., Longman Green, London, 1864, pp. 82–83.

Thus many labourers were from the agricultural classes and furnaces would often be closed during harvest. Their jobs were relatively steady and safe. Moreover the concentration of labour at the iron-making establishments meant a greater demand for agricultural products.[9] Equally, the waggonways linking iron-working sites improved communications, and eventually they and other highways were laid with crushed slag from the furnaces. But overall the devastation of Britain's woodlands by the infant iron industry must rank as one of the worst examples of environmental damage ever caused by the application of a chemical process.

## (b)    Replacement of charcoal by coke

One solution to the huge environmental and economic problems posed by deforestation was the discovery of an alternative reducing agent to charcoal. In 1709 a solution was found by an iron-founder at Coalbrookdale, Abraham Darby. As the place name suggests there was an abundance of coal in the vicinity, and others before him, such as Dud Dudley, had suggested replacing charcoal by coal. The difficulty is that many of the constituents of coal, including sulphur, render the iron useless. A similar problem had been found in the process of malting, where smoke and sulphur dioxide each make the resultant beer unpalatable. For some years maltsters had made coke by heating heaps of coal under wet ashes. Darby, who had once been apprenticed to a maltster, transferred the technology to iron-making. A result still visible to this day has given its name to the village of Ironbridge: a structure across the Severn constructed in 1779 by Darby's grandson Abraham III from cast iron blocks locked together in a style employed by workers in wood, with dowels and dovetail joints.

Replacement of charcoal by coke enabled still larger furnaces to be built since coke is not crushed so easily as charcoal and can bear a greater weight of furnace charge. Nevertheless the change was not an unmixed blessing. Coke-smelting was impracticable at first in areas like the English Midlands where (apart from the Severn) substantial water power was not available and where, in some cases, there remained plentiful supplies of wood. Thus Shropshire was still smelting with charcoal in 1788. Moreover coke, with a much smaller surface/weight ratio, does not burn so easily, requiring higher temperatures and so an even stronger air-blast. From the late 18th century the advent of steam pumping engines overcame this difficulty and coke now became a serious competitor of charcoal. Thus by 1788 there were 24 charcoal furnaces in England and Wales but 53 that consumed coke; by 1806 only 11 charcoal-burning furnaces remained in Britain. The last furnace to operate with charcoal was at Backbarrow in the Lake District, and this continued until 1920 (thereafter working with coke until closure in 1967).

[9] However it did not always work out like that in practice; at Carron ironworks in the 18th century the workforce consumed oats from Hull or Aberdeen as these were cheaper than the local produce (A. and N.L. Clow, *The chemical revolution*, Batchworth, London, 1952 [reprinted Gordon & Breach, Philadelphia, 1992], p. 339).

*The forehearth of the Old Furnace, Coalbrookdale, where in 1709 iron ore was first successfully smelted with coke*

*The world's first iron bridge (1779), made from cast iron but built on carpentry principles, with dovetail joints*

*The restored site of Abraham Darby's furnaces, Coalbrookdale*

*Three blast furnaces (1832, 1840 and 1844) at Blist's Hill, Ironbridge, operating until 1912 and excavated in 1973. They were charged from the bank at the rear*

*Sketch (1887) of the last charcoal-burning iron furnace in Britain at Backbarrow, Cumbria (ceased work in 1920)*

Location of iron furnaces now moved to coal-mining areas, and from the Weald to Yorkshire, Durham and the Black Country. In the latter they chiefly served the old-established nail industry at first, though in Birmingham the emphasis was on 'hollow-ware' – all manner of domestic pots, pans and kettles – and iron pipes.[10] Further north shipbuilding, railway construction and heavy engineering became the main consumers. The first furnace to be worked by coke in the north of England was founded by Isaac Cookson at Whitehill, near Chester-le-Street in County Durham before 1745. It used water power from Chester Burn, the vagaries of whose flow eventually led to closure before 1800. The problem for iron masters in the North East is that, although coal was abundant, suitable deposits of iron ore were not. In the valleys of the Derwent, Wear and Tyne some ironstone had been found, though, with one or two important exceptions, it was not enough to sustain lengthy operations. Generally speaking the iron ore supplies gave out long before the coal. Consett Iron Company is a good illustration, though it managed to survive until the 1980s. For coastal districts this was not too serious a problem since ore could be readily transported there by sea. By 1745 the Whitehill furnace was consuming iron from Robin Hood's Bay in North Yorkshire, brought in by sea and thence up the river Wear. The story of countless attempts to discover and exploit local ironstone deposits in Durham and Northumberland cannot be told here. Further south a number of small establishments were joined in 1840 by Bolckow and Vaughan at Middlesbrough, with financial backing from Joseph Pease. They, as their fellow iron masters further north had already discovered, were in need of ironstone from elsewhere. Their problem was solved by a chance discovery during the excavations for the Whitby to Pickering railway line, opened in 1835. In a cutting near Grosmont thick bands of a rich ironstone seam were uncovered and within a year the ore was being taken *via* Whitby to almost every iron-works on Teesside and further north. Even this was not enough for the hungry blast furnaces and further exploration of the Cleveland hills in North Yorkshire eventually (1850) led to the discovery of a 16 foot thick band of ironstone near Eston, on the northern scarp of the Cleveland hills and only 4 miles from the Tees. It was hailed by local patriots as 'inexhaustible', with an optimism rivalling that of Defoe. As a result of this discovery 'Teesside was virtually turned into a major iron-producing centre overnight'.[11] Indeed it became the world's leading producer of iron for the next 50 years, reaching its maximum in the 1890s but by 1914 with only one-third of the works still open.

Bolckow and Vaughan set up a blast furnace at Middlesbrough ('Ironopolis') in 1852, and by 1855 were supplying cast-iron pipes for the water mains of London. In 1853 the Bell Brothers set up an extensive iron-works on reclaimed land at Port Clarence. Here a howling wilderness, frequently flooded by the Tees, was converted into a stable site for the industry, protected by walls made from the slag waste-products. For its day it was a good example of conservation

---

[10] W.H.B. Court, *The rise of the Midland industries, 1600–1838*, Oxford University Press, London, 1953, p. 183.
[11] G.A. North, *Teesside's economic heritage*, Cleveland County Council, 1975, p. 19.

of land and materials. In northern Durham the Cleveland deposits were not so accessible and the famous Consett works of the Derwent Iron Company used local ores at first but, recognising a growing dependence on imported ironstone, in 1872 acquired a share in a company exploiting vast iron-ore deposits in Spain.

In Scotland the Carron iron-works, founded near Falkirk in 1759, was the first to use coke instead of charcoal. Because of the lack of local skills it had to employ labourers of all kinds from England, especially from Coalbrookdale. By 1793 it was said that the Carron and Clyde iron-works between them consumed as much coal as all the inhabitants of Edinburgh. Scottish iron-production was enormously increased by the rather belated recognition that the blackband ironstone, discovered in 1801 by David Mushet, had a high enough proportion of carbon for its own reduction. By 1842, 65 of Scotland's 88 furnaces were exploiting this mineral in one parish of Lanarkshire (Old Monkland).

Many improvements in the coke-fired blast furnace were introduced, including the use of limestone mixed with the ore to remove acidic oxides from the material (notably phosphorus). Few were more important than the invention in 1828 of a Glasgow gas-works manager, James Neilson. He showed that a hot blast of air (150 °C) can reduce coke consumption in a blast furnace by about a third. In fact the hot blast made it possible to use coal instead of coke, an advantage in Scotland where most coal deposits do not readily form coke. Meanwhile, however, other developments had taken place that increased still further the demand for iron.

## (c)    The puddling process

The large-scale manufacture of wrought iron, malleable and of great value in a multitude of situations where the brittle cast iron would be totally unsuitable, came about with a discovery in 1783. Henry Cort was born in Lancaster and became a Navy agent with little experience of metal-working. However, he owned a small forge at Funtley on the River Meon in Hampshire, a mile or two north of Fareham, and was under contract from the Royal Dockyard at Portsmouth to make iron hoops for ships' masts. Some water power was available from the small River Meon, which he diverted and dammed. The area was of little importance for iron-making, but it was very accessible to the naval market. Cort's achievement was to take an existing type of furnace, line it with haematite and use that to remove the excess carbon from cast iron. In his reverberatory furnace the iron bars were heated by hot gases from burning coal, deflected on to their surface by a downward-sloping roof; there was no direct contact between the fuel and the iron. Essentially the reactions were oxidations of the carbon by ferric oxide and by injected air:

$$3C + Fe_2O_3 = 2Fe + 3CO$$
$$2C + O_2 = 2CO$$

During the process the semi-fluid mass had to be stirred manually ('puddled').

*Cort's mill at Funtley. The water wheel was powered by the small River Meon, and drove rollers and tilt-hammer*

*Nasmyth's depiction of his steam hammer at work on a bar of wrought iron*

The iron was then withdrawn, rolled and hammered to extrude slag; if necessary the process was repeated. The puddling process enabled a furnace to produce 15 tons of wrought iron in twelve hours, compared with one ton by the traditional hammering method. In 1842 the invention of the steam-hammer by James Nasmyth gave an immensely valuable adjunct to the equipment already available. The result was a striking increase in growth rate of the British iron industry, so that by 1850 about 2½ million tons of iron were being produced each year. From the start puddled iron was used for making anchors but a multiplicity of other outlets soon became apparent.

*View and section of typical blast furnace of the 19th century*

There is little doubt that Cort's discovery initiated a mass movement of forges from the south of Britain to places with coal-bearing strata, just as the development of coke-fired blast furnaces had done. In South Wales iron-works established by immigrant Englishmen earlier in the century took up the puddling process with considerable zest from about 1790. This followed a visit to Funtley in 1787 by the partners in the great Cyfarthia iron-works, Richard Crawshay and James Cockshutt, the latter also being manager of the Wortley Ironworks near Sheffield where the new method was probably introduced at about the same time. In north east England Losh, Wilson & Bell, owners of a rolling mill at Walker-on-Tyne adopted the puddling process in 1833. One of their major products was wrought iron rails, widely introduced for railway track by John Birkinshaw of the Bedlington Iron Works in about 1820. Cort's process was soon adopted by Hawks & Co. at Gateshead (whose workers had a formidable reputation for truculence and pugnacity), the Derwent Iron Company at Consett, and others. One of the first activities of Bolckow and Vaughan at Middlesbrough was to set up a puddling furnace, in 1841. By 1850 there were about 300 such furnaces in the North East, and by 1864 a total of 646 (with 99 at the Derwent Iron Company).[12] Only with the later large-scale replacement of wrought iron by steel did the puddling process begin to decline.

[12] *Industrial resources* (ref. 8), p. 115.

The last puddling plant in the world, Walmsley's at Bolton, closed down in 1975.

## 3    The Coming of Steel

Iron of very high quality had been obtained in remote antiquity by heating and hammering wrought iron in the presence of charcoal, when the carbon particles diffuse into the metal to produce what we should call a steel (the cementation process). Such a product was in demand for tools and weapons required to have a sharp cutting edge. It was made in the Weald and in the North East, where Swedish bar-iron was imported;[13] Isaac Cookson was one manufacturer. The chemistry was entirely unknown but the result was spectacular. The story has been repeatedly told, but a brief account of the main stages may be helpful.

### (a)    Huntsman's crucible steel

Benjamin Huntsman was a Doncaster clock-maker in the 18th century who was not satisfied with the available steels made by the cementation process. Yet another person to take advantage of the newly available coke, with its high fuel potential, he tried heating some of the existing steel in a clay crucible, using coke as a fuel to achieve the high temperatures needed to melt the steel completely. Under these conditions small amounts of slag that had remained in the steel separated out and could be skimmed off. Also, a much more homogeneous distribution of carbon through the steel was accomplished, and the product had greatly superior properties for use in cutting tools. He perfected his process in Sheffield (to which he had moved to set up a steel manufactory) in the early 1740s. It was the beginning of that town's reputation as a world-class producer of high-grade steels and cutlery. But crucible steel was always expensive and a large-scale alternative was needed.

### (b)    The Bessemer converter

Henry Bessemer was a professional inventor who had devised a new type of gun barrel, for which he was experimenting with wrought iron. In 1856 the idea occurred to him of not allowing hot air to play upon the surface of hot cast iron, but rather of blowing it *through* the molten metal. He was rewarded by a spectacular display of flames and sparks at the mouth of the vessel, and it became clear that all carbon had been oxidised away from the original cast iron. No external heating was necessary as heat is generated by oxidation of silicon and manganese, whose oxides pass into a slag. Some of the iron is oxidised to $Fe_2O_3$ which then converts the carbon into carbon monoxide. As this is an exothermic reaction the temperature actually increased. The process was further improved by the addition to the molten steel of a calculated quantity of *Spiegeleisen*, a compound of iron, manganese and carbon. The manganese

[13] K.C. Barraclough, 'Swedish iron and Sheffield steel', *Hist. Tech.*, 1990, **12**, 1–39.

*A Bessemer converter*

removes any last traces of oxygen and the carbon brings the level up to what is required for a particular steel.

Ever the opportunist, Bessemer patented his process at once (1856), but trials by other people led to disastrous results. Too soon he realised that serendipity had played a considerable part in his experiments. Eventually it became clear that:

(a)  he had not lined his converter with a silica-rich refractory, and
(b)  his ores had an unusually low phosphorus content.

If a furnace lining is acidic (as silica) any phosphorus will not be removed and will remain in the iron with highly detrimental results to its properties. Given that there was little phosphorus in any case, and that what little there was would readily have separated from the iron, Bessemer was extraordinarily lucky. Those who attempted to work his process were not, for there was nothing to suggest a new kind of lining and most ores available in England were rich in phosphorus.

Nevertheless Bessemer had some success with non-phosphatic ores such as haematite, obtainable from Furness and Cumberland. He established his own factory at Sheffield two years later. One great advantage of Bessemer steel over wrought iron was a much greater suitability for railway lines. Rail made of the new mild steel lasted five times longer than its wrought iron counterpart, though it was much more expensive at first. There were experiments on it at Derby in 1857, a trial use on the High Level Bridge over the Tyne in 1862, and installation by the London & North Western Railway of a Bessemer plant at Crewe in 1864.

There remained the considerable problem of phosphatic iron ores. A solution was found by an amateur chemist, Sidney Gilchrist Thomas who suggested a *basic* lining to the Bessemer converter. By using calcined dolomite (a mixture of calcium and magnesium oxides) and adding limestone to the melt he succeeded in obtaining a steel that was free from phosphorus, which ended up in the 'basic slag' (a saleable by-product). In association with his cousin Percy Carlisle Gilchrist, chemist at the Blaenavon steel-works in South Wales, he had developed a workable process by the late 1870s. Persuaded by the management of Bolckow and Vaughan they moved to Middlesbrough and, on 4 April 1879, produced phosphorus-free steel from a Cleveland phosphatic ore. The product, known as Bessemer basic steel, marked a new era for steel production on Teesside and helped to continue the already swift decline in wrought iron production in favour of steel.[14]

Meanwhile another steel-making process was being developed at exactly the same time as that invented by Bessemer.

## (c) The open hearth process

Charles William Siemens was a young German with a flair for invention, employed in a succession of technical posts in England. At this time thermodynamics was slowly becoming recognised and problems of heat transfer and loss were being actively investigated. Siemens and his brother Frederick took out a patent for the utilisation of waste heat from furnace exhaust gases. The idea was immediately applied to iron extraction by a French iron master, Emile Martin, and the resultant process was sometimes incorrectly called 'the Siemens–Martin process', though this is to restrict it unnecessarily to one particular version.

In essence Siemens used a shallow open hearth furnace of the reverberatory type, heated by a gaseous fuel (such as producer gas), whose flames impinge directly on the charge while oxygen from the air removes the carbon. By means of a simple system of heat interchangers much of the heat in the escaping gases could be recovered and used to warm up the fresh intake, thereby saving fuel. This was important when it is recalled that the only heat required by the rival Bessemer process was that to melt the iron.

There were several advantages of the open hearth process. It could be used for scrap metal in almost any proportion (unlike the 5% maximum for Bessemer); as it was a much slower process (6–12 hours) it could be controlled more readily than the 20 minute pyrotechnics from a Bessemer converter; the iron did not absorb detrimental traces of nitrogen from an injected air blow; it could operate with ores that were low in phosphorus, the Bessemer basic process requiring at

[14] J.K. Almond, 'A century of basic steel: Cleveland's place in the successful removal of phosphorus from liquid iron in 1879, and development of basic converting in [the] ensuing 100 years', *Ironmaking & Steelmaking*, 1981, **8** (no. 1), 1–9.

least 1.7% of the element to generate the necessary heat; above all it was applicable to the phosphorus-rich ores so abundant in England, the phosphorus being removed by added lime flux.

Progress in England was at first slow, but in 1867 Siemens set up what was effectively a pilot-plant in Birmingham and, two years later, a much larger works in South Wales, the Landore–Siemens Steel Company. The Steel Company of Scotland was set up in 1871 to operate the Siemens process but the latter made its greatest mark on Teesside, where, as elsewhere, it had to compete with the modified Bessemer process. Again it took some time to get established. The firm of Dorman Long, founded at Middlesbrough in 1876 to make iron bars and angles for ship-building, moved from iron to steel in the early 1880s, replacing 122 puddling furnaces with seven open-hearth furnaces in order to make steel from imported haematite ores. Later they transferred to the Gilchrist–Thomas variation on the Bessemer process in order to use the local high-phosphatic ores. They became famous for bridge girders, most notably for the Sydney Harbour Bridge (1932). Bell Brothers at Port Clarence were also extensive users of the open-hearth process throughout the 1890s, opening new plant for the purpose in 1901. Faced with intense foreign competition the firm amalgamated with Dorman Long in 1923.

## (d)    Later modernisations

Many other developments were of an entirely non-chemical nature, such as casting machines and improved rolling mills of many kinds. Mechanisation of rolling mills in the 1860s meant that much heavier pieces could be handled. A reversing mill, enabling the metal to be passed to and fro through the rollers, was devised at Crewe by J. Ramsbottom of the LNWR in 1866. A continuous mill for wire rods was invented by G. Bedson at Manchester in 1862. The continuous wide strip mills, invented in the USA, made their first appearance in Britain at Ebbw Vale in 1938. A mechanically charged blast furnace, also devised in the USA, was introduced into Britain at Frodingham Ironworks, Scunthorpe, in 1904/5.

With the coming of relatively cheap electricity, electric power began to displace steam for much of the machinery involved in iron and steel production. In Britain this was surprisingly slow and was far from complete by 1939. The same is true of electric furnaces. An electric arc furnace was patented by William Siemens in 1878, though not used for steel until 1900. It was developed in the First World War, especially in Sheffield, for making steel from large amounts of iron scrap from the munitions industry. An induction furnace (invented by Ferranti in 1877 in Italy) was first employed in Britain on a production scale at Sheffield in 1927.

Important developments of alloy steels[15] took place in British industry. In 1868 Robert Forester Mushet, son of the redoubtable David, was producing crucible steel in the Forest of Dean and found that addition of tungsten gave air

[15] P.S. Bardell, 'The origins of alloy steels', *Hist. Tech.*, 1984, **8**, 1–29.

hardening properties, thus applying an Austrian discovery made some ten years previously. Seeking a metal that would harden in use as a tool, he found tungsten an ideal material for the heavier machine tools then being developed. Robert Hadfield was the son of a Sheffield steel maker, and in attempting to improve the firm's products experimented with added metals including manganese. This was quite apart from the established use of manganese as a deoxidant in the blast furnace. His manganese steel, with 12½% manganese, was not hard but immensely tough. It became widely used where there was great wear and tear, giving a five-fold increase of life to steel employed at railway crossings and junctions. The first major development of a ferromanganese industry was not, however, in Britain, but in France, in the 1860s. French railway companies exerted strong pressure for more and better rails, blast furnace design and conditions were good, and manganese ores could easily be imported from countries on the Mediterranean seaboard (such as Spain, Italy and Algeria).[16]

Another Sheffield invention was stainless steel, devised by Harry Brearley. In the course of experiments in 1913 on steel for rifle barrels he discovered that addition of chromium produced an extraordinary resistance to corrosion and staining in hardened steel. It became the first of an immense range of stainless steels used for cutlery, kitchen vessels and much else.

The advent of alloy steels has, since about 1910, enhanced the quality of domestic products and led to important advances in engineering. Yet, as P.S. Bardell has pointed out, there is another side to the picture. He writes:

> The desire for military superiority and the demands of warfare have been a constant stimulus to the search for and production of alloy steels throughout the past century.[17]

Countries with the appropriate mineral resources have thus gained immense political significance, such as Belgian Congo (cobalt), Rhodesia (chromium), Peru (vanadium) and New Caledonia which could supply chromium, manganese, cobalt and nickel. He argues that these deposits could even be a cause for war, as in Hitler's invasion of Yugoslavia (whose chromium was intended entirely for Britain) and the German military campaign for the manganese-rich Caucasus.

Finally mention should be made of the old process of 'pickling' iron to remove surface corrosion. This has traditionally used sulphuric acid (and, incidentally, accounted for a substantial consumption of that chemical). It continued to be applied to steel and inevitably added to the pollution of streams and rivers, though surely not so spectacularly as the 'picturesque example' from the USA. Here 'it is calculated that the iron sulphate flowing to waste from the American steel pickling plants contains sufficient iron to build 200,000 motor cars a year, not to mention the vast associated loss of sulphur!'

---

[16] E. Truffaut, 'The manganese industry in France in the 19th century', *Revue de Métallurgie, Cah. Inf. Tech.*, 1997, **94** (7/8), 955–966.
[17] Bardell (ref. 15), p. 22.

## 4   Chemistry and Iron Making

The iron and steel industries are often regarded as, at best, only a peripheral part of what is conventionally known as 'the chemical industry'. Often they are excluded altogether. Yet from what has been written so far it can be seen that there is a number of important ways in which chemistry was closely linked to the production of iron and steel. We can enumerate them thus:

### (a)   Overlapping personnel

From the late 18th century it was often the *same people* making (say) sulphuric acid and iron. There are dozens of examples of whom perhaps Garbett and Roebuck are the most prominent. Not only were they involved in the manufacture of acids in the English Midlands, but they also became partners in the Carron Iron Works in Scotland. Lord Dundonald dabbled with equal happiness in iron production and tar chemistry. Even the great Dr Joseph Black expressed opinions on iron and steel as well as fixed air. James Neilson, inventor of the hot blast, had been practising some kind of chemistry at Glasgow gas-works, while James Nasmyth of steam hammer fame pursued a strong amateur interest in chemistry as well as an interminable correspondence with the chemist Edward Frankland.

In mid-Victorian times this extensive overlap of interest is well seen in the Newcastle Chemical Society, whose founding president (1868) was Isaac Low-thian Bell, who became one of the most famous iron masters of the century. He had founded the iron-works at Port Clarence while still helping to run the Washington Chemical Works. His home, Washington Hall, was said to be alive with scientific and philosophical debate of the highest order. His father-in-law was Hugh Lee Pattinson, and his father, Thomas Bell, in the partnership Losh, Wilson and Bell, had simultaneously manufactured alkali and iron products at their Walker site. A later president (1879) was the chemical manufacturer R.C. Clapham, but he gave an address on 'The manufacture of steel from Cleveland iron'. Other members were concerned with iron-works at Spennymoor, Jarrow, Thornaby and Middlesbrough. All of this is not really surprising in view of a second consideration: chemical knowledge was useful, if not indispensable, to the extraction and fabrication of iron.

### (b)   Use of chemical knowledge

It is remarkable that many of the leading lights in British iron production had received some kind of chemical training, though at first formal qualifications did not exist. It has been well said of Roebuck that 'he had had such formal training in chemistry as was possible at that time'.[18] Consequently he set about quantitative studies of iron production at Carron. James Neilson, while at Glasgow gas-works attended science classes at the Andersonian Institution. An

---

[18] A. and N.L. Clow (ref. 9), p. 337.

interesting contrast exists between P.C. Gilchrist, the Blaenavon chemist, who had attended the Royal School of Mines in London where he must have learned some chemistry, and his cousin, S.G. Thomas, who although a police court clerk contrived to attend evening lectures in chemistry at the Birkbeck Institution (where, incidentally, he was prompted to enquire into the possibilities of adapting the Bessemer process to phosphatic ores by a chance remark by his teacher).

Such training was necessary if chemical knowledge was to be fruitfully applied to the industry. For centuries, of course, this was not the case but by the later 18th century the relevance of chemistry was increasingly obvious. Thus the ores from the Carron iron-works were sent to Birmingham for careful analysis, and trained analysts became in demand. This is well illustrated by the creation of an analytical laboratory at Crewe railway works (1864), whose duties included analysis of the materials for the production of Bessemer steel and of its products. Joseph Reddrop became chief chemist there in 1867 and developed the bismuthate technique for determination of manganese in steel. Analysis of steel rails was also a concern of Lowthian Bell, who tried to correlate their durability with their chemical composition.[19] By the 1870s many metal-producers were advertising for people who could competently analyse metals and ores.[20] Nor was it merely a matter of chemical analysis. As early as 1786 a paper by Berthollet and his colleagues showed that the differences between steels is chiefly due to their carbon content, though Guyton de Morveau had also come to the same conclusion.[21] The labours of Gilchrist and Thomas are perhaps the supreme example in the 19th century of the application of sound chemical theory to a practical problem of awesome dimensions. From then on many chemists wrote on iron constitution and metallurgy but the first major treatise was by I. Lowthian Bell, the culmination of many years' work at his Port Clarence laboratory, *The chemical phenomena of iron smelting*.[22] They were part of the chemical industry.

## (c)  Material transfer

By the late 18th century one product of the chemical industry was becoming useful for iron processing. This was sulphuric acid, or its derivative hydrochloric acid, employed for cleaning or 'pickling' metal, though it is impossible to say how much was used in the process. Other metals produced by chemical processes were used for alloying steel, notably tungsten, chromium and manganese. Oxygen, once proposed by Bessemer as an alternative to air in his process, was not in fact so used until after the Second World War, for the simple reason that that part of the chemical industry concerned with gas production

[19] Nevertheless Bell continued to believe in the superiority of wrought iron over steel for railway track, though his arguments were based not upon chemistry but upon apparent rates of wear and tear: D. Brooke, 'The advent of the steel rail', *J. Transport Hist.*, 1986, **7**, 18–31.

[20] C.A. Russell, N.G. Coley and G.K. Roberts, *Chemists by profession*, Royal Institute of Chemistry/Open University Press, Milton Keynes, 1977, pp. 97–98.

[21] J.R. Partington, *A history of chemistry*, vol. iii, Macmillan, London, 1962, p. 530.

[22] I.L. Bell, *The chemical phenomena of iron smelting*, Routledge, London, 1872.

was not until then in a position to produce enough from liquefied air. And of course in the laboratories of the large iron-works chemicals in plenty were in demand for analyses.

Nor was the debt in only one direction. Products from the furnaces such as basic slag were taken by chemical manufacturers for sale as fertilisers. This was the case in Middlesbrough from about 1887. Since many iron-works had coking ovens on site there were large quantities of by-product tar and some firms, such as Bell Brothers, exploited this tar as a source of other chemicals (though it has to be recognised that in this respect many British firms lagged behind the Continent). And most parts of the chemical industry became totally dependent on the availability of high-quality iron and steel for creation and maintenance of their manufacturing plant, everything from nuts, bolts and pipes to tanks, fractionating columns, and immense reaction vessels, often of stainless steel. It is therefore not surprising that on many sites chemical and iron-making plants operated in tandem, as witness the title of the Walker Iron and Chemical Company, or the existence of a blast furnace at the Seaham Chemical Company near Sunderland.

## 5  Copper in South Wales

### (a)  Early beginnings

Although the period during which copper was mined on a large scale in Britain was a comparatively brief one, confined chiefly to the 18th and 19th centuries, the exploitation of this mineral in Britain has a long history. Roman copper mining was concentrated in North Wales where ore deposits were shallow. Copper deposits were worked on a small scale in the Middle Ages, but most of the copper used in Britain during this time was imported. There is no record of significant copper production before the 1560s.[23]

The second half of the 16th century saw an intensification in Britain's exploitation of minerals. German capitalists, and hundreds of skilled workmen from Saxony, were invited by the Crown to develop a copper industry in the Lake District. Copper ore was mined and smelted at Brigham, near Keswick in Cumberland, under the auspices of the Company of Mines Royal. Letters patent were granted to Thomas Thurland and Daniel Hochstetter of Augsburg, who formed the company in 1568 with a further twenty-seven shareholders, to manufacture copper.[24] The sulphide ores used at Keswick were subjected to preliminary roasting to burn off excess sulphur, and then treated with nine horseloads of peat and five horseloads of 'stone coals' (a horseload was

[23] R. Chadwick, 'Copper: the British contribution', *Chem. Brit.*, 1981, **17**, 369–373; R.E. Glasscock, 'England circa 1334', in ed. H.C. Darby, *A New Historical Geography of England before 1600*, Cambridge University Press, Cambridge, 1976, 136–185 (172); G. Hammersley, 'Technique or economy? The rise and decline of the early English copper industry, *ca.* 1550–1650', *Econ. Hist. Rev.*, 2nd series, 1973, **26**, 1–27 (1); D.B. Barton, *A history of copper mining in Cornwall and Devon*, D. Bradford Barton, Truro, 1968, p. 9.

[24] M.B. Donald, *Elizabethan Copper: the history of the Company of Mines Royal 1568–1605*, Pergamon Press, London, 1955, pp. 7, 15, 24–25. A list of shareholders in the company is given on p. 242.

equivalent to 109 litres). Limestone was added as a flux and after smelting a matte or 'green stone' was run off. Subsequently, about eight days' recovery of matte was roasted with six peat fires, each hotter than the last, to produce 'copper stone' or 'black copper'. This was smelted once a month to give 'rough copper', and involved three separate smeltings with lead ore to extract the silver from the copper matte. This process of making copper at Keswick took eighteen weeks and five days.[25]

A smelter was in operation at Brigham until *ca.* 1640, while a second works at Aberdulais near Neath in Glamorganshire operated from 1584 to *ca.* 1603. Production from these two works could reach an estimated maximum of 100 tons a year, but was erratic and often fell significantly below this figure, an insignificant output by Continental standards. Other works were set up in Cornwall (*ca.* 1583–1603) and Staffordshire (*ca.* 1660–1670) but all other attempts to get copper in Britain between *ca.* 1610 and 1660 did not prosper. G. Hammersley argued that 'it was in essence a mining speculation which failed', explicable partly by the absence of strong demand for copper within the British economy, and partly by the economic advantages enjoyed by the Tyrolean, Hungarian and Swedish copper-mining industries.[26]

The environmental impact of the early-modern copper industry in Britain was therefore small: the smelters were high in fuel consumption, but records at Keswick suggest that charcoal and wood did not constitute more than 20% of the weight of fuel used after *ca.* 1570, peat and coal making up the remainder. At maximum production, the smelting of copper at Keswick could have consumed the produce of about 90 acres of woodland. Nevertheless, one historian has argued that a timber shortage contributed to the decline of the industry in the Lake District.[27]

For the remainder of the 17th century copper mining had become almost moribund in Britain, faced by cheaper supplies of Swedish copper dominating the European metal markets. The Society of Mines Royal began to concentrate on mining and smelting lead, and its monopoly on copper production was abolished by Acts of Parliament in 1689 and 1693. At the close of the century a new industry was founded based on copper smelters located at Redbrook on the Wye below Monmouth, and at Bristol, supplied by Cornish ore. The smelting works established by the Society of Mines Royal at Neath was considerably extended, and the first Swansea smelting works was established in 1717. By 1720, some 6000 tons of Cornish ore were being raised annually, this total being shared almost equally between the three groups of smelters. The Swansea works had access to ample cheap coal supplies, and was a day's sailing closer to Cornwall than Bristol or Redbrook. With these advantages, Swansea soon came to displace the other areas as the main smelting centre.[28]

The rise in the copper mining and smelting industries in the 18th century was

[25] *Ibid.*, pp. 182–215.
[26] Hammersley (ref. 23), pp. 1–2, 27; the quotation appears on p. 27.
[27] The reference is to J.U. Nef. See Hammersley (ref. 23), p. 19 and n. 1.
[28] Barton (ref. 23), pp. 11–12, 14–15.

*The old Forest Copper Works, Swansea, ca. 1730*

sustained by a big increase in market demand for copper.[29] Copper was used for the first time for British coinage, a market which absorbed 700 tons. Additionally, there was growing demand, both from within and from outside Britain, for a wide range of copper goods manufactured by 'battery', whereby items like pots and kettles were first cast and then hammered into form. Copper was also a constituent of brass (a copper–zinc alloy). Products containing copper included kitchen utensils, buttons, buckles, harness parts, guns and armour. By the 1780s the brass-using trades of the Birmingham area consumed about 1000 tons of copper annually. From the late 18th century, copper was also used for sheathing the hulls of naval and merchant vessels. Behind the success of this 18th century industry lay several important technical advances: the successful adaptation of reverberatory furnaces for burning coal, releasing the industry from its earlier requirement for charcoal, whilst gunpowder, and the development of steam pumps, facilitated mining at greater depths. Copper lodes were found below the upper veins of tin in Cornish workings, and the vastly increased exploitation of deep tin was another important factor in the opening up of Cornish copper-ore deposits. Specialist workmen from Germany and Sweden were also induced to operate in British copper works, lending their expertise to the enterprise. Output from the British copper industry grew rapidly, from under 1000 tons in 1712 to *ca.* 7000 tons at the beginning of the 19th century.[30]

Cornish mines were displaced from their pre-eminent position as suppliers of ore following the discovery of a vast deposit of shallow low-grade ore on Parys

[29] *Ibid.*, p. 15; R.O. Roberts, 'The development and decline of the non-ferrous metal smelting industries in South Wales' in W.E. Minchinton (ed.), *Industrial South Wales, 1750–1914: essays in Welsh economic history*, Frank Cass, London, 1969, pp. 121–160 (122).

[30] Barton (ref. 23), pp. 19–21, 27–28; Roberts (ref. 29), pp. 123–125.

Mountain in Anglesey in 1768. Until the 1790s, a flood of Welsh ore depressed the Cornish mining industry. By this date, Anglesey's vast open-cast mine covered an area over 17 acres, and had yielded 2.5 million tons of ore. The surviving Cornish mines were forced into making economies, concentrating primarily on the reduction of fuel consumption by the pumping engines. By 1775 John Smeaton had improved the Newcomen engine to a point where its inefficiency had been more than halved. The introduction of the Boulton and Watt engine into Cornwall achieved further efficiency gains and reduced coal consumption; for example, the Consolidated Mines installed five of the new engines in 1782, saving £11,000 annually on coal costs.[31]

## (b)  Copper mining in the 19th century and its environmental impact

In the 1790s, the supply of easily-worked ore from Anglesey declined, heralding a new era of expansion in the Cornish copper mining industry. Cornish mines now led the field in supplying the world's copper markets, and not until the 1820s did the industry face serious foreign competition, when the first really large shipments of foreign ore arrived at Swansea for smelting. The single parish of Gwennap at this time yielded more than 30% of the world's entire copper output, and its population doubled in the first three decades of the 19th century. The number and size of mines increased, and the output of ore from Cornwall and Devon steadily rose to a peak of 191,130 tons in the five-year period 1855–1860.[32]

The environmental impact of copper mining has received little attention in histories of the industry. Of chief concern to contemporaries was the effect of mining on the purity of local water-courses. In the early 1860s, the newly appointed Salmon Fishery Commissioners found that waste-water and slag from copper mines had damaged the fisheries of several rivers. In Cornwall, 'much apprehension was entertained that any steps taken for the purification of rivers would restrict the enterprise of the miners'. On the Tamar, the Devon Great Consols Copper Mine, at this date the largest British copper mine, employing 1200 men, was alone amongst the mines along the river in making attempts to purify mine waste-water by using settling pits to intercept the waste. Elsewhere, working mines, and mines newly fallen into disuse, continued to pour highly poisonous effluent into the rivers. On the whole, however, a far more serious threat to aquatic life was posed by lead mining.[33] After mid-century, copper mining in Cornwall steadily declined, slumping dramatically in 1868 as world copper prices fell below the break-even point of profitability. The county's whole economy was based on mining and its ancillary trades, and the collapse of the copper mines forced many miners to emigrate to mines in the Americas or Australia. The gradual shift of copper mining abroad displaced the

[31] Barton (ref. 23), pp. 26–27, 31–32, 39.
[32] *Ibid.*, pp. 50–51, 94 (Appendix II). See also R. Burt, P. Waite, and R. Burnley, *Cornish mines: metalliferous and associated minerals 1845–1913*, University of Exeter/Northern Mine Research Society, Exeter, 1987, p. xxxviii (Table 1).
[33] *Annual Reports of the Salmon Fishery Inspectors*, P.P., 1863 (84) XXVIII, pp. 61–64; 1867 (640) XVIII, p. 190; 1867–1868 (160) XIX, p. 836.

environmental impact of the industry to new areas. Beginning in the 1820s, exporters of ore, and later of 'regulus' (part-smelted ore), emerged in South America, where smelting took place on a small scale and was high in timber consumption. After the mid-1860s, new exporters began to play an important role: particularly North America, southern Africa, Spain, and Portugal.[34] Overall, the history of the exploitation of copper minerals was one characterised by a pattern of 'mining progressively leaner ores from increasingly large individual deposits'.[35] For example, average English copper ore yielded from 9.27% copper in 1800 to 6.56% copper in 1870–1885, whilst the average for ore from the United States was only 2.5% in 1906. The lower-grade ores, occurring in huge disseminated ore-bodies, pushed production towards larger-scale and more capital-intensive mining enterprises. Overall, the contribution of Britain to world copper mineral production fell from 35.5% in 1850–1859 to 1.9% in 1900–1909.[36]

### (c) The rise of copper smelting and its environmental impact

By the mid-18th century the smelting of copper in Britain was already largely concentrated in the western part of the South Wales coalfield; in 1750, the Swansea area was responsible for 50% of total British production. The bulky fuel requirements of the industry, about three tons of coal per ton of copper smelted, weighted the locational advantages to the coal-field, rather than the ore-mining areas like Cornwall, lacking in coal.[37] This remained the case up to the 1880s, when, on average, smelting techniques required two tons of coal to smelt one ton of copper ore, coal accounting for about 45% of total smelting costs.[38] Additionally, by locating in the western, coastal areas of the South Wales coalfield, smelters had access to a range of coals: anthracite, sub-bituminous, coking and gas coal. The sustained growth of the industry in the 19th century was fuelled by new markets for copper, including copper boilers and tubes for steam engines; copper vats for brewing, distilling and dyeing; copper rollers for textile printing; and copper wires and plates for telegraph and telephone equipment. In 1857, it was estimated that there were 600 furnaces employing 4000 workers in the Llanelli to Port Talbot area. By this period, the output of the British copper industry had increased to *ca.* 22,000 tons, reaching its peak in the late 1880s and early 1890s, with South Wales possessing about 90% of Britain's copper smelting capacity.[39] The British smelting industry benefited from the lag between the development of copper mining and the establishment of copper smelting overseas, allowing the Swansea area to become the principal world export market for copper ore. After the 1820s,

---

[34] E. Newell, '"Copperopolis": the rise and fall of the copper industry in the Swansea district, 1826–1921', *Business History*, 1990, **32** (3), 75–97 (78–79); Barton (ref. 23), p. 95 (Appendix II).

[35] C. Schmitz, 'The rise of big business in the world copper industry 1870–1930', *Econ. Hist. Rev.*, 1986, 2nd series, **39**, pp. 392–410 (398–399).

[36] Burt *et al.* (ref. 32), p. xxxviii (Table 1).

[37] Roberts (ref. 29), pp. 125–127.

[38] Newell (ref. 34), p. 75.

[39] Roberts (ref. 29), pp. 122, 125, 135.

*One of 13 copper smelting works to appear in the Lower Swansea Valley by 1850, the Bristol Company Copper Works in action in 1813*

British copper ore imports were made from an increasing diversity of sources, especially Cuba, Chile and Australia, later supplemented or superseded by supplies from North America, South Africa, Venezuela, and elsewhere. In the late 1840s the industry also began importing regulus on an increasing scale.[40]

The 'Welsh process' which characterised the copper smelting industry of South Wales, and differentiated it from its European counterparts, was based on the utilisation of a wide range of ores of varying chemical composition (copper sulphide ores: $Cu_2S$, $CuFeS_2$, and carbonate/oxide ores: $CuCO_3$, $CuO$, $Cu_2O$), but most notably high-grade sulphide ores, using coal-fired reverberatory furnaces. Newell and Watts have recently discussed the chemistry of this process of copper refining in some detail.[41] The Welsh process was a four-stage operation involving calcination of the ore to reduce its sulphur and arsenic content, smelting by melting the calcine with a silica-based flux, converting the resulting 'matte' or 'regulus' from $Cu_2S$ into $Cu$, and finally fire-refining to produce copper of 99.5% purity. For carbonate/oxide ores the initial calcination stage was sometimes omitted, the ores being smelted immediately, but with the addition of sulphur to the charge. Large amounts of sulphur and arsenic were released during the first calcination stage, making this the chief cause of the environmentally damaging 'copper smoke' pollution. In addition, lead, cadmium, antimony, and traces of hydrofluoric acid were present depending on the ores being used and smelting conditions. The main chemical species emitted from copper smelting works were $CO_2$, $CO$, $SO_2$, $SO_3$, $H_2SO_3$, $H_2SO_4$,

---

[40] Newell (ref. 34), pp. 78–79.
[41] E. Newell and S. Watts, 'The environmental impact of industrialisation in South Wales in the nineteenth century: "Copper smoke" and the Llanelli Copper Company', *Environment and History*, 1996, **2**, 309–336 (318–322).

$H_2O$, $As_2O_3$, HCl, S, $Sb_2O_3$, Pb and Cd. The principal chemical reactions of the smelting process producing copper smoke were oxidation of coal to oxides of carbon and of copper pyrites to sulphur dioxide, followed by conversion of the latter to sulphurous and sulphuric acids; further reactions included:

$$FeAsS \longrightarrow FeS + As(g) \ (700\,^{\circ}C)$$
$$FeS_2 \longrightarrow Fe + S_2$$
$$2Sb_2S_3 + 9O_2 \longrightarrow 2Sb_2O_3 + 6SO_2$$
$$2As_2S_3 + 9O_2 \longrightarrow 2As_2O_3 + 6SO_2$$

and evolution of hydrogen halides by liberation from mineral halides with sulphuric acid.

The damaging effects of these emissions resulted in measures to exclude copper smelting from the borough of Swansea as early as the 1760s, a measure which merely displaced the industry beyond the borough boundary. Litigation for damages against the industry dates back to at least the 1820s, damages commonly being alleged to crops, livestock, and more controversially, to public health.[42]

By the mid-19th century, the effects of copper smoke in the Swansea region were all too obvious to all. As one eye-witness put it, the appearance of the affected areas was 'extraordinary': 'The immediate district of Swansea, which is subjected to the direct and concentrated influence of the copper smoke, is entirely denuded of vegetation; the hillsides have not a blade of grass upon them, but are converted into a mass of debris of gravel and stones'.[43] Further away from the immediate source of the pollution, evidence of rabbits, sheep, and horses dying or failing to thrive after having grazed the copper- and arsenic-contaminated pastures mounted. A recent study of the Llanelli Copper Company during the 1860s, a company responsible for *ca.* 15% of copper production in South Wales at this time, has estimated that heavy metals (As, Sb, Pb, Ag, Cu) were deposited within 12 km of the smelting works in concentration of up to 10–15 $\mu$g m$^{-3}$, which is 2–3 times the maximum levels measured in the vicinity of modern smelters or modern cities. This implies widespread contamination of crops, grazing land, and water supplies, a finding which is supported by studies of heavy metal contamination of stream sediments in the area.[44] This study also suggests that $SO_2$ and $SO_3$ concentrations reached 100 $\mu$g m$^{-3}$ over 10 km away from the Llanelli Copper works, five times greater than peak $SO_2$ concentrations in the modern city, and implying the incidence of respiratory problems in the local population.[45] Considering the emissions from the copper industry in Llanelli and Swansea as a whole, it is probable that the effects would have been felt over much of South Wales, as well as the North Devon and

[42] R. Rees, 'The South Wales copper smoke dispute, 1833–1895', *Welsh History Review*, 1980–1981, **10**, 480–496 (482, 483 and ref. 13, 484–486).

[43] *Report from the Select Committee of the House of Lords on Injury from Noxious Vapours, 1862*, P.P., 1862 (486) XIV, p. 63. See also p. 67.

[44] Newell and Watts (ref. 41), p. 310.

[45] *Ibid.*, pp. 331–332.

*A graphic impression of atmospheric pollution from two copper smelting works (Vivian & Sons, and Pascoe Grenfell & Sons) in Swansea about 1860. It is roughly the same position as the previous picture, 47 years earlier*

Somerset coasts. The industry additionally left a legacy of toxic copper and arsenic slag in the lower Swansea valley. It has been said of this area that: 'Nowhere in . . . Britain is there a more dismaying example of man creating wealth while impoverishing his environment'.[46]

Despite the scale of the pollution, it is apparent that sections of the local community were prepared to tolerate it. Following the unsuccessful indictment brought against the copper smelters, John Henry Vivian and Sir Richard Hussey Vivian, tried at Carmarthen in 1833, for example, Swansea was 'diffused [with] the greatest joy . . . manifested by ringing of bells and firing of cannon throughout the day'.[47] In the 1840s, Swansea's surgeon and registrar was prepared to argue that copper smoke had prophylactic properties, and was a beneficial check to the spread of epidemic diseases. Neither the Swansea board of health, nor the local courts took effective action to curb the copper industry's emissions. Community response was in general to support jobs over clean air. One landowner's attempt to serve an injunction against the Rio Tinto company in Cwmavon in the 1890s, for example, was met with a petition of 3000 signatures for its withdrawal.[48]

Perhaps because pollution from the copper industry was on so vast a scale, there were early attempts by some of the captains of the industry to remedy the problem by drawing on chemical expertise. In 1821, John Henry Vivian under-

[46] J. Barr, *Derelict Britain*, Penguin, Harmondsworth, 1969, p. 79; E.M. Bridges, *Healing the scars: derelict land in Wales*, University College of Swansea, Swansea, 1988, pp. 20–21, 60–61.
[47] Rees (ref. 42), p. 487.
[48] *Ibid.*, p. 493.

took a series of experiments to try to reduce the noxious effects of copper smoke, assisted by Richard Phillips of Swansea, Michael Faraday and Sir Humphry Davy. Vivian also contributed to a prize of £1000, offered by the Royal Institution of South Wales, for a solution to the smoke problem. By 1832 Vivian's experiments had cost him £12,000.[49] The most successful of these was the construction of long flues from the calciners through which the smoke was passed and treated with water before emission from high chimneys. This process was said to be successful in removing the arsenic, hydrofluoric acid, arsenious acid and sulphuric acid, but failed to eliminate the sulphur dioxide. Later, his son, Henry Hussey Vivian, arranged for long flues to be attached to the smelting furnaces, which by simply cooling the furnace gases, was successful in preventing some of the heavy metal emissions. Finally, in the 1860s, Vivian's Hafod works adopted the Gerstenhofer furnace. The German chemist Moritz Gerstenhofer discovered that by using lead-lined flues and condensing chambers, he could convert sulphurous gases into sulphuric acid and, with further processing, into superphosphate manure.[50] Despite the claim that the Gerstenhofer furnace could cleanse the copper smoke and increase profits, other copper manufacturers were reluctant to adopt it, preferring instead to rely on the simple device of a tall chimney to carry the offensive gases away from the neighbourhood.[51] Despite his pioneering efforts to reduce copper smoke pollution, Henry Hussey Vivian was a prominent opponent of proposals to introduce government inspection of copper smelting works, and to regulate the emissions of the copper industry in South Wales. He argued that where people of a locality did not complain, no inspection was required and none of the improved processes necessary;[52] the local population of South Wales, as already discussed, was more tolerant of copper smoke than their counterparts in Lancashire and the North East. Ultimately, the legislature accepted the representations of the copper smelting industry that no technical solution was available to supply an economic remedy for their pollution problems: the industry was exempted from inspection and regulation until the Alkali Act of 1906, and was thereby deprived of the technical assistance of the Government chemists of the Alkali Inspectorate which might have led to a reduction in emissions.[53]

## (d)   The chemical industry and the growth of the 'wet-copper' industry

Sulphuric acid manufacturers had long been interested in the recovery of copper from the copper pyrites cinders which, from the mid-19th century, formed the waste-product of the industry's raw material source of sulphur. The 'wet-copper' process was a method developed for the extraction of copper from low grade sulphide ores. In this process, the burnt or calcined ore was mixed with

[49] *Report of the Royal Commission on Noxious Vapours, 1878*, P.P., 1878 (2159, 2159-I) XLIV, pp. 464–465.

[50] Rees (ref. 42), pp. 483–484, 491. See also Newell and Watts (ref. 38), pp. 312–315, for a discussion of the limitations of the technological solutions available.

[51] *Royal Commission*, 1878 (ref. 49), pp. 496–497; Rees (ref. 42), p. 492.

[52] *Royal Commission*, 1878 (ref. 49), p. 496.

[53] Newell and Watts (ref. 41), p. 312.

salt and the mixture roasted. The product was afterwards extracted with water, and the copper in the resulting solution was precipitated by the addition of iron:

$$CuCl_2 + Fe \longrightarrow FeCl_2 + Cu$$

The precipitated copper, contaminated with iron oxide and other impurities, was afterwards converted into regulus (partly smelted ore), which could then be smelted to produce marketable copper. The whole cycle of roasting, extraction and preciptitation was patented by William Henderson in 1858, and first introduced at his sulphuric acid works near Llanelli. Subsequently the process was adopted at sulphuric acid works in St. Helens, Widnes, Tyneside, and Glasgow, and became a common adjunct of sulphuric acid manufacture.[54] By 1889, *ca.* 26% of the copper produced in Britain was derived from this source, at a time when British mines contributed only 1.54%. The importation of copper pyrites, which was only 179,225 tons in 1863, had risen to 596,774 tons in 1887, supplied chiefly by Spain and Portugal. By 1887, Tyneside's five copper manufacturers were producing an estimated 9700 tons of copper annually, almost a 14-fold increase since 1863. These developments within the chemical industry helped to shift production of copper away from Swansea. The increased use of this new method of copper extraction also owed much to the development of open-cast pyrites mining in southern Spain and Portugal by the British Rio Tinto and Tharsis companies from the 1860s.[55]

By the 1870s, a groundswell of complaints against this new branch of the chemical industry was developing. Damage to agricultural land, as a result of emissions from 'alkali' and copper works in Tyneside and St. Helens, was brought to the attention of the government Alkali Inspectorate. The 1871 report of the chief Alkali Inspector, R. Angus Smith, found that the greater part of the injury sustained to vegetation, and the worst pollution of the atmosphere in many areas, was caused by the emission of 'sulphur acids', not hydrochloric acid gas, the only gas controlled by Act of Parliament (Alkali Act, 1863). Amongst other industries chiefly responsible for the emission of sulphurous and sulphuric acids, Smith recommended that 'wet copper' works be brought under the jurisdiction of the Alkali Acts, and that the definition of 'noxious gases' be extended to include sulphuric and sulphurous acids, nitric acid, hydrogen sulphide and chlorine. From 1st March 1875, under the Alkali Act, 1874, 'wet copper' works were bound to use 'best practicable means' to prevent the discharge of sulphurous and sulphuric acid gases. Meanwhile the 'dry' copper smelting industry in England and Wales continued to be unregulated.[56]

[54] W.A. Campbell, *A century of chemistry on Tyneside 1868–1968*, Society of Chemical Industry, London, 1968. Forty-six joint producers of sulphuric acid and copper using the wet process between 1867 and 1914 have been identified: J.B. McIntyre, 'The role of the wet process in the growth of the pyrite industry' (unpublished PhD thesis, Nottingham University, 1975), p. 146; quoted in Newell (ref. 31), p. 96, n. 51.

[55] G. Gatheral, 'Copper', *Rep. Brit. Assoc. Adv. Sci.*, 1889; local guide, pp. 121–127; pp. 123–125; Newell (ref. 31), p. 87.

[56] Alkali Act, 1874: *Statutes, Public and General*, 1874 (37 and 38 Vict.) c. 43; Royal Commission, 1878 (ref. 46), p. 7.

Environmental damage, in the form of the destruction of hedges and trees in the areas affected by this new branch of the chemical industry, formed the background to this legislation. John Pattinson, Public Analyst for Newcastle-upon-Tyne, testified in 1877 to the great extent of the damage done to vegetation on the banks of the Tyne, expressing the opinion that the copper works on Tyneside, established for ten years, had caused more injury than any of the alkali works in the area.[57] Neighbouring landowners reported destruction of crops and failure of sheep and cattle to thrive, and this was clearly reflected in the depressed value of agricultural land in the vicinity. The greater part of the damage around St. Helens was also attributed to copper smoke. By the early 1860s there were six to eight large copper smelting works established there, probably two of which were based on the 'wet process'. At this time, the combination of alkali and copper smoke had produced a stark impression on the surrounding landscape. One witness described it as 'one scene of desolation. You might look round for a mile, and not see a tree with [much] foliage whatever . . . three-fourths of the trees are totally dead'.[58]

## (e)  Decline of the British copper industry

Britain's pre-eminence in copper smelting was inevitably challenged, paralleling shifts in world copper-mining. Decline set in as copper smelting capacity increased in countries where new sources of copper ore had been opened up, particularly South America, Australia, the United States, Spain, Africa and Canada. Concentrates and blister copper came to South Wales in increasing quantities between the 1850s and 1890s. The increase in world supplies of copper, and falling prices, caused the collapse of copper mining in Cornwall, Cuba and Chile, Swansea's established sources of ore. These closures helped to shift copper smelting to the United States, where copper mining was expanding rapidly. A temporary stay of execution for Swansea's copper industry was provided by the discovery of new deposits of high grade sulphide ore in Newfoundland and southern Africa, ensuring the continuance of copper smelting in South Wales to the end of the 19th century. From the late 1860s diversification into a range of by-products, sulphuric acid, superphosphate fertilisers, arsenic, cobalt, antimony and nickel, also assisted its survival.[59]

New technologies also contributed to the shift of copper production away from the Swansea area. Electrolytic refining methods, developed by George and Henry Elkington, enabled the economic extraction of precious metals from the copper ore and produced copper of high purity and electrical conductivity. Copper produced by this method, first used commercially in 1869, met the expanding market of the electrical industries. Fire-refined copper could not

[57] *Royal Commission*, 1878 (ref. 49), pp. 324–325, 19, 21.
[58] *Select Committee*, 1862 (ref. 43), pp. iii–iv.
[59] Roberts (ref. 29), p. 142; Newell (ref. 34), pp. 90–91; R. Chadwick, 'A history lesson on nineteenth century metallurgy', *J. Birmingham Metall. Soc.*, 1957, **37**, pp. 583, 591–592.

meet the standards of purity required, and competition from American producers led many British smelters to specialise in fire-refined copper for brass manufacturers and the engineering and shipbuilding industries. Increasingly exposed to international competition, the copper smelting industry of South Wales declined rapidly in the early years of the 20th century, and ceased altogether in 1921.[60] It took another forty years before the environmental legacy of the copper and other departed metal refining industries of the Lower Swansea Valley, in the form of slag heaps, contaminated soils and watercourses, began to be fully addressed. On 16 August 1961 the Lower Swansea Valley Project was inaugurated to examine the valley floor and tackle the physical problems of redevelopment with the aim of bringing the area back into normal social and economic use. In the 1960s and 1970s the area was the focus of a large-scale, publicly-funded reclamation project.[61]

## 6 Three Other Non-ferrous Metals

We shall now glance briefly at three other products of the metal extractive industries which, in their different ways, have had profound effects on society and the environment. All have been known for centuries.

### (a) Lead

This soft and malleable metal, with a low electrode potential, is one of the seven metals known in antiquity. In Britain it has been extracted at least since Roman times, occurring widely, with large deposits in Derbyshire, eastern Cumbria and southern Scotland. By the mid-19th century Britain was the largest lead producer in the world. The metal was formerly required extensively for water pipes, drains and cisterns, for roofing purposes and for printers' type. To these specific functions must be added the manufacture of lead shot and (more recently) of bullets. In the early 19th century vast quantities were needed for the 'lead chambers' of sulphuric acid manufacture and for the production of lead compounds (of which tetraethyllead became the most important and probably the most environmentally undesirable). Lead has also been converted into its carbonate, oxides and acetate for a variety of uses, especially in the paint industry (though white lead has now been superseded by the non-toxic titania, $TiO_2$). Lead chromate was a very popular yellow pigment, being used for such diverse purposes as a paint for coach panels (as in that used by Queen Adelaide and in some early railway carriages) and, most reprehensibly, as an additive to sweets.

Lead has been obtained from its chief ore, galena (PbS), by aerial oxidation. The ores are ground, washed and then submitted to a two-stage roasting in air, the second stage involving a raising of temperature and the addition of lime:

---

[60] Newell (ref. 34), pp. 92–93.
[61] Barr (ref. 46), pp. 79–156; Bridges (ref. 46), pp. 20–21. See also J.H. Beynon and D. Betteridge, 'The rise and fall of copper: a Swansea chronicle', *Chem. Brit.*, 1979, **15**, 340–345.

First stage:

$$2PbS + 3O_2 = 2PbO + 2SO_2$$
$$PbS + 2O_2 = PbSO_4$$

Second stage:

$$PbS + 2PbO = 3Pb + SO_2$$
$$PbS + PbSO_4 = 2Pb + 2SO_2$$

A slag is formed from the lime which contains more lead, and this is recovered by smelting in a small blast furnace.

In Derbyshire smelting was introduced in the 16th century on something like an open forge, the reverberatory furnace being brought in by the London Lead Company about 1747 and completely replacing its rival by 1780. A smelting works near the Cromford Moor Mine near Wirksworth was operated by Joseph Wass until 1880, remains of one furnace being still visible. He also owned the last smelter-plant in Derbyshire, at Lea, which continued well into the 20th century. The Scottish works were centred on Leadhills and Wanlockhead in the bleak and remote terrain of the Lowther Hills on the borders of Lanarkshire and Dumfriesshire. In Scotland coal was first used for lead smelting in 1727. Steam power was early introduced to Wanlockhead (1782) but not to Leadhills (1842), though it is not clear when the reverberatory furnace made its first appearance. Lead smelting continued in the area until 1928. The extraction of lead in north east England was centred on the area around Alston, where plentiful remains are still recognisable today. The industry was early aware of the importance of chemical analysis, and the London Lead Company built assay laboratories at three of its sites in 1833. In the following year lectures were given to assayers and assistants by J.F.W. Johnson, Reader in Chemistry at Durham, and in later years by Thomas Richardson, at Newcastle. The scientific investment was soon repaid by a discovery in 1833 by Hugh Lee Pattinson, Assayer at the Alston Moor Lead Company. He devised a new method of purifying lead by melting in a crucible and skimming the crystals that formed at the top. Since one of the impurities was silver the method was of more than academic significance. Successive generations of Pattinsons and Cooksons continued extraction of lead, setting up later establishments on Tyneside.

It has, of course, long been recognised that lead and its compounds pose a peculiarly nasty hazard to health. In the Scottish area in the middle of the 18th century most smelters were said to die insane (though the curious reason was advanced that this resulted from abstinence from spirits while at work!). Later visitors remarked on the healthy appearance of the workforce, and an unusually high consumption of milk (now known to reduce the effects of plumbism) may have been responsible. Pollution of water-courses below a smelter was soon recognised, though often attributed to other undesirables such as arsenic.[62]

[62] A. and N.L. Clow (ref. 9), pp. 378–379.

*Remains of lead-smelting flues at Rookhope in Weardale The twin flues were smaller than usual, but were carried from the mill by a series of stone arches (one of which remains as shown here), and then turned half-left up the hill for a distance of 1.5 miles to a chimney at the top*

The most obvious pollution was of the air, as smoke belching from lead furnaces was visible for miles and unpleasant. Cattle and sheep were poisoned by grazing on the contaminated land near a smelter, the so-called 'belland ground'. Moreover its lead content rendered it expedient on economic grounds to remove as much as possible. Whether for economic or environmental reasons the effect was to induce manufacturers to resort to the most elaborate and costly forms of 'condensation'. Dry condensation was not very successful in ordinary chimneys. One approach was to pass the gases through wire gauzes just immersed in water, and this was moderately successful. The only alternative was to have flues of an immense length, suggested by Bishop Watson in the late 1770s. These were introduced into Derbyshire in 1777 at Middleton Dale. Impressive remains exist of zigzag flues climbing their way up a hill to a chimney at the top (Bonsall and Alport-by-Youlgreave). Nowhere in Britain was there a longer flue than that of the London Lead Company at Allendale, Cumberland, which was lengthened between 1845 and 1850 to over 8000 yards. The noxious gases thus had a journey of nearly 5 miles in which to deposit their objectionable burden. In the lead mills owned by the MP Wentworth Blackett Beaumont (later Lord Allendale) it was reported that the horizontal flues were masonry structures, 8 feet high, 6 feet wide and extending to an aggregate length of nine miles.[63] Many flues, and a few chimneys, remain in the north Pennine area.

The workers themselves were, of course, in need of protection from the fumes, so most smelt mills had an abundance of ventilation, so much so that while workers were over-heated in front their backs were continuously cooled by the hurricane of cold air sweeping in through wide arches behind them. Thus in the cool North East many workers wore thick shawls round their necks as some form of protection. This was the case at Rookhope, dating back to 1752 but rebuilt in 1884, but not at Blackton, built 1820 with rebuilt furnaces from 1862 and a superior ventilation system that avoided blasts of cold air on the workers' backs. Similar hazards were experienced by those who worked in the production of lead pigments.

Two environmental legacies of lead smelting remain with us to this day. It may be noted that washing of ores was 'not carried out to the same degree of perfection as in other countries',[64] and that much lead therefore remained in waste tips to poison the immediate environment for the indefinite future. Secondly, the need to transport ores, coal and peat fuel and particularly the finished product led to the creation early in the 19th century of a new network of roads in the North East, jointly funded by the companies and the Admiralty, for their Greenwich Hospital Estates which owned most of the land.

## (b)  Tin

Tin is a metal that appears to resist corrosion and therefore remains bright under most circumstances. It has been known since antiquity and, when alloyed with copper, gives rise to bronze. It is almost entirely found in nature as the oxide $SnO_2$, or cassiterite. Substantial deposits of this mineral in Cornwall have brought a succession of acquisitive visitors, from the Phoenicians in 500 BC onwards. For centuries tin mining and smelting had been an important part of the economy of western Cornwall, the 'stannaries', or mining communities, possessing special rights and privileges. At the end of the 18th century demand increased with the growth of tin-plating and the discovery of Britannia metal, with 90% tin; much metal was also exported. By the mid-19th century Cornwall produced nearly half the world's tin, though by 1900 the industry had suffered an almost total collapse in face of cheap imports from the Far East.

Tin is extracted by reduction of the ore with anthracite in a reverberatory furnace:

$$SnO_2 + 2C = Sn + 2CO$$

and purified by liquation (when the fusible tin separates from a dross containing copper, iron, arsenic *etc.*), and then by 'poling', when greenwood poles are used

---

[63] *Industrial resources* (ref. 8), p. 140.
[64] S. Muspratt, *Chemistry, theoretical, practical and analytical*, Mackenzie, Glasgow, n.d. [c.1860], p. 462.

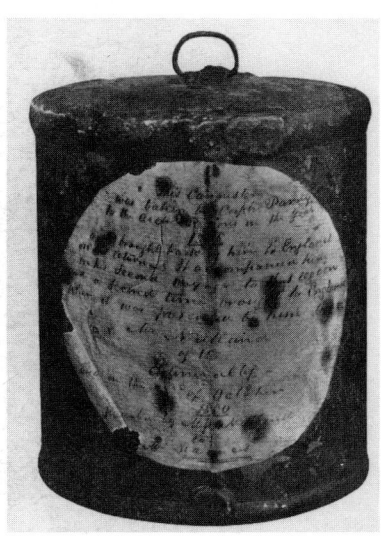

*A very early 'tin can' (1824), made by the London firm of Donkin, Hall and Gamble*

to stir the molten metal and the reducing gases liberated remove further impurities.

By 1800 the chief use of tin was coming to be as a protective coating for iron. Tin-plating had been known for well over a century in Saxony but attempts to introduce it to Britain in the 17th century had been unsuccessful. The process was conducted by pickling and cleaning an iron sheet and immersing it in molten tin, followed by rolling. It was first successfully established in South Wales where it had been operated at least as early as 1760, one of the earliest works being at Kidwelly. Ready access to Cornwall across the Bristol Channel, combined with abundant supplies of coal and iron, maintained a Welsh supremacy for the process. By the late 19th century, however, British production began to decrease and stagnate. Partly this was due to spectacular growth of a rival industry in the USA and partly to the inherent conservatism of the Welsh tinplate makers and their reluctance to innovate.[65]

Although the discovery that food, heated in a sealed container, can be kept indefinitely seems to have been made by a Frenchman, Nicholas Appert, the application of the principle in a tin can was a British invention. Patented in 1810 the tin can was made in large quantities in Bermondsey for use by the Royal Navy. The social and environmental consequences of this simple application of a chemical product are clearly immense. Some tin-plating was carried on in the Midlands, but for heavier and more durable items. The tinning of steel was apparently introduced by Siemens at Landore in 1870.

[65] W.E. Minchinton, 'The tinplate maker and technical change', *Explorations in Entrepreneurial History*, 1954–1955, **7** (2), 1–11.

*Seaton Carew works, Co. Durham, in 1919: the zinc smelter is on the left and the sulphuric acid plant (using sulphur dioxide from the smelter) on the right*

### (c)  Zinc

Zinc was first known to humanity as brass, an alloy with copper and formed by smelting mixed ores of the two metals. The element itself was brought to Europe from the Far East in the 17th century, made in a Swansea copper works about 1720, and first produced on a large scale in Britain by William Champion in 1740 at Bristol which became, for a century, the world's largest supplier.

In the traditional method of extraction the ore, zinc blende (ZnS), was first oxidised by air to the oxide which was then reduced by coke to metallic zinc:

$$2ZnS + 3O_2 = 2ZnO + 2SO_2$$
$$ZnO + C = Zn + CO$$

The method remained essentially unchanged until the recent use of electrolysis which now accounts for over half the world's zinc production.

The production and use of zinc generated some interesting environmental problems. The great importance of the older method is that it was generally used in conjunction with the production of sulphuric acid, thus utilising the waste sulphur dioxide. Hence many plants had zinc smelters and lead chambers side by side (as at Seaton Carew near Hartlepool). The coupling of zinc smelters with vitriol plants was environmentally and economically sensible, though even then the piles of solid waste products were considerable.

Where the effluent sulphur dioxide was not used for sulphuric acid production, but allowed to escape into the atmosphere, the nuisance received the predictable reactions. This was true even in a remote and sparsely populated area like Tindale Fell in Cumbria, site of the only zinc works in the north of England, opened in 1845. Within a few years complaints were received that the effects of the smoke were 'worse each year and . . . they were much injured by it'. So reluctant – or unable – were the owners to eliminate this nuisance that, twenty-five years later, compensation of £65 *per annum* was being paid for damage to vegetation and to farm animals. When, in 1895, the lease expired it was not renewed as the landowners, the Earl and Countess of Carlisle, opposed pollution on their estate. This is an interesting and somewhat surprising case of members of the landed aristocracy placing environmental protection above economic gain, though they were known also for being ardent social reformers and supporters of the temperance movement.[66]

The most important use for zinc was introduced in 1837 with the invention of protective zinc coatings on iron. The latter was dipped in molten zinc ('galvanising'), coated by zinc dust at 350 °C ('sherardising') or plated electro-lytically. Disposal of waste liquors after pickling the iron prior to galvanising caused considerable problems. Several manufacturers in the late 19th century applied for permission to dispose of the waste into the local sewer. This generally received an unfavourable response. Thus in 1889 a West Bromwich firm, Ash and Lacy, met with the response that such waste would corrode the iron sewer pipes. The Corporation did not accept the company's suggestion that the waste could be treated with limestone to neutralise the acid. The outcome of this disagreement remains uncertain. A few years later, Wolverhampton Corporation were in contention with the firm of John Lysaght over pollution of a local brook with the same material. The firm's response to their refusal was to relocate their galvanising works at Newport in Monmouthshire.

## 7  Metals and Electricity

### (a)  The coming of electric power

One of the most remarkable features of the chemical industry, in Britain or elsewhere, is its enormous consumption of energy. Very few products can be made that do not, in at least one stage of their production, require strong heating. Where high pressures are additionally involved the energy demands are often very great. In an island 'built on coal' this might not have mattered too much, except that any profligate expenditure of fossil fuel reserves is environmentally unacceptable today. Even in the years before the Second World War there was a growing awareness of the problem, though it was rarely expressed in those terms. Towards the end of the 19th century fresh hopes were raised with the new availability of virtually unlimited electric power at the touch of a switch. With the creation of central electricity generating stations, and an efficient means of distribution, the chemical industry could contemplate manufacturing possibilities undreamed of before.

By the middle of the 19th century it was possible to build a dynamo capable of

[66] J.K. Almond, 'Zinc production at Tindale Fell, Cumbria', *J. Hist. Metall. Soc.*, 1997, 11 (1), 30–38.

supplying electricity to a small grid. Apart from a few isolated cases of electric illumination there was not much demand until the sudden appearance of the incandescent electric light bulb, invented almost simultaneously in 1878 by the American inventor Thomas Edison and the Sunderland pharmacist Joseph Swan. It was in north east Britain that several more important developments took place. Following several attempts to supply a local grid from dynamos powered by reciprocating steam engines a Newcastle engineer Charles Parsons devised a steam turbine in which the rotor was made to rotate far faster than was possible than in the older system of cranks and connecting-rods. For the British chemical industry the first important applications of electricity also took place on Tyneside, appropriately for one of the two or three areas in Britain that could legitimately be called the cradle of its chemical industry.[67] Two 75 kW Parsons turbo-alternators were installed in 1890 at the Forth Banks power station of the Newcastle and District Electric Lighting Company. Two larger generating stations were subsequently built at Neptune Bank, Wallsend (1901) and Carville (1904). The subsequent spread of electric power generators need not concern us here, but by 1914 the triumph of the steam turbine was virtually complete.

Of course there remained the environmental cost of coal depletion, since all early power stations were coal-burning. And there was also an economic cost in that electric power was relatively expensive in the UK as compared with some countries overseas, most notably Canada and the United States. For since the 18th century a parallel development had been taking place in other parts of the world, and concerned the evolution of a water turbine, *i.e.* one driven by large volumes of falling water. By 1850 some simple turbines were replacing the traditional water wheels in mills of various kinds, especially in North America. The use of turbines to generate electricity was first developed in countries with large water resources, as the USA, Canada, Norway and Switzerland. At Niagara Falls two turbines generating 500 h.p. each were installed in one of the earliest large-scale hydroelectric generators. In Britain more effort went into improving water-wheels, the first example of hydroelectric power being a small 9 h.p. turbine installed in 1880 by W.G. Armstrong at his home *Cragside* in Northumberland. It was intended to provide electric light for his picture gallery, but soon *Cragside* became the first house in Britain to be lit entirely by electricity. The environmental advantages of this renewable source of power are obvious, though it was early recognised that they may be accompanied by visual pollution, an observer of the scene at Niagara complaining that 'it has been spoiled because it has lost its setting – the environment of nature undisturbed'.[68] Whatever may be the pros and cons of that environmental debate the lack of huge water-drops in the UK precluded hydroelectric power from becoming a significant public energy source until Scottish developments after the Second World War.

[67] See W.A. Campbell, *A century of chemistry on Tyneside*, Society of Chemical Industry, London, 1968, pp. 56–58; and *idem*, 'Early electrochemistry on Tyneside', Symposium volume *The history of electrical engineering*, Institution of Electrical Engineers, London, 1976, pp. 44–49.

[68] P. Lewis, *The romance of water-power*, Sampson Low, London, n.d., p. 93.

Meanwhile, as indicated above, power transmission over moderately wide areas had been introduced, though anything like a national grid had to wait until 1927. With the opening of the power station at Wallsend in 1901 it 'was delivering electricity over what was probably the most widespread interconnected power system in the country'.[69] Yet again Tyneside was in the forefront of progress, and indeed saw also the first major attempt at railway electrification when in 1904 the North Eastern Railway electrified 37 route miles of track in North Tyneside in order to compete with yet another innovation resulting from the new power source: the electric tramway.

Nor was the chemical industry slow to take advantage of the new situation. Already there had been the invention of *electrothermal* techniques where furnaces were raised to a high temperature by passage of electricity through the mix to be heated. Such was the case with the invention in 1888 of a new method for producing phosphorus from calcium phosphate, sand and coke. The patent by Readman, Parker and Robinson of Wolverhampton was applied at Wednesfield by the Phosphorus Company, later acquired by Albright and Wilson and operated until 1920 at their site at Oldbury. Thereafter, enticed by the relative cheapness of hydroelectric power, production was exiled to Canada.[70] For similar reasons calcium carbide manufacture had a brief and chequered career in Britain (Chapter 10). Nevertheless the prospect of cheap electricity on Tyneside attracted the firm of Thermal Syndicate who from 1906 have used electrothermal methods to manufacture from fused silica such products as crucibles, tubing, Vitreosil globes and so on.

However, these products (phosphorus, carbide and silica) are not metals, and it was in metal production that electricity was most fruitfully applied to industrial chemistry. The techniques employed were not *electrothermal* but *electrolytic*. Perhaps the two most important examples are sodium and aluminium.

## (b)  Sodium

Although sodium metal was first isolated by electrolysis (Davy, 1807) the cost was prohibitive on an industrial scale, and for much of the 19th century demand was not great. However, the known ability of sodium to displace aluminium from its chloride (see below) had been developed on a small scale at the Washington Chemical Company (Co. Durham) in about 1860, where R.S. Newall experimented on the properties of aluminium wire. However large-scale production remained elusive, waiting only for a cheap and reliable source of sodium. This was discovered in the USA by H.Y. Castner, who modified an older process by reducing sodium hydroxide with a mixture of iron and carbon (1879). He brought the process to England in 1886, building a plant at Oldbury to supply sodium to the nearby aluminium manufacturer, Webster Crown

---

[69] F. Atkinson, *Industrial archaeology of north-east England*, David & Charles, Newton Abbot, 1974, vol. i, p. 156; an original turbine from the earlier Forth Banks site is now displayed at the (dismantled) Stella Bank power station (*ibid.*, vol. ii, p. 263).

[70] It returned to England in 1934 when imports became too expensive (Key Industry Duty).

Metal Co. Within three years a rival firm, the Aluminium Alliance Company, was making its own sodium by Castner's reductive method, at Wallsend-on-Tyne.

There matters might have rested had not an alternative route to aluminium been invented that did not require sodium at all: the electrolytic reduction of aluminium oxide. Fortunately for the indefatigable Castner the glut of sodium about to appear on the marker could be disposed of in a new way. The South African gold rushes towards the end of the century required sodium cyanide for extracting the gold, and Castner had, in 1894, patented a route to this desirable chemical by allowing sodium, carbon and ammonia to react in a single-stage process. It was rumoured that the salt was sold as 130% *potassium* cyanide! In Castner's case production helped to create demand and now he was able to devise an improved way to meet that demand. As reliable electric power supplies were becoming available the answer might lie in electrolysis.[71]

Castner reverted first to Davy's electrolysis of caustic soda, in which the copper cathode was separated from the anode by an iron gauze diaphragm. Such cells were in use in England from their introduction in 1890 until 1952. The early stages of development were dogged by various difficulties, which Castner was persuaded arose from impurities in the soda. These could be avoided, he thought, by preparing the caustic soda electrolytically from brine. One problem was that the sodium hydroxide was instantaneously formed at a cathode, thus mixing with the salt solution. This was solved by using a mercury cathode, for under these conditions sodium is discharged, enters the cathode and forms an amalgam. On contact with water it slowly yields sodium hydroxide. By constructing a cell that was rocked on an eccentric, Castner was able to achieve a continuous production of sodium hydroxide, the mercury being allowed alternately to take up the sodium and then to discharge it away from the solution being electrolysed.[72]

A different kind of amalgamation was needed when Castner discovered that his use of the mercury electrode (1892) coincided with a similar discovery by a German chemist, C. Kellner. To avoid litigation the two joined forces and the Castner–Kellner Alkali Company was launched in Britain in 1895. They commenced operations at Runcorn two years later, having failed to interest Brunner Mond or the United Alkali Co. in their processes. In 1906 they established a factory at Wallsend, where the relatively cheap electricity enabled Castner's new process for making sodium to replace his older one, though as before in the service of the local aluminium industry.

Environmental problems raised their head at an early stage of the Castner–Kellner cell. A correspondent in the *Liverpool Courier* feared there would be '130 tons of mercury sent into the atmosphere every year; enough to salivate every living thing for miles around.' The same concern motivated James Hargreaves and Thomas Bird to develop a mercury-free cell with asbestos diaphragms instead, each having 'declined to have anything to do with a

[71] D.J. Adam, 'Early industrial electrolysis', *Educ. Chem.*, 1980, **17**, 13–14, 16.
[72] T. Wallace, 'The Castner sodium process', *Chem. Ind.*, 1953, 876–882.

business likely to injure his fellow men, no matter what profit it might bring'. The Hargreaves–Bird cell was exploited in the General Electrolytic Alkali Company founded at Middlewich in 1899 and restyled the Electro-Bleach and Byproducts Company in 1915. The works survived until 1927, then being part of ICI. In fact fears of mercury poisoning were groundless, as became apparent in the next 50 years.[73]

In 1924 J.C. Downs at Niagara succeeded in devising a method for electrolysis of fused sodium chloride, thus producing equimolecular parts of chlorine and sodium. The Downs process was taken up by ICI in 1937 and now constitutes the main route to metallic sodium.

It remains to mention the uses to which this sodium metal was put, once the metal was no longer needed to make aluminium, and once the gold rush was over. Apart from its importance in sodium lights and in the production of sodium compounds as the cyanide, peroxide *etc.*, the chief outlet for the metal since 1923 has been the production of tetraethyllead, the 'anti-knock' constituent of motor fuels:

$$4Na + Pb + 4C_2H_5Cl \longrightarrow Pb(C_2H_5)_4 + 4NaCl$$

The well-known additive to 'leaded' petrol, it has been the subject of fierce environmental debate, as the exhaust gases must contain lead compounds and these are known (or believed) to have a detrimental effect on brain-growth in young children. Also, to eliminate the lead from the cylinder a further substance is added, 1,2-dibromoethane (which ensures most lead is expelled as a bromide). The modern industry for extracting bromine from sea-water, as at Amlwch in Anglesey, depends on this outlet. But injection of bromine compounds into the atmosphere has largely unknown effects but is highly likely to help in the destruction of the ozone layer.

For these and other reasons successive governments have favoured the use of unleaded petrol. The environmental dilemma, however, is that production of high octane unleaded petrol, though beneficial in other ways, is wasteful of energy in that additional refining losses are involved in its manufacture. It was recognised as long ago as 1973 that elimination of leaded petrol, together with other measures involving purification of exhaust gases, could raise petrol consumption by up to 20%.[74] In one sense this is the oldest environmental dilemma of all: material gain *versus* energy loss.

## (c) Aluminium

Aluminium is an important and versatile metal.[75] The first industrial method in Britain for extracting aluminium involved, as we have seen, reduction of the chloride with sodium. The double chloride $AlCl_3.NaCl$ was first employed (at

[73] D.W.F. Hardie, *A history of the chemical industry in Widnes*, ICI, London, 1950, pp. 191–193, 197.
[74] *Energy: efficiency in use*, Shell information brief, Royal Dutch/ Shell Group, 1973.
[75] G.B. Kauffman, 'Aluminium: its discovery, properties, uses and production', *Cahiers d'Histoire de l'Aluminium*, 1989, **5**, 38–52.

*Inspection of construction at Foyers in the early 1890s (Lord Kelvin third from left)*

Oldbury) but later at Wallsend a mixture of fused cryolite ($Na_3AlF_6$) and salt was employed – a hazardous operation where men stood on boxes to manipulate large balls of sodium into the molten mixture. However, they were paid unusually well, with a 56 hour working week and an annual week's holiday with pay.[76]

The electrolytic process for aluminium began in 1886 with independent discoveries by C.M. Hall in the USA and P.L. Héroult in France (whose lives coincidentally spanned the same period of 1863–1914). The mineral cryolite, found almost exclusively in Greenland, had the ability when fused to dissolve aluminium oxide, the solution then being capable of electrolysis to yield aluminium at the cathode. Aluminium oxide is a very abundant mineral, and was usually in the form of bauxite, which needed some preliminary purification, an effective method having been discovered by an Austrian chemist, K.J. Bayer, about 1888. In addition to the electric current consumed by the actual electrolysis much is also used in maintaining the bath at the necessary temperature of about 1000°C. At first production was possible only in the USA and Switzerland (in the late 1880s) where hydroelectric power was available. Britain had to wait until 1894 when the British Aluminium Company was formed to exploit it.[77] It became British Alcan Aluminium in 1982.

Within two years aluminium was being produced at the company's first smelter, at Foyers in Scotland, on the banks of Loch Ness and powered by hydroelectricity generated from the Falls of Foyers. Alumina was shipped to it from Northern Ireland, *via* the Caledonian Canal. The plant, though small, was able to supply the total world demand for aluminium. It continued to operate until 1967. Two others followed, also dependent on hydroelectric power derived from Scottish rivers: at Kinlochleven in Argyllshire (1907) and at Fort William

[76] Campbell (ref. 6).
[77] Anon, *The history of the British Aluminium Company Limited, 1894–1955*, BAC, London, 1955.

*Pipeline construction at Foyers (1896)*

(1929). Together they used nearly 100,000 h.p and obtained the water from 100,000 acres. Much more recently they have been joined by another at Invergordon in Ross and Cromarty, this time drawing electricity from the public supplier, the North of Scotland Hydroelectric Board (1971). To put the UK production in a larger perspective it is salutary to note that, in 1938, the British output was 30,000 tons, only about 5% of the world's production

The environmental consequences of all this activity need some investigation. First, with regard to the uses of the product, aluminium has little or no directly adverse effect on the environment. Although it is extremely reactive, in practice a sheet of the metal freshly exposed to the air becomes immediately coated with a very thin, invisible but tenacious film of oxide. Further reaction then becomes very difficult. It is, however, corroded by sea-water, unless exposed first to heat treatment. The metal's low density has made it available for an immense range of uses, ranging from kitchen-ware to aircraft. The first aeroplane with an all-aluminium frame was made in 1920, and since then it has often been employed as an alloy with magnesium ('magnalium'), or another with copper and magnesium ('duralumin'). Its aeronautical value began to be appreciated in the First World War when it was also used as fuse-caps, bottles and mess-tins for troops in the trenches, and in gearboxes and crank cams for military vehicles. In the Second World War demand was even greater, with 80% used in light alloys for aircraft construction and a tenfold increase in production of forgings, especially for propeller blades and aircraft engines. Consumption of electro-lytically produced aluminium grew from 78,000 tons in 1939 to 208,000 tons in 1943 (its peak). With the fall of France in 1940 supplies of French bauxite dramatically ceased, though efforts were made to import it from the Gold Coast and many other places. Raw metal obtained from Canada was always at risk

*View down the pipeline to Kinlochleven smelter*

from being torpedoed in the Atlantic. Accordingly Britain began an intensive programme of recycling, a process particularly suitable for aluminium, having been tried on a small scale in the inter-war years. The peak wartime production of recycled metal amounted to 110,000 tons.

The ability of aluminium to reflect heat has made it invaluable in both storage and packaging, products including milk-bottle tops, kitchen foil and storage containers for petrol, milk and so on. An excellent conducting power combined with the lightness of aluminium led to trials in overhead electric cables at the turn of the century. In combination with a central core of steel wire it gave cables of such strength that the supporting and unsightly pylons could be placed much further apart. This has been almost universal in Britain since the end of the Second World War. At least from the point of view of the visual environment this has to be an advantage.

In other ways aluminium represents a classic case for environmental enquiry.[78] To bring dramatic industrial development to the most unspoilt and spectacular scenery in Britain, the Scottish Highlands, seemed to many the height of irresponsibility. The worst fears would have been confirmed by a visit to Kinlochleven during the construction years 1904–1907, an area hitherto

[78] See ALFED Fact Sheet no. 15, *UK aluminium industry: aluminium and ecology*, Aluminium Federation Ltd., Birmingham, 1992; *Environmental issues: the aluminium industry*, European Aluminium Association, Brussels, [*c.* 1994]; and *A commitment to continual environmental improvement*, Alcan Aluminium Ltd., Quebec, 1995.

*The Lochaber smelter (1929), powered by water from Loch Treig conveyed by a tunnel through Ben Nevis*

devoid of houses and even roads. Between 2000 and 3000 workmen were employed, living in 'shanty towns' in the hills. Add to this the creation of an 86 ft (26 m) high dam across the Blackwater River, so creating a storage reservoir of 3930 million ft$^3$ (372 million m$^3$), then the largest in Europe. If this were not enough a similar drama was to re-enacted in 1924 with an even larger smelter and hydroelectric plant at Lochaber, Fort William.

Yet this is a one-sided picture. Once construction was complete the 'shanty towns' vanished, as they are wont to do with large engineering projects. Their sites were returned to nature, suitable land was made over to sheep farming and (since the 1940s) a long-term programme of re-afforestation has begun. Moreover much of the immense engineering project is completely invisible. For the Lochaber works the water is conveyed from Loch Treig to the power house by a tunnel, 15 ft (4.6 m) in diameter and 15 miles (24 km) long through the heart of the Ben Nevis mountain range. In human terms an area of chronically high unemployment provided work for nearly 1000 people even before the third large plant came on stream (at Invergordon). Two new communities have been created, with social and other amenities. It has brought prosperity to an area in a way that is reminiscent of that conferred on Shetland by the oil industry.

As for materials, the one thing that cannot be claimed for aluminium is

*Electrolytic cell at Lynemouth aluminium smelter*

scarcity; it is the most abundant metal on Earth, being about 8% of the earth's crust. The one rare material involved in the extraction is cryolite, but this is hardly consumed (the small amounts of fluorine-containing gases having in recent years been recovered from the effluent). Moreover, as we have seen, the metal is very readily recycled, a fact that has been utilised in the UK since about 1920. In one other major respect the aluminium smelting enterprise is out-standingly successful on environmental criteria. Though highly energy-dependent its four earliest smelting plants drew on energy sources that are all renewable. They used hydroelectric power based on exceptional rainfall and mountainous terrain. For that reason also there is little atmospheric pollution.

Perhaps unfortunately this tradition was broken in 1972 when a smelter was opened at Lynemouth, Northumberland, drawing energy from its own power station that was fired by coal. This made sense for the local economy, providing over 600 jobs in an area of high unemployment and using locally produced fuel. Very recently (1990s) strenuous efforts have been made to achieve high levels of environmental performance.[79] However the long-term wisdom of choosing a non-renewable energy source remains to be seen. Even so, few branches of the modern chemical industry can claim such an enviable record of environmental probity as that of aluminium production.

---

[79] Alcan, *Annual environmental report, Lynemouth smelter and power station*, London, 1996.

Chapter 11    ***Chemical Industry and the Quality of Life***

N.G. COLEY AND S.A.H. WILMOT

## 1  Introduction

In tracing the development of the British chemical industry and its relationship
with society and the environment, we have tried to show both the benefits it has
brought and some of the disadvantages. The contributions which chemical
products have made to the quality of life, both positive and negative, have been
sketched. The 'quality of life' is a notion which covers a very wide range of
issues. Besides a feeling of progress and general well-being, it includes important
social concerns such as standards of health-care, safety and comfort, the
possibilities of travel, communication and leisure activities, diet, dress, books,
magazines and newspapers, and so on – in fact all of the trappings of modern
life. It is a concept almost impossible to define as the expectations of every
individual and every society are different, but we easily recognise when the
quality of our own lives is good, or poor, improving or declining. During the
20th century the concept has been expanded to include a consideration of the
well-being of other species besides our own. Indeed, the quality of life for
humans is recognised as intimately bound up with the welfare of the broader
natural environment.

It is a long standing myth that social problems can largely be solved by
applying science and technology to deliver material goods thought to be
essential to 'progress' (however ill-defined that term). The chemical industries
have been constantly engaged in this process, inventing new compounds and
improving old ones, creating a real dependence upon chemical products in many
ways. Despite this fact the popular perception of 'chemicals' is broadly negative.
They are widely regarded as inherently nasty, poisonous and dangerous. The
latter view tends to be promoted by the media, especially the popular press
where the word *chemical* is often preceded by the adjective *dangerous*, even when
quite innocuous substances are involved – or even where they are of entirely
natural origin! As William H. Brock has commented in concluding his recent
history of chemistry: 'while the repair of the chemical industry's image, and of
chemistry itself, is a performance challenge for the 1990s, it is clearly also one of
historical awareness'.[1]

[1] W.H. Brock, *The Fontana history of chemistry*, Fontana Press, London, 1992, p. 662.

## 2   Environment: the Big Issue

### (a)   Some general considerations

Until the 1960s the idea of progress based on scientific and technological advances was widely accepted and the general benefits of science, including industrial chemistry, went largely unquestioned. Public confidence in the ability of science to improve the quality of life remained strong. New products were, and still are, despite increasing scepticism, promoted using the popular perception of science and technology as the engine of progress. Advertising often relies on scientific or pseudo-scientific ideas and terms which tend to create a direct, if sometimes spurious, link between science, technology and the quality of life. This is especially true of new chemical products. New products for home decorating, household cleaning or laundry are presented with scientific backing, while extravagant therapeutic claims are made for proprietary medicines on the basis of scientific or pseudo-scientific tests. These techniques have served to maintain a generally beneficent image for science and technology. On the other hand there is undoubtedly a negative side to this industry where social costs and benefits are perhaps more starkly opposed than in most others. Against the social benefits brought by chemical industry must be placed the fact of environmental pollution and the dangers resulting from potential and actual chemical accidents. These also have their effect on the quality of life, and policy makers are constantly engaged in maintaining an equilibrium between the social and economic advantages, and the disadvantages of the chemical industry's activities.

Nineteenth century industrialists were aware of health and environmental issues, both for their workers and for those people living in the vicinity of their works. As a body, however, industry generally reacted to public criticism rather than acting positively to improve the environment. Sensitivity to public opinion was sufficient to induce manufacturers to make emissions to the atmosphere just sufficiently clean to escape adverse criticism. In some quarters the pressure brought by the Alkali Acts (beginning in 1863) to restrict emissions of certain scheduled gases was also sufficient to prompt the beginning of a rhetoric of wider social and environmental responsibility.[2] However, as late as 1957 the main concern voiced at a Clean Air Conference was the avoidance of 'justified complaints by the public'. The speaker, Sir Ewart Smith of ICI, observed:

> The main objective must be to make any discharge to atmosphere sufficiently innocuous, having regard to local conditions, topography and climate, for the inhabitants of the area to suffer no significant or avoidable discomfort from dirt, corrosion, loss of sunlight or injury to health.[3]

[2] S. Wilmot, 'Pollution and Public Concern: The response of the chemical industry in Britain to emerging environmental issues, 1860–1901', in E. Homburg, H. Schroter, and A.S. Travis (eds.), *The chemical industry in Europe, 1850–1914: industrial growth, pollution and professionalization*, Kluwer Academic Press, Dordrecht, Holland, 1998; J.A.G. Drake (ed.), *Integrated pollution control*, Royal Society of Chemistry, Cambridge, 1994; R.M. Harrison (ed.), *Pollution, causes, effects and control*, Royal Society of Chemistry, Cambridge, 3rd. ed., 1996.
[3] 'Air pollution in the chemical industry': report on Clean Air Conference organised by the Institution of Mechanical Engineers, 19–21 Feb., 1957, *Chem. Age*, 1957, **77**, 369.

Such admirable sentiments were nevertheless by today's standards both parochial ('inhabitants of the area') and restricted entirely to human beings (who might otherwise complain). The idea of cleaning up the environment for its own sake and to safeguard the quality of life for future generations was beginning to be articulated, but it is only in the last 20 years or so that this has become an overriding concern. This is due largely to pressure from groups such as Friends of the Earth, Greenpeace and other environmental organisations, including government agencies and political parties. It may also be part of a wider disenchantment with science and technology for wholly other reasons.

At this point a *caveat* is in order. Some of the more evident effects of industrialisation on the environment have been caused only in part by the chemical industry. For example, limestone, a primary raw material for making alkalis and glass, is also quarried for agriculture, building and cement manufacture. Timber is also felled for construction purposes, and until the 19th century vast quantities of wood were used for building ships and buildings. In the 18th century wood was also required for making charcoal, used in smelting iron as well as other metals. Coal was used mainly as a fuel, until the second half of the 19th century when the newly established organic chemicals industry used the products of coal-tar distillation as a primary feedstock; sand was required for mortar in building as well as in glass manufacture. The mining and quarrying of all these natural materials would have occurred without industrial chemistry, though it is true that large quantities have long been extracted for use in the chemical industries and there is a tendency to regard all extractive industries as fundamental to chemical manufacturing.

Public fears that finite raw materials and fossil fuels are being used up at an increasing rate that will ultimately deprive future generations of supplies contribute to negative images of chemical industry. Industrial pollution, in which the chemical industries play a part, adds to public disquiet. Chemical industries have been, and still are, responsible for massive environmental damage and pollution, not least the damage caused to the land on which chemical plant is sited.[4]

However, the history of the British chemical industry also reveals the tremendous past effort that has gone into the reduction of waste products. In order to become more efficient small chemical works were often clustered together and their operations were interrelated so that the waste products of one chemical process could form the primary materials for another. Where waste and pollution could be reduced, all the industries involved benefited economically. For example glass, soap, copperas and alum works were often found close to one another and to dye works and tanneries. All engaged in mutual exchanges and traded with each other in the course of their manufacturing operations. Glass makers could use soap boilers' waste which consisted mainly of calcium carbonate formed when soda was causticised:

$$Na_2CO_3 + Ca(OH)_2 = 2NaOH + CaCO_3$$

[4] See A. Markham, *A brief history of pollution*, Earthscan Publications Ltd., London, 1994; R.E. Hester and R.M. Harrison (eds.), *Contaminated land and its reclamation*, Royal Society of Chemistry, Cambridge, 1997.

Some glass makers may have obtained their soda by mixing common salt with wood ashes:

$$2NaCl + K_2CO_3 = 2KCl \downarrow + Na_2CO_3$$

Potassium chloride, less soluble in water than sodium carbonate, crystallises out and can be separated off and sold for alum manufacture. Waste from copperas production could be used in the tanneries.

From this it is clear that early chemical works often grew up as inter-dependent groups of small firms each making a different product and each depending to some extent on waste discarded by their neighbours. Ideally all the by-products of each chemical process would form the starting point for the manufacture of another useful product, but the history of chemical manufacturing is littered with waste products of little or no commercial value. Solid waste has often been dumped to form unsightly spoil-heaps while gases have been allowed to escape into the atmosphere and liquids have polluted the nearest river. Although litigation against chemical manufacturers who caused such pollution dates from the 1830s, national legislation limiting emissions was not introduced until 1863, since when relatively little has been done to reduce the problems until the 20th century. As awareness of pollution and ecological damage has grown, and the toxicity and variety of chemical effluents has been increasingly recognised, more stringent controls have become essential with consequent increases in production costs.

The environmental debate surrounding the chemical industry in Britain has its origins in the conspicuous air pollution produced by the alkali industry in the 19th century. Public anxiety was aroused both by emissions from factory chimneys, and by gases released by chemical reactions in the soda waste deposits which surrounded many alkali-producing areas. The targeting of the alkali industry by government regulation at local and national levels reflected both the unprecedented scale of local environmental devastation caused by the industry, and the apparent ready availability of technologies for the reduction of this major source of pollution. The lack of agreed appropriate technologies for tackling emissions in other industries, and the treatment of 'noxious vapours' and smoke pollution as two separate problems for the purposes of legislation and regulation, exempted other industrial polluters from the first phase of pollution legislation. This was in spite of the fact that the environmental damage resulting from industries like copper-smelting, brick-making and pottery-production was on an equivalent scale.[5]

River pollution engaged the attention of the British public from the early 1840s, but debate was not primarily focused on industrial contaminants. The

[5] R.M. MacLeod, 'The Alkali Acts administration: The emergence of the civil scientist', *Victorian Studies*, 1965, **9**, 85–112; *Report of Select Committee of the House of Lords on the Injury from Noxious Vapours*, P.P, 1862, XIV, pp. 3–5; E. Newell, '"Copper smoke": A case of acid rain in the Industrial Revolution', unpublished paper, Nuffield College, Oxford, 1992; A. Voelcker, 'On the injurious effects of smoke on certain building stones and on vegetation', *J. Soc. Arts*, 1864, **12**, 146–153 (150–151); R.E. Hester and R.M. Harrison (eds.), *Air quality management*, Royal Society of Chemistry, Cambridge, 1997.

main challenge facing Victorian Britain was the alarmingly high death rate in its major cities, and this gave a greater urgency to the problem of pollution from human, animal, and plant wastes. Industrial pollution *per se* was perceived as a lesser danger to health. Of the groups opposing river pollution in Britain, it was the fisheries lobby which focused greatest attention on industrial contaminants. However, their concern was more with the damage done to salmon fisheries by mining and tin production at river head-waters, and the paper-making industry was considered to be the worst polluter of down-stream stretches of river in Britain. Whilst the chemical industry was responsible for some locally spectacular river pollution, stimulating specific legislation for the treatment of alkali wastes, its role in this sphere attracted little national public controversy. On the whole, public perception of the chemical industry's contribution to river pollution was relatively weak. Additionally, public concern over the effects of marine dumping of chemicals was slow to emerge.[6]

The chemical industry faced significant public opposition soon after the establishment of the first alkali works in Liverpool in 1823. The alkali manufacturer E.K. Muspratt commented that hardly an assize passed without an action for nuisance or damage in the interval between the industry's foundation and the passing of the first Alkali Act in 1863. Pressure was exerted both by landowners in areas affected by air pollution, and by some urban authorities on behalf of wider public and trade interests. In general, large agglomerations of chemical industries polluted the air with more impunity than isolated chemical factories, and this remained true after the implementation of the Alkali Act in 1863. In addition to the piecemeal pressure of litigation exerted by private individuals and local authorities, the 1870s saw the growth of more organised attempts by sections of the public to get local and national government action to suppress chemical pollution. Five anti-pollution organisations were founded between 1873 and 1876, based in Warrington, Liverpool, Northumberland, Manchester, and London.[7]

However, the strength of the 19th century environmental lobby should not be exaggerated. For example, moves to tighten regulations on the chemical and other industries were often opposed by employees, who were prepared to mobilise on behalf of their employers. In South Shields, when Cooksons, the glass and lead manufacturers, were facing litigation for damage to farmland, the workmen held a grand procession to the Town Hall to enlist public sympathy for their employers.[8] In addition, some local authorities appear to have been more tolerant of the air pollution caused by chemical and other industries in

[6] A. Wohl, *Endangered lives. Public health in Victorian Britain*, J.M. Dent, London, 1983, pp. 234, 238–239; *Letter to First Lord of the Treasury by the President of the Sanitary Association and Vice-President of the Fisheries Protection Association relating to the pollution of streams*, P.P., 1864, L, p. 329.

[7] E.K. Muspratt, 'President's address', *J. Soc. Chem. Ind.*, 1886, **5**, 401–414 (407); A.E. Dingle, 'The monster nuisance of all: Landowners, alkali, manufacturers, and air pollution, 1828–1864', *Econ. Hist. Rev.*, 1982, **35**, 529–548; MacLeod (ref. 5); Minutes of the Manchester and Salford Noxious Vapours Abatement Association, 1876–1891, Manchester Central Library Archives: M 126/6/1/1: Minutes for Nov. 2nd, 1876.

[8] W.A. Campbell, *The chemical industry*, Longman, London, 1971, p. 37.

their areas than other local authorities. In Swansea, as discussed in Chapter 10, the local board of health, the courts, and the community were generally supportive to the copper smelting industry, notwithstanding the immense local environmental impact; this was symptomatic of the priority accorded by many to employment over clean air. It should also be stressed that the environmental concerns of Victorian Britain differed substantially in character from those of the environmental campaigners of the 20th century. The Victorian anti-pollution debates reflected the influence of a broad-based campaign to improve both the health and amenities of British urban life. Mainstream debate, however, did not develop the pollution issue into a more general anti-industrial rhetoric. Damage to property or amenity was often of more concern to anti-pollution campaigners than the possible effects on health. Additionally, whilst the public worried about arsenic or lead contamination of water or food supplies, there is little evidence to suggest that a widespread fear of chemical compounds had emerged in Britain by this period.

Nevertheless there is evidence that public opposition to pollution from chemical factories was significant in shaping the development of the 'chemical industry' as a body. The founding of organisations to represent the interests of chemical industry was at least in part a response to pressure for legislation. The first of these groups was the Widnes Alkali Association, founded in 1859, 'for the purpose of rendering the condensation of muriatic acid compulsory'. The most important of later interest groups was the Society of the Chemical Industry, founded in the same year that the Alkali and Works Regulation Act (1881) came into force. The Society's President in 1886 acknowledged that it was understood when the society was first inaugurated that:

> legislation on the control of the emission of noxious vapours or the pollution of rivers . . . could be more wisely considered and guided by a society such as ours, containing both scientific and practical chemists, than by any individual, however eminent, or by the individual action of each industry affected.[9]

The 20th century began with considerable optimism about the potential contribution of chemistry and the chemical industry to the improvement of material wealth and the quality of life, encapsulated by Du Pont's 1935 slogan 'Better Things for Better Living Through Chemistry'. After the Second World War there was more public uncertainty about the direction in which industrial technology was marching. The role of science in two world wars, the persistence of inequalities in material wealth, and the pollution and risks emanating from industrial processes helped to generate the beginning of an anti-technology movement. Some commentators would argue that this fear of technology has become so entrenched in western culture that society is increasingly moving towards the irrational.[10] However, social surveys demonstrate that public

[9] E.K. Muspratt, 'President's address', *J. Soc. Chem. Ind.*, 1886, **5**, 401–414 (402).
[10] Brock (ref. 1), pp. 656–658; H.D. Crone, *Chemicals and society: a guide to the new chemical age*, Cambridge University Press, Cambridge, 1986, pp. 4, 237–238; M. Kranzberg, 'One last word – technology and history: Kranzberg's Laws', *Technology and culture*, 1986, **27**, 544–560 (552); D. Cardwell, *The Fontana history of technology*, Fontana Press, London, 1994, p. 508.

*Environmental damage caused by a working chemical plant: the manufacture of phosphoric acid at Oldbury (1951) in a plant opened in 1924*

opinion on environmental issues cannot be taken for granted, that the 'general public' does not exist as a homogeneous entity, and that hazards are seen in specific contexts. What might cause great concern to an outsider may be accepted by local inhabitants as a result of tradition, repeated exposure to the hazards in question, or the financial benefits in terms of employment.[11]

Public concern about the chemical industry has tended to focus on the possible risks associated with industrial accidents, and the dumping of toxic waste. The explosion at ICI's nylon works at Flixborough, Lincolnshire, in 1974, and the release of methyl isocyanate by the Union Carbide plant at Bhopal, India, in 1984, which resulted in 2500 deaths, received widespread media coverage and raised the public perception of the risks associated with chemical manufacture. Aside from spectacular disasters such as these, the problem of disposing of hazardous chemical waste has brought the chemical industry into prominence in environmental debates. It has been argued that more than any other issue, hazardous waste poses major political problems for the chemical industry. The development of an international trade in toxic waste has added to political sensitivities.[12]

---

[11] D. Smith and A. Irwin, 'Public attitudes to technological risk: the contribution of survey data to public policy-making', *Trans. Institute of British Geographers* (n.s.), 1984, **9**, 419–426 (423–424).
[12] D. Smith, 'Beyond the boundary fence: decision making and chemical hazards', in J. Blunden and A. Reddish (eds.), *Energy, resources and environment*, Hodder & Stoughton/Open University, London, 1991, pp. 267–291 (281, 286–287).

Before the mid-1970s the local effects of hazardous pollutants were the major focus of scientific concern about chemical use. Few scientists would have placed low-level emissions of non-toxic, unreactive chemicals high on the agenda, or suspected that they could have devastating global effects in a relatively short period of time. It is now recognised that we can alter the atmosphere and climate with unprecedented speed, and that we may already have done so inadvertently.[13] The concerns of environmental campaigners have also shifted in emphasis in the course of the century. The 1960s saw the growth of wide-ranging fears about the possible risks to human health of chemical exposure, including exposure to chemicals released by nuclear explosions. In the 1970s, the fear of resource depletion began to dominate environmental debates, coupled with the emergence of a 'new environmentalism' characterised by a sense of humanly induced global environmental crisis.[14] In the 1980s and 1990s attention has increasingly focused on the possible effects of pollution on global climate change.

In the next few pages we glance at several major environmental issues that have become particularly important in the 20th century.

## (b)  Agriculture, agro-chemicals and biodiversity

Since the 18th century the chemical industries have played an important role in helping farmers to maintain an adequate supply of basic foods such as grain and dairy products. This depended on improvements in agriculture arising partly from the development of improved breeds of animals and crops, partly from increased yields due to well-balanced feeding and soil fertility maintenance, and partly as a consequence of the protection of crops and animals from diseases and the ravages of insect and microbial pests. In all of these respects the chemical industries have played a part. From the Middle Ages it has been understood that arable land requires regular regeneration. With the relatively light usage of earlier times it was found sufficient to allow the land to lie fallow for a season. Various minerals and organic manures were used to improve the land during the 16th and 17th centuries, but as the population increased and the need for more intensive cropping grew in the 18th and 19th centuries, it was realised that other measures were necessary. The use of organic waste, guano, animal manure, and sewage as fertilisers was followed by the recognition that natural fertilisers were still insufficient and that industrially prepared fertilisers would be necessary to maintain the rate of cropping required to feed growing populations. The first manufactured fertiliser, calcium superphosphate, was made on an industrial scale by John Bennet Lawes at Deptford about 1840. The process involved heating minerals containing calcium phosphate with 70%

---

[13] S. Ross and R. Blackmore, 'Atmospheres and climatic change', in R. Blackmore and A. Reddish (eds.), *Global environmental issues*, Hodder & Stoughton/Open University, London, 1996, pp. 129–191 (186).

[14] S. Brown, 'Humans and their environment: changing attitudes', in J. Silvertown and P. Sarre (eds.), *Environment and Society*, Hodder & Stoughton/Open University, London, 1990, pp. 238–269 (257).

sulphuric acid. Perhaps the most important developments have been the industrial methods of 'fixing' atmospheric nitrogen to form ammonia, nitric acid and their compounds. These processes have been discussed in Chapter 5 and their relation to other chemical processes has been explored. As agriculture became mechanised and took on the aspect of an industry, financial gains and losses became more important and the need to supply the market with high-quality food became imperative. Consequently the results of organic chemical research and synthesis were vital to the success and even the survival of farming. To control the ravages of pests and diseases, compounds containing arsenic were introduced, as were naturally formed substances such as pyrethrum; then followed the synthetic pesticides and herbicides. Pesticides and other agronomic inputs have been of major benefit to world agriculture. The world's population has grown geometrically since 1950 and food supply has more than kept pace.[15]

The contribution of the chemical industries to maintaining the adequacy, reliability, and quality of food supplies is perhaps the most significant of their contributions to the quality of human life. In the future, only rising yields will permit the non-agricultural use of land and the protection of biodiversity in our wilder habitats. Modern crop varieties are dependent on high agronomic inputs, and current yields could not be sustained without them. Chemicals designed to control diseases, pests and weeds have opened up new opportunities for farming and crop breeding. Without effective weed control, for example, it would not have been possible to introduce higher-yielding dwarf varieties of cereals. Use of these chemicals also gives the farmer greater flexibility over the timing of crop sowing. Before the introduction of herbicides, weeds often dictated what crops were grown. Herbicides have also significantly reduced labour costs in agriculture. For example, one hectare of cotton crop in 1933 required 200 human-hours per year to grow and harvest using manual hoeing and picking, but by 1985 this requirement had fallen to only 50 hours as a result of chemical weeding, defoliation, and mechanical picking. It has also been possible to reduce soil erosion and degradation, and to crop steeper slopes safely, herbicides having made the adoption of minimum tillage techniques possible.[16]

In England, the average wheat yield has more than doubled since 1950, while the record yield is more than five times as high. One estimate suggests that varietal improvement accounted for 45, 38 and 23% of increases in national yield for wheat, barley and oats, respectively, between 1947 and 1983. In other words, over the past fifty years, approximately half of the increase in cereal yields can be attributed to plant breeding and half to the improvement of agronomy and management, including the use of fertilisers and chemical crop protection. A global survey in 1967 suggested that between 24 and 46% of the world's staple grain crops were lost as a result of the ravages of insects, diseases and weeds. It has been estimated that the elimination of all pesticide use in the

---

[15] L.T. Evans, *Crop evolution, adaptation and yield*, Cambridge University Press, Cambridge, 1993, p. 34.

[16] G.R. Conway and J.N. Pretty, *Unwelcome harvest*, Earthscan Publications Ltd., London, 1991, p. 81; Evans (ref. 15), p. 28.

USA would reduce crop and livestock production by around 30%; in the UK potato yields would be reduced by an estimated 42% and cereals by 45%.[17]

Insecticides have also made a substantial contribution to human health. DDT, for example, had its first major use among troops in controlling typhus in Naples in 1943–1944. DDT appeared to many to be an outstanding insecticide owing to its toxicity to a wide range of insect pests, its low mammalian toxicity under normal circumstances and its unprecedented persistence. By utilising DDT and other powerful organochlorine insecticides, the insect carriers of malaria, yellow-fever, sleeping sickness, dysentery, typhus and bubonic plague were greatly reduced, and thereby millions of lives were saved. At the same time major insect destroyers of crops, particularly locusts, were controlled. The immediate benefits of these chemicals were therefore considerable. When the effects of DDT on wildlife became clear, western industrialised nations banned it because they could afford alternative insecticides, and also in response to a powerfully orchestrated political campaign in the USA. India continued to use it because it was not economically feasible to change to less persistent insecticides, and because its use in agriculture was secondary to its role in disease prevention. According to the World Health Organisation, the use of DDT in the 1950s and 1960s in India cut the incidence of malaria from 100 million cases a year to only 15,000, and reduced the number of deaths from 750,000 to 1500 per year.[18]

Damage to wildlife must be balanced against the advantages to agriculture and to human health. The massive increase in the use of chemicals in agriculture since 1945 has not been without some environmental problems (see below). However, improvements in efficiency and selectivity of insecticides, fungicides and herbicides have helped to minimise their unwanted effects. Application rates have declined through three or more orders of magnitude since the turn of the century. Methods of application have improved in parallel with the introduction of low and ultra-low volume sprays. With smaller doses, the risks of environmental contamination have been reduced. Additionally, more recent pesticides have been developed which have a reduced potential for accumulation and increased degradability.[19]

Many have come to date the rise of the modern environmental concerns from the publication of Rachel Carson's book *Silent Spring* in 1962.[20] Carson had been on the staff of the US Fish and Wildlife Service until her retirement in 1952.[21] She took as her central theme the indiscriminate use of pesticides, fungicides and herbicides in modern farming, and warned of ensuing environmental damage. Carson linked these developments with the growth of the chemical industries during the Second World War, and the sudden increase in the production of synthetic chemicals with insecticidal properties. Prior to the

[17] H.H. Cramer, *Plant protection and world crop production*, Bayer, Leverkusen, 1967; quoted in Evans (ref. 15), pp. 344–345.

[18] Conway and Pretty (ref. 16), p. 105; Kranzberg (ref. 10), p. 546.

[19] Evans (ref. 15), pp. 345–347.

[20] A. Dobson, *The green reader*, Andre Deutsch, London, 1991, p. 25.

[21] J. Sheail, *Pesticides and nature conservation: the British experience 1950–1975*, Clarendon Press, Oxford, 1985, p. 87.

1950s, the pesticides available to farmers were derived either from plants (for example, pyrethrum, nicotine, and derris), or were simple inorganic chemicals. A new era of chemical pest control began with the introduction of dinitro-*ortho*-cresol (DNOC) as a weed-killer in the 1930s. Synthetic plant growth regulators, MCPA and 2,4-D, were developed during the Second World War, enabling the farmer to kill weeds chemically without damaging crops. The 1940s also saw the introduction of organochlorine insecticides, as DDT and $\gamma$-BHC. Carson depicted a future in which the enormous biological potency of these new synthetic chemicals outstripped the capacity of humankind to control the consequences:

> In the less than two decades of their use, the synthetic pesticides have been so thoroughly distributed throughout the animate and inanimate world that they occur virtually everywhere. They have been recovered from most of the major river systems and even from streams of ground-water flowing unseen through the earth. Residues of these chemicals linger in soil to which they may have been applied a dozen years before. They have entered and lodged in the bodies of fish, birds, reptiles, and domestic and wild animals so universally that scientists carrying on animal experiments find it almost impossible to locate subjects free from such contamination. They have been found in fish in remote mountain lakes, in earthworms burrowing in soil, in the eggs of birds – and in man himself. For these chemicals are now stored in the bodies of the vast majority of human beings, regardless of age. They occur in the mother's milk, and probably in the tissues of the unborn child.[22]

*Silent Spring*, most of which had been published in instalments in the *New Yorker* in June and July 1962, captured the public imagination and became emblematic of a wide-ranging fear of chemical exposure amongst environmental commentators and pressure groups. A British edition of the book was published in 1963, and though its appearance shaped subsequent political debate on pesticide pollution, concern about the issue in Britain did not originate with it.

During the 1950s, evidence had been accumulating which suggested a causal link between pesticides and damage to flora and fauna. In May 1950 the Nature Conservancy received the first complaints concerning damage to flora caused by the spraying of roadside verges with 2,4-D (2,4-dichlorophenoxyacetic acid) and MCPA (2-methyl-4 chlorophenoxyacetic acid). The Nature Conservancy was concerned about the possible effects of spraying on bird life and fauna resulting from the loss of natural habitat. As a result of their interventions, the mid-1950s saw the beginning of attempts to regulate roadside spraying to limit the damage to wildlife. At the international level, the potential dangers of increased pesticide use were discussed as early as 1949. The International Technical Conference on the Protection of Nature identified the destruction of non-target species, the disruption of normal predator–prey relationships, and the jeopardising of pollination as possible side-effects. In Britain, concern was focused on the effects of pesticides on human health: seven workers died of DNOC poisoning between 1946 and 1950. Regulations were introduced to

[22] R. Carson, *Silent spring*, Hamish Hamilton, London, 1963, p. 13.

protect agricultural workers using sprays with the passing of the Agriculture (Poisonous Substances) Act, 1952. The early 1950s also saw the beginning of worries about pesticide residues in food which resulted in the setting up of an inter-departmental Advisory Committee on Poisonous Substances used in Agriculture and Food Storage in 1954.[23]

The 1950s also saw the accumulation of circumstantial evidence suggesting that pesticide use was indeed a threat to wildlife. The autumn of 1952, for example, saw an unprecedented number of reports of birds and mammals being found poisoned after crops were sprayed with the organophosphorus insecticide schradan. The terms of reference of the Advisory Committee on Poisonous Substances were extended to include reporting on any damage caused by chemicals in commercial use. The Advisory Committee reached an agreement with industry for a voluntary notification scheme whereby new chemicals, or new formulations, would not be marketed without providing the Ministry of Agriculture with details of their properties and break-down products. The information supplied was reviewed by a panel of scientists selected for their expert knowledge of pesticides. In 1961, a set of detailed recommendations of measures to protect wildlife was incorporated in an appendix to the notification scheme. Between 1961 and 1973, over a hundred trials were carried out by manufacturers on the effects of new pesticides, and of new applications of pesticides already in commercial use. A number of new products were refused registration on the grounds that they represented an unacceptable hazard to wildlife.[24]

Seeds had long been treated with fungicides with beneficial results to agriculture. Problems began when seed-dressings began to incorporate insecticides. After 1955, farmers were encouraged to make greater use of the newly available chemicals aldrin, dieldrin and heptachlor for this purpose. These compounds were particularly effective against the most serious insect pest of winter wheat, the wheat bulb fly. The toxicity of these chemicals to birds and mammals was known, but as the seed dressed with the chemicals was to be buried in the soil, manufacturers and agriculturists did not anticipate problems. Evidence of heavy mortality among wood pigeons and pheasants was reported in the spring of 1956. Further reports of exceptional mortality were received by the Nature Conservancy in 1959. A wide range of bird species were affected, including birds of prey. The British Trust for Ornithology and the Royal Society for the Protection of Birds set up a joint committee to collect further evidence in 1960 and 1961. Gradually a picture emerged of serious effects on seed-eating bird species, and on species eating insects or carrion in which residues had accumulated.

Unexplained deaths of unprecedented numbers of game birds and foxes were also causing concern to those involved in field sports. Circumstantial evidence, eventually confirmed by post-mortem analyses, pointed to the foxes having been poisoned by eating pigeons which had themselves eaten dressed seed. These

---

[23] Sheail (ref. 21), pp. 7–15, 20–23.
[24] Sheail (ref. 21), pp. 29–36.

reports of bird and fox deaths attracted widespread press coverage. As a result of government investigations into these reports, a voluntary ban on the use of aldrin, dieldrin and heptachlor as spring seed-dressings took effect in 1962. The final demise of these insecticides was facilitated by the introduction of two new organophosphorus insecticides, chlorofenvinphos and carbophenothion.[25]

It is salutary to reflect that the organophosphates were first developed as nerve gases for offensive use in war. They attack the central nervous system, producing paralysis and death. Modified as insecticides some of these compounds have been highly effective, especially in cases where insects have developed a resistance to other insecticides. However, they have caused great concern due to the worrying neurological effects on agricultural workers using them to spray crops.

About 200 pesticides were in common use by the 1960s, 90% of which were synthetic organic compounds. During the 1950s, damage to wildlife appeared to be localised and without long-term effects on species populations. The lack of major poisoning incidents, the improvement of regulations governing pesticide use, and the trend to less toxic chemicals added to a general complacency. However, by the early 1960s warnings about pesticide use had suddenly gained credibility.[26] The Association of British Manufacturers of Agricultural Chemicals (ABMAC) published a commentary on *Silent Spring* welcoming the book for stimulating the public to take greater care in the use of pesticides, but stressed that the dangers of these chemicals should be balanced against the pressing problems of world health and food supplies.[27]

Prior to the *Silent Spring*'s publication, the Nature Conservancy had been contending that more serious long-term side-effects to the use of pesticides existed and required study. Surveys carried out in the USA were beginning to show the effects of insecticide spraying on birds and other fauna, caused by the persistence of organochlorine compounds. Persistence made it easier for these chemicals to be transferred to other organisms (including humans) and environments, and their toxicity was cumulative. There were also worries about the incidence of residues in soils and their potential effects on soil fauna that might lead to a decline in soil productivity. By the mid-1960s Britain had its first documented example of a significant decline in a wildlife species since the introduction of organochlorine insecticides. The peregrine falcon population had fallen to under 45% of its pre-war level. Post-mortem studies on this and other raptorial and fish-feeding species recorded high levels of organochlorine residues. These studies underlined the need for manufacturers and policy makers to consider not simply the oral toxicity of a pesticide, but also the cumulative effects of small doses, effects on species reproduction, and the movement of pesticides through the environment. A ban on most uses of aldrin, dieldrin and heptachlor took effect in 1965.

---

[25] Sheail (ref. 21), pp. 58–84, 195. Despite significant improvements, modern compounds (including carbophenothion) still cause damage to wildlife; see Conway and Pretty (ref. 16), p. 81.
[26] Sheail (ref. 21), pp. 19, 58, 227–229.
[27] *Silent Spring: a review and commentary*, Association of British Manufacturers of Agricultural Chemicals, London, 1963, quoted in Sheail (ref. 21), pp. 91–92.

## (c)    Global issues: ozone depletion and global warming

*Silent Spring*'s concern was chiefly with the effects of persistent pesticides in the food-chain and did major service in emphasising the value of an ecological perspective.[28] Perhaps the most striking development of the 20th century has been the increase of concern about the global ecological effects of chemical use. In 1974 CFCs, which had been in use since the 1930s as refrigerants, and subsequently as propellants and solvents, were linked to the destruction of the ozone layer. In 1984 scientists reported the discovery of a hole in the ozone shield over Antarctica, resulting from a massive loss of stratospheric ozone that occurs each September/October. Fears about depletion of the ozone layer, first raised in connection with the effect of air travel on the stratosphere in the early 1970s, had already been projected as a human threat in the form of a probable increase in skin cancer resulting from the likely increase in ultra-violet radiation reaching the surface of the Earth. Newspapers carried headlines such as 'Aerosol cans could endanger all human life, scientists say' and 'Sprays are linked to skin cancer threat'. The issue was placed on the international political agenda, and on 16 September 1987 the Montreal Protocol was signed, the first international agreement to restrict the release of substances into the atmosphere deemed damaging to the global environment. Since the late 1970s ozone losses have been observed in mid-latitudes of both hemispheres as well as in the polar regions, and the issue continues periodically to fuel public imagination.[29] The debates about ozone depletion signalled a major shift in international attitudes towards the global environment; as Roger Blackmore has encapsulated it, it was shocking to recognise that damage to the remote and pristine environment of Antarctica, or increased incidence of human skin cancer, could result from such everyday activities as getting rid of an old fridge, or using a deodorant.[30]

Global warming is another issue that has underlined the view that the atmosphere is a complex and finely-balanced system which may be changed dramatically by the everyday use of chemicals in modern industrial societies. There is a balance between the solar radiation received by the Earth and the long-wave radiation that it emits. Some of this long-wave radiation is absorbed by atmospheric gases, the most important of which is water vapour. Concern grew in the 1980s that atmospheric concentrations of other radiatively active, or 'greenhouse', gases were increasing as a result of human activities. Such increased concentrations have the capacity to enhance the natural greenhouse effect, and cause significant changes in the world's climate with potentially disastrous consequences for many human populations. The effect of individual gases on radiation balance can be calculated from a knowledge of the atmospheric concentration of the gas and its rate of increase, the relationship between concentration and radiation absorption for that gas, and the extent to which its absorption bands overlap those of other gases. Estimates of the relative

---

[28] Sheail (ref. 22), p. 95.
[29] R. Blackmore, 'Damage to the ozone layer', in Blackmore and Reddish (ref. 13), pp. 70–71, 102, 117.
[30] *Ibid.*, p. 123.

contribution of the major 'greenhouse' gases to global warming during the 1980s suggested that $CO_2$ contributed about 50%, and that $CH_4$ and the major CFCs were the most important of the other gases. Of the industrial contribution to greenhouse gases, the most important resulted from fossil fuel burning ($CO_2$ and $N_2O$ production), and from the release of CFCs and their substitutes. The two most important chlorofluorocarbons in terms of global warming were $CFCl_3$ (CFC-11) and $CF_2Cl_2$ (CFC-12). Other major anthropogenic sources of 'greenhouse gases' included agriculture (including changes in land-use, biomass burning, fertiliser application, rice paddies, domestic ruminants) and deforestation.[31] And of course these gases are also generated by entirely natural means, such as volcanic activity.

Although the causes and consequences of global warming remain the subject of controversy, the issue has been successfully placed on the international agenda as a global threat without parallel. In 1996, the Intergovernmental Panel on Climate Change (established in 1988), concluded that global warming was already detectable. The first week of December 1997 saw the gathering of delegates from 166 nations in Kyoto, Japan, to consider what action to take against the pollution implicated in global warming. Uncertainty exists as to the likely effects of global warming on world climatic systems: they may change in a gradual, predictable way or they may not, and the changes will not be easily reversible.[32] However, the economic, political, and social obstacles in the way of reaching an international solution to the problem of greenhouse gas production are immense, given both the dependence of modern society on fossil fuels and the difficulties involved in changing current land-use practices.

Also on the international political agenda is the conservation of biodiversity world-wide. The maintenance of biodiversity is of practical significance to humankind. Genetic diversity in crop species is desirable for maintaining resistance to pests and diseases. Species diversity provides us with innumerable products, including medicines, food and industrial products. Ecosystem diversity, amongst other things, ensures the cycling of water, carbon, minerals and other materials, the regulation of watersheds and the stabilisation of soils. It is also important for the regulation of climate at micro- and macro-levels. Biodiversity is also increasingly valued on aesthetic and philosophical grounds. There are several ways in which chemical use in modern society may conflict with the maintenance of biodiversity: the application of fertilisers and pesticides results in the erosion of habitat for plants and wildlife in areas given over to intensive agriculture. Leakage of chemicals from agricultural areas may cause disruption in neighbouring terrestrial and aquatic environments. Industrial chemicals and chemicals present in domestic sewage and refuse can endanger the reproductive success of wildlife through the pollution of riverine and marine environments. Atmospheric pollution, particularly as oxides of sulphur and nitrogen, has caused devastating effects on vegetation when deposited as acid rain. Finally, human-induced global climate change may have disastrous

[31] M. Ashmore, 'The green house gases' [diagram], *Trends in ecology & evolution*, 1990, **5**, 296–297.
[32] G. Lean, 'It may be our last chance', *Independent on Sunday*, 30 November 1997, p. 30.

consequences for the functioning of ecosystems adapted to different climate conditions.[33]

## (d)   Health and diet

In regard to personal hygiene and health the chemical industries play a central role. The products of the pharmaceuticals and fine chemicals industries (discussed in Chapters 6 and 7), such as toiletries, cosmetics, shampoos and perfumes, have made a vast improvement in the quality of life for ordinary families. Synthetic perfumes are widely used in washing powders, air-fresheners, polishes, insecticides *etc*. The chemist has been able to supply synthetic substitutes for plant- or animal-derived essential oils, and in consequence has succeeded in overcoming the prohibitive cost of natural perfumes. The perfume and cosmetics industries are close to the much larger drugs industry, since both produce substances for personal use which must be carefully tested to ensure safety. The chemical industries have contributed to the rise of public health standards equal to those made by the progress of medical knowledge. Their contributions to safeguarding standards of purity in water supply have already been discussed. Additionally, new drugs, anaesthetics, antiseptics and antibiotics have played a crucial role in improving medical treatments and in reducing the risks of infection. In some cases the chances of curing formerly incurable diseases have also been increased, whilst analgesics have helped to reduce pain and suffering. Hospitals have become safer for patients and life expectancy in the community generally has been extended. This is an undoubted contribution to the improvement of quality of life, and there is also every hope that innovation will provide further improvements in the future. Yet most people give all the credit for these advances to medicine, not realising that without the products of chemical research and industry they could never have been made.

However, the late 20th century has seen growing anxiety about carcinogens in the environment. In 1901 infectious diseases were the major killers, and tuberculosis alone killed twice as many people as cancer. Public perception of cancer was not acute, even the risks of smoking tobacco being little appreciated and given little official recognition until the 1960s. Since then there has been a significant increase in the number of people suffering and dying from cancer. Over 130,000 men and women died of cancer in England and Wales in 1980 and the disease was the cause of about 22 in every 100 deaths.[34] Whilst some commentators in the 1980s argued that there had been a real rise in cancer rates in all ages, others argued that the rise in cancer deaths reflected merely the decline in infectious disease mortality which resulted in more deaths from diseases of old age. The result of this change, it was argued, was that cancer had become a more prominent cause of death than it was formerly, but the risk of

[33] C. Clubbe, 'Threats to biodiversity', in Blackmore and Reddish (ref. 13), pp. 192–237 (193, 217–218, 226–233).

[34] L. Doyal, K. Green, A. Irwin, D. Russell, F. Steward, R. Williams, D. Gee and S.S. Epstein, *Cancer in Britain: the politics of prevention*, Pluto Press Ltd., London, 1983, p. 8.

death due to cancer was still not high, nor had the risk increased, with the sole exception of tobacco-related respiratory cancer.[35]

In 1978 the political temperature of the debate was raised when Samuel Epstein, Professor of Occupational and Environmental Medicine at the University of Illinois, published *The Politics of Cancer*, in which, supported by estimates produced by US Federal experts, he argued that at least 20–40% of cancers were work-related, and that industrial pollution and the chemicals used in consumer products were significant causes of cancer. Epstein downplayed the role of lifestyle factors, arguing that the contribution of smoking had been exaggerated, and that the role of diet, except in the specific case of deliberate or accidental carcinogenic food additives, was negligible. In this analysis, the solution to the cancer problem lay not solely in health education, but in the identification and regulation of chemicals in the workplace, in consumer products, and in the wider environment. Epstein's argument, and his analysis of the evidence, was challenged by the cancer epidemiologist Richard Peto in an article published in *Nature* in 1980. Peto dismissed the estimate of the relative importance of occupational causes of cancer as wildly exaggerated, and restated the lifestyle theory of cancer causation, stressing that smoking-derived and fat-associated cancers accounted for more than half of all cancer deaths. Peto's argument was developed in *The Causes of Cancer* (1981), which Peto co-authored with Richard Doll. Their analysis suggested that occupational factors caused less than 5% of all cancers, pollution 2%, smoking 30% and diet 35%. This remains a matter of considerable dispute, and more information is required about environmental pollution before definitive conclusions about the relative importance of different causes of cancer can be reached.[36]

Whilst the overall contribution of industrial pollution to cancer deaths is disputed, and disagreements exist about the historical trends in cancer rates in all age groups, there are certain instances in which the relationship between occupational exposure to chemicals and cancer has been well established. Some of these relationships are so well established that certain cancers are designated 'prescribed diseases', entitling the victim to possible compensation. Occupational cancers granted this status in Britain include skin cancer caused by exposure to soot, tar and mineral oil (prescribed in 1921); lung cancer caused by exposure to gaseous nickel compounds (1949); bladder cancer caused by aromatic amines in chemical dyestuffs and in rubber production (1953); mesotheliomas caused by exposure to asbestos (1969); adenocarcinoma of the nose caused by exposure to wood dust (1969); angiosarcoma of the liver resulting from exposure to vinyl chloride monomer (1976); and cancer of the nasal cavity contracted in the manufacture and repair of footwear (1979).[37]

[35] Crone (ref. 10), pp. 85–86; R. Peto, 'Distorting the epidemiology of cancer: the need for a more balanced overview', *Nature*, 1980, **284**, 297–300.

[36] S.S. Epstein, *The politics of cancer*, Anchor Press, New York, 1979; R. Doll and R. Peto, *The causes of cancer*, Oxford University Press, Oxford, 1981; S.S. Epstein and J.B. Swartz, 'Fallacies of life-style cancer theories', *Nature*, 1981, **289**, 127–130.

[37] Doyal *et. al.* (ref. 34), p. 17.

Additionally, concern exists about the possible effects of carcinogenic substances in consumer products.

The chemical industries also give support to a large food manufacturing industry. For instance, preserves, pickles and soups are manufactured on a very large scale and are sold in glass jars and bottles, or in cans and packets. Fruit, vegetables, meat and fish are cooked or heated before being canned. Bread is made on an industrial scale and is now usually wrapped in thin polythene, or other plastic material. Dairy products such as milk, butter and cheese are processed and packaged in vast quantities. Margarine is manufactured from vegetable oils, while cornflakes and other cereals are manufactured from grain. In all of these examples, and in many others, the products of the chemical industries are involved, from glass to sheet steel and tin-plate, from culinary preservatives and colourings to cellophane, polythene and other packaging materials. One of the most difficult problems for farmers and the food industry is the question of preserving fresh fruit and vegetables in good condition over long periods in warehouses and during transport. Various methods such as bottling and canning are used along with a range of chemical preservatives in the manufacture of foods with the aim of prolonging their 'shelf life'.

Substances in food emanating from methods of food production (pesticide use, additives to cattle feed), food processing (colourings, flavourings and preservatives), and from food storage (contamination from plastic containers and food wrappings) have all come under suspicion. Yet, when examined in detail, the evidence does not suggest that chemical residues in food are a high risk to human health. A study of DDT and the organochlorines, the most persistent and therefore probably the most hazardous pesticides over the long term, exemplifies this. Their cumulative effect on bird populations in the 1960s was dramatic and has been well documented. It seemed reasonable to assume that human beings might be similarly at risk. However, experiments on laboratory mammals have shown that, whilst at very high dosages DDT causes death, there is little evidence of chronic effects. Rats and mice fed diets high in DDT for several generations have shown no effects on reproduction, or on survival of progeny. Occupational exposure also suggests there is little risk to humans. Workers received heavy exposures as spray operators during the 1950s and 1960s, yet there is no epidemiological evidence of damaging effects. DDT and some of its derivative compounds are found in almost all people throughout the world, most being exposed to levels 200–1000 times less than those occupationally exposed. Most epidemiological studies have shown no relationship between the amounts of DDT stored in the body and cause of illness or death. The World Health Organisation published a report on DDT in 1979 and concluded that the 'safety record of DDT is phenomenally good'.[38]

At the very low concentrations in which they occur, there is also no evidence that chemical additives in foodstuffs cause cancer in humans. The main evidence of risk is derived from the feeding of laboratory animals high concentrations of chemicals; some toxicologists in North America and Britain doubt the relevance

[38] Conway and Pretty (ref. 16), pp. 93–98.

of feeding studies involving the incorporation of unrealistically high levels of test substances in the diet. On the other hand, many naturally occurring contaminants of food are extremely dangerous. Aflatoxins, for example, produced naturally by mould, are a class of chemicals with extremely toxic and carcinogenic properties. Chemical food additives have reduced some of the naturally occurring food risks by improving preservation and preventing the partial spoilage of stored foods by micro-organisms. Nitrites, for example, extensively used in the curing and preservation of meat, kill the organism that produces *botulinum toxin*, one of the most poisonous substances known.[39] The case of nitrites exemplifies the complexity involved in balancing risks to human health. Whilst studies have raised concerns about the safety of nitrites in respect of their potential carcinogenicity (after reacting with foodstuffs, or in the stomach, to form nitrosamines), their use for the protection of the population against botulism will continue until a safer substitute can be found.[40]

Turning to pesticide residues in food, the level of exposure of the general population to residues in Britain is assessed by the Total Diet Study which began in 1966. Foodstuffs are purchased from a wide variety of retail outlets and the levels of residues estimated after preparation and cooking in the usual way. The food is divided into 20 food groups and the daily intake is then computed, based on the contribution of these groups to the typical diet as found in national food surveys. The study has revealed the steady decline in organo-chlorine intake since 1966. In all cases the residues have been found to be below the maximum 'acceptable daily intake' (ADI) as set by the Food and Agriculture Organisation and the World Health Organisation. Because they are less persistent, organophosphate residues in food have always been lower than organochlorines. Much higher pesticide residues have been detected in some foods, including home produce, due either to incorrect pesticide application, or to post-harvest treatment of crops, particularly the use of fumigation during storage. However, the overall hazard presented by pesticides in Britain and other industrialised countries appears to be low. In developing countries, where daily intakes of pesticide residues in total diets are often very high and in some cases exceed the ADI, the overall risk is considerably higher.[41]

The question of the possible risks to humans from food contaminants derived from plastic food wrappings has recently attracted additional interest as part of a wider debate about the effects of the breakdown products of plastics and synthetic chemicals on the wider environment. Research has revealed that a number of diverse man-made chemicals possess oestrogenic and in some cases anti-androgenic activity. In other words, such chemicals have been found to act as sex hormones capable of developing female characteristics in the bodies of males, or of interfering with the normal functioning of male reproductive systems. The effects of what are now known as environmental 'endocrine disrupters', or more specifically oestrogen-mimicking chemicals, first generated concern during the 1970s. Among the most recent compounds found to

[39] Editorial, 'Sweet reason', *Lancet* 1977, 634; Crone (ref. 10), pp. 91–96.
[40] Doyal *et. al.* (ref. 34), pp. 94–99; Crone (ref. 10), p. 96.
[41] Conway and Pretty (ref. 16), pp. 129–136, 145–147.

demonstrate oestrogenic activity are the phthalates, a large group of chemicals widely used throughout the world in the manufacture of plastics. As a consequence these are among the most common aquatic micropollutants. Since then the number of studies of incidences of reproductive disorders and abnormalities among wildlife and human populations has steadily risen. It has become one of the most controversial environmental issues for many years. A direct relationship between oestrogen mimics and effects in humans has not been established to date. Studies of wildlife in the aquatic environment, however, have demonstrated the potentially powerful effects of these substances on fish and aquatic reptiles. Reassuringly, there has been no evidence of oestrogenic compounds in raw water reservoirs prior to treatment for public water supply.[42]

Research into British rivers was stimulated by the discovery of hermaphrodite roach in stretches of river close to sewage outfalls. More research is required before the implications of these findings can be fully interpreted. The natural occurrence of hermaphroditism in coarse fish in Britain and the overall ecological implications of hermaphroditism on fish population levels remain to be determined. However, studies of effluents from sewage treatment works in Britain have demonstrated that most if not all are oestrogenic to fish. In three rivers (the Lea, Arun, and Aire) downstream stretches have also been shown to be oestrogenic. The chief effect on the fish is the inhibition of testicular growth, a phenomenon attributed largely to the presence of alkyl phenolic compounds in the sewage effluent. The source of these chemicals is likely to have been detergents and industrial surfactants which degrade during sewage treatment. However, many other oestrogen-mimics are also likely to have been present in the sewage effluent.[43] These and other studies of oestrogenic activity represent another well-publicised example of the possibility that synthetic chemicals will behave in quite unexpected ways when released into the environment, with potentially undesirable and wide-ranging effects on wildlife and human populations.

## 3   Consumer Benefits from the British Chemical Industry

Despite these problems, many of which have only been recognised in the last three or four decades, the chemical industry can claim to be one of the most important contributors to improvements in the quality of life. In order to get some idea of the truth of this we might pause for a moment to consider what a world without its products would be like. Consider first the great variety of products from the chemical industries in common everyday use. They range from manufactured foods like margarine and cornflakes to silicon chips; they provide materials for clothing, carpets and furnishings, and building and

---

[42] J.E. Harries, D.A. Sheahan, S. Jobling, P. Mathiessen, P. Neall, J.P. Sumpter, T. Tylor, and N. Zaman, 'Estrogenic activity in five United Kingdom rivers detected by measurement of vitellogenesis in caged male trout', *Environmental toxicology and chemistry*, 1997, **16**, 534–542 (534–535). For a list of studies of the effects of oestrogenic chemicals on wildlife, see their refs 2–8.
[43] *Ibid.*, pp. 539–540.

decorating materials which enhance the attractiveness of domestic housing, both outside and inside. Detergents, silicone waxes, cleaning agents, washing powders and antiseptics help to reduce the physical effort of housework while improving domestic hygiene. The chemical industries also contribute substantially to agricultural productivity and food distribution. Artificial fertilisers and synthetic hormones have increased yields; insecticides and herbicides have reduced pests; sprays preserve crops after harvesting and retard the ripening of fruit to ensure transport in good condition to distant destinations. The chemical industries have also contributed to improved standards of health by the invention and manufacture of new drugs to prevent and to treat diseases, and by contributions to the food industry which have led to a healthier, more varied and interesting diet. They have helped to make more time available for leisure pursuits by increasing the speed and thoroughness of domestic cleaning and at the same time they have brought leisure activities, especially sports like tennis and golf, which were once the prerogative of the middle classes, within the reach of working people by the introduction of cheaper, often stronger and lighter, synthetic materials for the manufacture of affordable equipment. Without the products of the chemical industry the quality of life would be much poorer in all these and many other ways. For most ordinary people a world without modern chemical products would be full of drudgery and heavy physical labour. A much greater dependence on natural resources would bring the prospect of scarcities. Given the size of modern populations it would also be a world in great danger from hunger and from epidemic outbreaks. In fact we are envisaging a return to the conditions of three or four hundred years ago.

Thus we cannot ignore the huge benefits conferred by the chemical industry, though an immense pall of ignorance obscures much of the picture. It has often been remarked that many chemical products, especially heavy chemicals, are used only within industry and remain unknown to the public. Since the late 19th century, however, new chemical manufactures have led to more direct contacts between the chemical industries and the public in regard to products such as paints, plastics, wall and floor coverings, horticultural products, cosmetics, drugs and medicines.

Some of these products were invented, developed and manufactured specifically for the retail market. The trend began with celluloid and the various kinds of rayon, or artificial silk, but it grew rapidly with the introduction of bakelite and ebonite, synthetic rubber, new dyes, drugs, cosmetics and a wide variety of other materials purchased by consumers. These products are continually modified as a result of new discoveries and in response to public demand. In some cases very specific criteria are set and the success of a new product is judged in relation to the extent to which it meets the users' requirements. As improvements are made in the variety and quantity of their products, the chemical industries themselves are also steadily changing; their processes becoming more streamlined and automated, their products more sophisticated.

Poly(vinyl chloride) (PVC), well known as a cheap plastic material manufactured on a very large scale, is a good example here. Its manufacture involves a range of technical processes from the preparation of vinyl chloride, through

*Mouldings in parkesine (1861–1868)*

polymerisation and various methods of conversion to useable, attractively coloured plastic materials. Each stage of the process must be carefully controlled. Many different grades of PVC are made to satisfy a wide range of demands and the variations depend on such factors as molecular weight, particle size, the proportion of added plasticisers, dyes or pigments, and other such technical factors, none of which are of the slightest concern to the consumer. What matters to the user is that it is durable, non-toxic, attractive, a good electric insulator, fire-resistant and so on; these are the criteria the chemical manufacturer is required to meet and when this is done successfully the product may be considered to fulfil the demands made of it.

Sometimes a new material introduced for one purpose has been used in ways unforeseen by its inventor, and as a result has had unexpected social consequences. Perhaps the most striking example of this is celluloid. Invented in the 1860s by Alexander Parkes as a substitute for ivory, 'parkesine' had the great disadvantage of being not only highly inflammable but also rigid and non-plastic. When its rigidity was modified by treatment with camphor, the new product which resulted, celluloid, introduced about 1870, found numerous domestic uses. At about the same time the possibilities of producing photographic images were being investigated, initially using glass plates which were inconvenient and cumbersome. When it was found that celluloid could be made into a flexible film that would carry the chemicals necessary to produce a photographic image, the portable camera and the film industry were born. The celluloid film made the Kodak box-camera possible and gave a great boost to personal and documentary photography. About the same time it was realised that by taking a series of pictures of moving objects and then projecting them at a suitable speed a moving image could be produced and the cinema was born. This development produced important social effects in the 1920s and 1930s as cinemas superseded the old music halls.

Films, especially American films, served to raise people's notions of the quality of life and to implant new ideas and aspirations. No single advance in the chemical industry has had quite the same social impact. Thus celluloid film ushered in new art-forms and made possible the growth of industries far removed from the chemical industry itself, in fields entirely different from those for which the material had first been introduced. Moreover, this new material helped to enhance the quality of life for millions of people. The chemical industries have provided other new materials for leisure activities which have facilitated the development of a leisure industry for the masses, and the flourishing of popular culture: for example, ebonite for gramophone records, fibreglass and carbon fibre for fishing rods, tennis racquets and golf equipment. Many of these products were cheaper than the natural products which they replaced and consequently they helped to bring many leisure activities, formerly the preserve of the better-off, within the range of ordinary working people.

Nowhere has the impact of the chemical industry on the quality of life been more marked than on the domestic scene where comfort, hygiene, safety and attractiveness have all been enhanced by chemical products. Home furnishing and decoration have been immensely improved using chemical products. After the introduction of bakelite (phenol–formaldehyde resin) about 1910, plastics went through a period of rapid development. Urea– and thiourea–formaldehyde resins, methyl methacrylate polymer (perspex) and other compounds brightened and modernised homes, and with new detergents the hygiene and ease of cleaning of kitchen and bathroom worksurfaces and floors was improved. These developments have occurred during the present century though not always smoothly as new products sometimes turned out to have undesirable faults. Early plastics suffered from a variety of problems. Plastic work surfaces tended to 'craze', becoming unsightly and unhygienic, providing breeding-grounds for germs; PVC tended to harden and crack in the cold and become deformed in contact with heat; early nylon would not absorb moisture and so was unsuitable for underwear. Chemists were ultimately successful in correcting many of these faults, but the public perception of plastics as ersatz, 'cheap and nasty' substitutes, was slow to change. Only quite recently have plastics come to be widely regarded as materials in their own right rather than as cheaper substitutes for natural materials such as wood, metal or glass.

Domestic detergents, derived from industrial de-greasing agents such as teepol, were introduced to improve cleaning by removing grease and fats left after cooking. They were used in washing-up liquids to produce sparklingly clean crockery and glassware, and in washing powders developed to replace soap. But these new products were not always welcomed by the housewife. When, in 1938 the soap manufacturers, Procter and Gamble, introduced their new detergent washing powders 'Drene' and 'Dreft' in competition with 'soap flakes' for washing delicate fabrics they did not sell, and it was only after

*Fashion advertisement for Enka Rayon (1952)*

American troops brought nylon stockings to Britain and nylon underwear became available that Procter and Gamble found themselves leaders in a new market.

New polishes have also been developed based on silicone waxes for cleaning modern surfaces. These produce a harder smear-free surface with less effort. The production of foam rubber and plastics for use in upholstery has also widened the variety of choice of home furnishings. Again initial problems were experienced with some of the innovatory new materials. New fire-risks, including fierce and rapid burning, and the production of poisonous fumes, led to a number of fatal accidents. Research into fire-retardants to overcome this problem has progressed and regulations have restricted the use of certain materials in furnishings, whilst fire-resistant fabrics and upholstery have been developed.

Fabrics have also been transformed by the chemist. Beginning with rayon at the end of the 19th century, new fibres and fabrics have continued to be invented. Whilst rayon is composed of reconstituted cellulose fibres, nylon, terylene and the acrylics are truly synthetic in that they are fabricated from raw materials, notably oil-based compounds, which are not originally fibrous. Their composition and properties have been chemically constructed on the basis of the theoretical study of the properties of large molecules and they aim to copy in synthetic form protein molecules such as those of silk or wool. New man-made fibres have produced lighter and easier-care fabrics, helping to reduce the labour of domestic laundry, in addition to promoting greater choice and comfort.

Nevertheless, these man-made fibres are not identical to any natural ones and they require special dyes and dyeing techniques. This has resulted in a vast extension of the chemistry of dyes and dyeing which has produced a wide range of new colours and methods of applying them to fabrics. Synthetic dyes for fabrics and carpets and coloured fillers for plastics used in floor-coverings, work-surfaces and domestic utensils have all played a part in improving the quality of life in homes and workplaces. These developments have brightened homes, offices, hotels and other public buildings, as well as multiplying the choice of clothing fabrics available to the consumer. Clothes and fashions have appeared in all imaginable colours and endless shades. The psychological effects of the dyestuffs industry on the quality of life are difficult to quantify. The inventions of the colour chemist raise the spirits rather than causing tangible physical effects as so many other chemical products do.

Turning to a very different field we find that the contribution of the chemical industries to modern communications is also immense. The paper industry is heavily dependent on chemical products and methods. The variety of types and qualities of paper is very great, and some forms such as newsprint are made in vast quantities – a major industry. Chemistry also supplies the various inks and dyes used in printing, as well as the chemicals and other materials needed for photography, from the film to the chemicals used to develop the photographs, and the inks needed to print them. We have already discussed the role of the chemical industries in the production of films for the cinema, but today this is also extended to radio and television and to recording apparatus both for sound and for vision. The chemical input to radar systems, the telecommunications industry, and the burgeoning computer industry at all levels is incalculable. All these things have become essential to modern life, for the running of the economy and for leisure pursuits. It would be difficult to envisage life without most of them – international markets depend on rapid communication for example – and the measure of how far they have improved the quality of life may be found in the reluctance with which we would be willing and able to give them up.

The rise of chemical industry has exercised a powerful influence on developments in communications and on the urban infrastructure. Primary raw materials, whether imported or indigenous, required transport to chemical works from the ports, mining areas and other centres of production. Similarly, the industry's products had to be transported from the factories to points of distribution and use. Canals, linking industrial sites with each other and with navigable rivers and ports, provided the main method of transporting heavy goods up to the 1830s. They were slow, required constant maintenance and could become frost-bound in winter, but they provided valuable links for commerce during the 18th century. After about 1830 they were gradually superseded by the railways which could carry the same heavy loads in larger quantities much faster. Roads were also improved from the late 18th century and coastal shipping increased. Communications and the means of transferring heavy goods between chemical factories and the ports improved greatly during and since the 18th century, due in part at least to the demands of chemical manufacturers. The early 'cottage industries' using primitive chemical techniques were sited near the source of their raw materials and were geographically scattered, for example, the kelpers in the Scottish highlands and in parts of

Ireland, or the iron founders of the Forest of Dean and the Sussex Weald; the modern chemical industries have agglomerated in urban areas. Ports such as London, Bristol, Liverpool, Newcastle, and Glasgow attracted chemical industry to their vicinity owing to the ease of importing raw materials and exporting the finished products. The chief centres of British chemical industry from the mid-18th century developed in the urban areas around Glasgow in Scotland, Newcastle on Tyneside, Widnes and Warrington on Merseyside, the Black Country north of Birmingham, the East-end of London, and Bristol. Each of these areas supported a range of chemical and associated industries. Besides chemical factories there were glass houses, iron foundries, metal works, soap boilers, alum and copperas manufacturers, dyers, brewers, tanners and many others more or less closely related to chemical manufacture. Expansion was stimulated by the demands of industrialisation.

The availability of transport is an important aspect of the quality of life which is heavily dependent upon the chemical industry. From oils used as fuel and for lubrication, to steel and aluminium for construction, all forms of transport use many of the products of the chemical industries. Plastics, new materials, foam rubber and insulation materials are used in the construction of vehicles of all kinds. In road building coal-tar and bitumen are used for surfacing, as is cement and reinforced concrete. Steel is used both for body-work and for engines in vehicle construction. Aluminium alloys reduce weight, especially important in the aircraft industry, while paint, rubber, plastics and man-made fabrics are used in vehicle interiors. On the railways steel is used in large quantities for rails, engines, wheels and exterior coachwork. Special alloy steels are also used for points and other heavy duty uses. Diesel and lubricating oils are needed, and for electric railways copper conductors and plastic insulators. In sea and air transport too, a whole range of chemical products are required. It is clear that the chemical industries supply the materials which make all of these forms of transport possible.

While chemical products have improved many aspects of transport, there are still problems associated with the bulk transport of the raw materials for chemical industry and finished chemical products themselves. The illustration shows a horse drawn dray about 1925, loaded with chemical products for local delivery – a far cry from modern bulk transport by road or rail. It is the heavy tankers thundering through towns and villages or along the motorways that make the general public most aware of the importance of the chemical industry. This is especially true when in an accident chemical substances are spilled in quantity on to the carriageway and reports of the accident appear in the local or national press. But it is the disastrous consequences of major oil-spills causing serious ecological damage and the difficult problems of containing the oil slick and of cleaning up its effects on beaches, rocks and marine life which make front page news. It is ironic that the means of correcting the problems caused by crude oil in such disasters depend on the use of detergents and oil dispersants, which are themselves products of the oil-based chemical industry.

Until the middle of the 20th century coal was the most important fuel, both for domestic and industrial heating, for steam traction on the railways and at

*Typical local delivery for small quantities of chemicals (1925): pollution-free transport?*

sea, and for the generation of electricity. The burning of raw coal is both inefficient and highly polluting, but it continued in general use in Britain until the late 1950s. For some industrial purposes it was already clear that coke was preferable and in the smelting of metals it was essential. For domestic heating, lighting and cooking, coal-gas was often used. Until the 1950s there was a large and diverse coal-gas industry producing 'town gas' for domestic and industrial use. The by-products of this industry – coke, coal-tar and ammoniacal liquor, all found industrial uses, although coal-tar was regarded as a nuisance until it was realised that it contained a large number of organic compounds, mainly aromatics, which could be used as the basis for the manufacture of dyes, drugs, explosives and many other organic chemical products. From the mid-20th century petroleum and natural gas have taken over from coal both as fuels and to provide the necessary feedstocks for the organic chemical industry. Oil refining is itself a large-scale industry based on chemical technology; its products form the basis for the manufacture of synthetic organic chemicals.

Some of chemistry's most important contributions to improvements in health and the quality of life are those to food safety. Since the early 19th century it has been recognised that there is a need for regulation of quality and hygiene in the handling and sale of food. Since the first Public Analyst was appointed about 1872, this service has grown and by the use of chemical and microbiological testing has carefully monitored the safety and quality of foods and the water supply. In the 19th century the analysis of water samples was an important source of income for many chemists in Britain, and many hundreds of such analyses were carried out even before the Institute of Chemistry established its qualification for public analysts in 1899. Ever since then chemists have endeavoured to improve the quality of water supply and maintain high standards of purity. This is an area where the chemical industries have made substantial improvements by introducing methods of chemical analysis and testing, rather than new products, and, although just as important, the involvement of the chemical industries is often less evident to the consumer.

Chemical methods of retarding the ripening of fruit have also helped to improve the quality and variety of diets, though the most widely used method of food preservation is by refrigeration. Here again the chemical industry has been involved in developing safe, effective refrigerants. This technology has also enabled a domestic revolution in food storage. Traditional methods of home preservation had included salting and pickling, but most fresh foods had to be consumed quickly to avoid spoilage. Domestic refrigerators began to be available in the 1920s, and now most modern homes will possess one in addition to a freezer to preserve foods for longer periods than was possible hitherto. In recent times the chemical industry has had to devise substitutes for the compounds formerly widely used as refrigerants, the chlorofluorocarbon compounds (CFCs), after it was discovered that these compounds were damaging the ozone layer in the upper atmosphere.

The chemical industries are also closely involved in establishing and maintaining standards of public safety in factories, offices, shops and hotels, in public places such as sports halls and stadia, swimming pools, theatres and cinemas; even in our homes. Fire risks are investigated by chemical methods, and fire-resistant materials have been produced and regulations introduced to control their use. Smoke alarms, non-slip materials for floors, improved insulators for electrical systems, new refrigerants to remove the hazards of CFCs, and many other safety measures reduce the risks of accidents, and all in various ways contribute to improvements in the quality of life.

## 4 Conclusion

The development of a new chemical product is hardly ever easy; chemical manufacturers have to wrestle with many intractable problems quite unknown to the general public. First among these is the *inexorability of natural laws*. In considering chemical manufacturing it must be remembered that no matter what is being produced, all chemical reactions are governed by the laws of conservation of matter and energy. Neither is destroyed during any chemical change, both represent money to the chemical manufacturer and there is a balance which accounts for all the reactants and all the products. Energy is usually exchanged in the form of heat, sometimes absorbed but more often evolved during the reaction.

In an efficient process a high percentage of the initial reactants are transformed into the required product, but there are always by-products of the reaction and the importance of saleable by-products, and of methods of conserving evolved heat for re-use, cannot be over-emphasised. This has been mentioned in one way or another in earlier chapters whenever the question of economic viability has arisen. The chemical manufacturers' attempts to avoid pollution are often driven by economics. To allow waste products to escape represents financial loss, but the introduction of measures to prevent pollution is also expensive and if the cost is too high the price of the product may become prohibitive, making the whole process unprofitable. This too has to be borne in mind, in keeping with our treatment of specific operations in the chemical industry. As the nature of the chemical industries changes, the waste products also change and new methods to prevent pollution and clean-up the environment are constantly required. Whenever these are not successful the 'quality of life' is reduced and there is always a balance to be struck between introducing new products and developing new ways to prevent pollution.

No manufacturing process can ever be wholly free from pollution and a compromise must be made between the desirable state of zero pollution and the best degree of clean-up that is achievable.

Then there are questions of *scale-up*. Devising efficient methods of satisfying these criteria and of manufacturing the product on a large scale poses many problems. Once the theoretical aspects have been resolved and a workable laboratory method has been devised it becomes necessary to scale-up to manufacturing proportions. This cannot be done in one stage; it requires a stepwise approach. A process which works well enough in the laboratory using small quantities of pure reactants under easily controlled conditions of temperature and pressure may not work on a larger scale without modification.

The first stage is to make a small scale model using a few litres of material, then a larger scale set-up using 20 or 30 litres. A semi-commercial model may then be built before moving to the full-scale commercial plant. At each stage new problems may be encountered. Owing to the increasing mass of material being used, the process may slow down, stop or even take a different direction. It may be that a catalyst or some traces of foreign substances are needed, or even that a complete new process must be devised. Heat evolved in the process may be enough to require heat exchangers, or the product may need to be continuously removed from the mixture to keep the process going. The development of these new techniques and the plant necessary to operate them takes time, skill, imagination and not a little luck.

A third problem is that of *disposal* of a substance when its useful life has ended. Even when a new product has been successfully manufactured and marketed, this problem remains. It is particularly important in the case of packaging, large quantities of which are constantly thrown away. Aside from the difficulties of disposal of such urban and industrial detritus, and its environmental impact, this represents a colossal waste of potentially re-usable material. Recycling has not always been a realistic option, but in recent years, faced with growing mountains of urban waste, efforts have been made to find economically viable methods of recycling waste paper and cardboard, glass and metals such as the aluminium cans which are discarded annually in their millions.[44] Some of the most intractable problems arose in the case of plastic

---

[44] N.G. Coley, 'Removing mountains, industrial waste and the environment', in M. Fetizon and W.J. Thomas (eds.), *The role of oxygen in improving chemical processes*, Royal Society of Chemistry, Cambridge, 1993, pp. 123–130.

containers which resist degradation in the landfill sites used for the disposal of urban waste. Efforts have been made to produce bio-degradable plastics, but more recently the possibilities of recycling plastic waste have been explored.[45]

The production of new materials is the most obvious way in which the chemical industries have improved the quality of life, but it is not the only way. Less visible to the public, yet just as important to the quality of life, are the service industries which depend on chemical products and methods. Of these the most important is the water industry where chemical and microbiological analysis and quality control maintain the purity and safety of public water supplies. The water industry also operates a continuous check on pollution in rivers, streams, and reservoirs, and the quality of effluent flowing into rivers from sewage works is carefully monitored by chemical and other tests. The oil and natural gas industries also maintain analytical procedures to avoid pollution of the atmosphere. Similar controls are maintained in the nuclear industry to ensure that radioactivity is kept to prescribed levels around nuclear power stations and processing plants. The chemical industries also have a big part to play in controlling atmospheric pollution from internal combustion engines, partly by the refinement of fuels, partly by the design of catalytic converters and partly by the chemical testing of engine emissions. In all these cases chemical products in the form of pure reagents are employed, but what is more important is the application of chemical methods of analysis and testing, nowadays based on electrical, optical and electronic methods and equipment.

If we have shown that modern life would be impossible without the methods and products of the chemical industry, it has also become evident that we must come to terms with the unfortunate consequences of chemical manufacturing. The 20th century has been exceptional for the economic, material and social benefits derived from the chemical industry, but also for the attention that has been given to solving the problems of pollution. It has become increasingly clear that we need to put as much effort and ingenuity into eliminating or correcting the chemical industry's faults as we put into the manufacturing processes themselves. As this effort is not economically rewarding in immediately obvious ways, regulations to ensure a cleaner environment must be enforced with adequate legal sanctions and we must all be prepared to pay dearly for the benefits gained. The recent past is exceptional as an age in which concern for the environment has become an important social and political issue. Fears for the future of the planet, whether or not they are well-grounded, have driven environmental movements to demand higher standards of protection. Equally, the products of the chemical industry have become much more varied and the potency of many has become greater, making the need for concern and action imperative. Considerable efforts are now being made to repair earlier environmental damage and to avoid future problems.

Foremost among environmental issues today is the problem of air quality in urban areas and the means of monitoring and improving it are presently under

---

[45] W. Hoyle and D.R. Karca (eds.), *Chemical aspects of plastics recycling*, Royal Society of Chemistry, Cambridge, 1997.

active consideration. Another environmental issue of current concern is the reclamation of contaminated land on urban sites previously occupied by gas-works and chemical plants. In 1965 fifty people died and a hundred were made seriously ill in Japan due to mercury poisoning caused by eating fish caught in Minimata Bay. It was found that a chemical firm had discharged mercury compounds into the sea and these had been converted by micro-organisms into highly toxic methylmercury which was then concentrated in fish subsequently eaten by the victims. Accidents of this kind led to demands for closer control and adequate regulations. Today the discharge of toxic waste into rivers and the sea is a topic of concern in both scientific and political circles as effective methods of prevention and clean-up are sought.[46]

These and related issues have been reviewed repeatedly in recent years and are commonly featured in the media, but while it is easy to identify the problems and their consequences, to propose standards and aims, it is much more difficult to put effective measures for solving the problems in place. Nevertheless, the British chemical industry is at least trying to meet the required standards and to exceed them wherever possible. The old image of the chemical industry as the great polluter is slowly but surely changing as a keener awareness of these environmental issues and the potential for avoiding and correcting them grows within the industry.

---

[46] J. Tapp, J.R. Wharfe and S.M. Hunt (eds.), *Toxic impact of wastes on the aquatic environment*, Royal Society of Chemistry, Cambridge, 1996.

# Index of Persons

Except for living persons, full dates are given whenever possible. Otherwise the symbol *fl.* ( = *floreat*, flourishes) indicates approximately the time of their activity. Page numbers in *italics* indicate illustrations.

# Subject Index

Principal references only are given. Page numbers in *italics* indicate illustrations.